Ioannis Emmanouil

Idempotent Matrices over Complex Group Algebras

 Springer

Professor Ioannis Emmanouil

Department of Mathematics
University of Athens
157 84 Athens
Greece
E-mail: emmanoui@math.uoa.gr

Mathematics Subject Classification: 16S34, 18G, 19A31, 19D55, 20C07, 20J05, 46L10, 46L80

Library of Congress Control Number: 2005930320

ISBN-10 3-540-27990-3 Springer Berlin Heidelberg New York
ISBN-13 978-3-540-27990-7 Springer Berlin Heidelberg New York

Springer is a part of Springer Science+Business Media
springeronline.com
© Springer-Verlag Berlin Heidelberg 2006
Printed in The Netherlands

Typesetting: by the authors and TechBooks using a Springer LATEX macro package

Cover design: *design & production* GmbH, Heidelberg

Printed on acid-free paper   SPIN: 11529439   41/TechBooks   5 4 3 2 1 0

στους γονείς μου

Preface

The study of idempotent elements in group algebras (or, more generally, the study of classes in the K-theory of such algebras) originates from geometric and analytic considerations. For example, C.T.C. Wall [72] has shown that the problem of deciding whether a finitely dominated space with fundamental group π is homotopy equivalent to a finite CW-complex leads naturally to the study of a certain class in the reduced K-theory $\widetilde{K}_0(\mathbf{Z}\pi)$ of the group ring $\mathbf{Z}\pi$. As another example, consider a discrete group G which acts freely, properly discontinuously, cocompactly and isometrically on a Riemannian manifold. Then, following A. Connes and H. Moscovici [16], the index of an invariant 0th-order elliptic pseudo-differential operator is defined as an element in the K_0-group of the reduced group C^*-algebra C_r^*G.

The idempotent conjecture (also known as the generalized Kadison conjecture) asserts that the reduced group C^*-algebra C_r^*G of a discrete torsion-free group G has no idempotents $\neq 0, 1$; this claim is known to be a consequence of a far-reaching conjecture of P. Baum and A. Connes [6]. Alternatively, one may approach the idempotent conjecture as an assertion about the connectedness of a *non-commutative space*; if G is a discrete torsion-free abelian group then C_r^*G is the algebra of continuous complex-valued functions on the dual group $\widehat{G}$, which is itself a compact and connected topological space. Even though the complex group algebra of a group G is a much simpler object than the corresponding reduced group C^*-algebra, the idempotent conjecture for $\mathbf{C}G$ remains still an unproved claim when G is an arbitrary torsion-free group. The latter problem has attracted the attention of ring theorists since the middle of the 20th century.

On the other hand, reformulating a theorem of R. Swan [70] about projective modules over integral group rings of finite groups in terms of the Hattori-Stallings rank, H. Bass stated in [4,5] a conjecture about the trace of idempotent matrices with entries in the group algebra of a group with coefficients in a suitable subring of the field $\mathbf{C}$ of complex numbers. An immediate consequence of the validity of the conjecture is the equality of various Euler characteristics that can be defined for groups. Furthermore, as shown by B. Eckmann [20], Bass' conjecture is related to the freeness of certain induced finitely generated projective modules over the von Neumann algebra of the group.

This book provides an introduction to the study of these problems for graduate students and researchers who are not necessarily experts in the field. Our aim is to show the unified character of the conjectures mentioned above and present the basic elements of an area of research that has recently experienced a revival, in view of its close relationship with deep geometric problems. At the same time, we hope that this book will become a valuable aid to the experts as well, as it collects and presents in a systematic way basic techniques and important results that have been obtained during the past few decades.

The pace of the book is suitable for independent study and the level of the presentation not very demanding, assuming only familiarity with the techniques of Algebra and Analysis that are usually covered during the first year of graduate studies. Moreover, in order to facilitate the reader, we have decided to include a few Appendices that detail some of the tools used in the main text. On the other hand, we have restrained ourselves from using some of the more advanced techniques that may be employed in the study of these problems, such as refined tools from K-theory.

In the first chapter, we fix the notation used in the rest of the book and properly formulate the Bass' and idempotent conjectures. As a warm-up, we prove the idempotent conjecture for torsion-free ordered groups and introduce the Strebel-Strojnowski class, providing some additional examples of groups satisfying the idempotent conjecture.

In Chap. 2, we present the simplest examples of groups that satisfy Bass' conjecture, namely the abelian groups and the finite ones. We put the conjecture in a geometric perspective, by relating it, in the abelian case, to the connectedness of the prime spectrum of the group algebra. Using some basic representation theory, we establish the equivalence between Bass' conjecture for finite groups and Swan's theorem on integral representations, which served itself as the primary motivation for H. Bass to formulate the conjecture.

In Chap. 3, we study idempotent matrices with entries in complex group algebras by reduction to positive characteristic. This technique was pioneered by A. Zaleskii [75], in order to complement a result of I. Kaplansky [38] on the positivity of the canonical trace. Using the action of the Frobenius operator in the positive characteristic case and then lifting the result to **C**, we prove two theorems of H. Bass and P. Linnell describing some properties of the support of the Hattori-Stallings rank of an idempotent matrix.

In Chap. 4, we present another method that may be used in the study of the idempotent conjectures, which is of homological nature. We define cyclic homology and relate it to the K-theory and the Hattori-Stallings rank. The nilpotency of Connes' periodicity operator in the cyclic homology of group algebras suggests the definition of a class $\mathcal{C}$, which provides us with many interesting examples of groups that satisfy the idempotent conjectures.

In the last chapter, we study idempotent matrices with entries in the reduced group C^*-algebra of a discrete group and prove the integrality of the canonical trace, in the cases of abelian and free groups. In the abelian case, this follows from the connectedness of the dual group, whereas the free group case is taken care of by considering a free action of the group on a tree. We construct the center-valued trace on the von Neumann algebra of a group from scratch (i.e. without appealing to the general theory of finite algebras) and study its importance in K-theory. In particular, we prove the result of B. Eckmann on the freeness of induced finitely generated projective modules over group von Neumann algebras.

The five Appendices at the end of the book summarize the results from Algebra, Number Theory and Analysis that are needed in the main text.

At this point, I would like to acknowledge the intellectual debt owed to the mathematicians whose work and ideas build up this book; in particular, to H. Bass, A. Connes, B. Eckmann, I. Kaplansky, P. Linnell and A. Zaleskii.

Athens, Greece   *Ioannis Emmanouil*
June, 2005

Contents

1

Introduction

1.1 Preliminaries

We assume that the reader is familiar with the basic elements of Algebra and Analysis that are usually covered during the first year of graduate studies. In particular, we assume some familiarity with the basics of group theory (structure theorem for finitely generated abelian groups, rank of abelian groups, solvable groups), ring theory (unique factorization in commutative rings, tensor product and its relation to torsion, flatness) and field theory (field extensions, Galois theory). Prerequisite notions also include the basics of topology (completion of metric spaces, the Stone-Weierstrass theorem, connectedness and compactness) and functional analysis (completeness, orthogonality and bases of Hilbert spaces).

For future reference, we record the basic results from Algebra and Analysis that are needed in the sequel in §1.1.1 and §1.1.2. We give the details of the construction of the K_0-group of an algebra R in §1.1.3, by means of finitely generated projective modules and idempotent matrices, and explain the way that traces on R induce additive maps on $K_0(R)$ in §1.1.4.

1.1.1 Basic Notions from Algebra

In this first subsection, we discuss about a variety of topics, including extension and restriction of scalars in module categories, the order structure on the set of idempotents of a ring and record certain basic facts about group rings.

Unless otherwise specified, all rings will be associative and unital and all ring homomorphisms will be assumed to be unit preserving. In the same way, all modules will be understood to be left modules. If R is a ring, then $U(R) \subseteq R$ will denote the group of invertible elements and $\mathrm{nil}\, R \subseteq R$ the subset (ideal, if the ring R is commutative) of nilpotent elements. Given a ring R, the set $R^{op} = \{r^{op} : r \in R\}$ can be endowed with a ring structure, by letting $r^{op} + s^{op} = (r + s)^{op}$ and $r^{op} \cdot s^{op} = (sr)^{op}$ for any $r^{op}, s^{op} \in R^{op}$; as such, R^{op} is called the opposite ring of R. Then, a right R-module is simply a (left) R^{op}-module.

I. EXTENSION AND RESTRICTION OF SCALARS. If $\varphi : R \longrightarrow S$ is a ring homomorphism then S can be regarded as a left (resp. right) R-module, by letting $r \cdot s = \varphi(r)s$ (resp. $s \cdot r = s\varphi(r)$) for all $r \in R$ and $s \in S$. For any R-module M the abelian group $S \otimes_R M$ can be viewed as a left S-module, where $s \cdot (s' \otimes x) = (ss') \otimes x$ for all $s, s' \in S$ and $x \in M$. The resulting functor from the category R–Mod of left R-modules to the category S–Mod of left S-modules is referred to as the extension of scalars along φ. On the other hand, given φ, any left S-module N can be regarded as a left R-module by restriction of scalars along φ, i.e. by letting $r \cdot x = \varphi(r)x$ for any $r \in R$ and $x \in M$; the resulting R-module will be denoted by N'. With this notation, there is a natural isomorphism of abelian groups

$$\lambda : \mathrm{Hom}_S(S \otimes_R M, N) \overset{\sim}{\longrightarrow} \mathrm{Hom}_R(M, N') \, ,$$

which is defined by letting $\lambda(f) : M \longrightarrow N'$ be the map $x \mapsto f(1 \otimes x)$, $x \in M$, for any $f \in \mathrm{Hom}_S(S \otimes_R M, N)$.

For later use, we record the following simple property of the restriction of scalars functor.

Lemma 1.1 *Let $\varphi : R \longrightarrow S$ be a ring homomorphism, N a left S-module and N' the left R-module obtained from N by restriction of scalars.*

(i) If $\{s_i : i \in I\}$ is a set of generators of S as a left R-module and $\{x_j : j \in J\}$ a set of generators of the left S-module N, then $\{s_i x_j : i \in I, j \in J\}$ is a set of generators of the left R-module N'.

(ii) If the left R-module S is free with basis $\{s_i : i \in I\}$ and the left S-module N is free with basis $\{x_j : j \in J\}$, then the left R-module N' is free with basis $\{s_i x_j : i \in I, j \in J\}$.

(iii) If the left R-module S is projective and the left S-module N is projective, then the left R-module N' is projective. □

II. THE ORDERING OF IDEMPOTENTS. If R is a ring we denote by $\mathrm{Idem}(R)$ the set of idempotent elements of R, i.e. we let

$$\mathrm{Idem}(R) = \{e \in R : e^2 = e\} \, .$$

It is clear that $0, 1 \in \mathrm{Idem}(R)$. A non-trivial idempotent of R is an idempotent e with $e \neq 0, 1$. A ring homomorphism $\varphi : R \longrightarrow S$ maps idempotents of R to idempotents of S and hence restricts to a map

$$\mathrm{Idem}(\varphi) : \mathrm{Idem}(R) \longrightarrow \mathrm{Idem}(S) \, .$$

Two idempotents $e, f \in R$ are called orthogonal if $ef = fe = 0$; if this is the case, the element $e + f \in R$ is also idempotent. It is easily seen that the following two conditions are equivalent for two idempotents $e, f \in R$:

(i) $ef = fe = e$ and

(ii) the element $f - e \in R$ is an idempotent orthogonal to e.

We define a relation $\leq$ on the set $\mathrm{Idem}(R)$, by letting $e \leq f$ if either one of the equivalent conditions above is satisfied. We now state the basic properties of that relation:

Proposition 1.2 *Let R be a ring and $\leq$ the relation defined above on the set $\mathrm{Idem}(R)$. Then:*

(i) The relation $\leq$ is an order on $\mathrm{Idem}(R)$.

(ii) For any $e \in \mathrm{Idem}(R)$ we have $0 \leq e \leq 1$.

(iii) For any $e \in \mathrm{Idem}(R)$ the element $e' = 1 - e \in R$ is an idempotent, such that $e \wedge (1 - e) = 0$ and $e \vee (1 - e) = 1$.[1] We shall refer to e' as the complementary idempotent of e.

(iv) If $e, f \in R$ are two idempotents then $e \leq f$ if and only if $f' \leq e'$; here, f' (resp. e') is the complementary idempotent of f (resp. of e).

(v) If $e, f \in \mathrm{Idem}(R)$ are two commuting idempotents then the elements $ef, e + f - ef \in R$ are also idempotents; in fact, we have $e \wedge f = ef$ and $e \vee f = e + f - ef$.

(vi) If $e, f, g \in \mathrm{Idem}(R)$ are three commuting idempotents then we have $e \wedge (f \vee g) = (e \wedge f) \vee (e \wedge g)$.

Proof. The verification of the properties of an order for $\leq$ is a routine exercise, whereas assertion (ii) is an immediate consequence of the definitions.

(iii) First of all, we note that for any idempotent $e \in R$ the element $1 - e \in R$ is also idempotent. Moreover, if $x \in \mathrm{Idem}(R)$ is such that $x \leq e$ and $x \leq 1 - e$, then $x = xe = x(1 - e)e = x0 = 0$. In view of (ii) above, we conclude that $e \wedge (1 - e) = 0$. On the other hand, if $y \in \mathrm{Idem}(R)$ is such that $e \leq y$ and $1 - e \leq y$, then $y = ey + (1 - e)y = e + (1 - e) = 1$ and hence $e \vee (1 - e) = 1$, in view of (ii) above.

(iv) It is easily seen that the equalities $ef = fe = e$ are equivalent to the equalities $(1 - f)(1 - e) = (1 - e)(1 - f) = 1 - f$.

(v) If $e, f \in R$ are two commuting idempotents, then the element ef is idempotent as well. Since $eef = efe = ef$ and $fef = eff = ef$, we conclude that $ef \leq e$ and $ef \leq f$. Moreover, if $x \in \mathrm{Idem}(R)$ is such that $x \leq e$ and $x \leq f$, then $xef = xf = x$ and $efx = ex = x$, i.e. $x \leq ef$. It follows that $e \wedge f = ef$. Since the complementary idempotents $e' = 1 - e$ and $f' = 1 - f$ are also commuting, we have $e' \wedge f' = e'f'$. Invoking (iv) above, we conclude that $e \vee f = (e' \wedge f')' = (e'f')' = 1 - (1 - e)(1 - f) = e + f - ef$, as needed.

(vi) This is an immediate consequence of the formulae established in (v) above. □

An ordered set $(X, \leq)$ is a Boolean algebra if it satisfies the following conditions:

---

[1] If x, y are elements of an ordered set X, we denote by $x \vee y$ (resp. $x \wedge y$) the supremum (resp. infimum) of x and y, whenever that exists.

(BAi) There exist two elements $0, 1 \in X$, such that $0 \le x \le 1$ for all $x \in X$.

(BAii) Any two elements $x, y \in X$ have a supremum $x \vee y$ and an infimum $x \wedge y$.

(BAiii) For any $x \in X$ there is an element $x' \in X$, such that $x \wedge x' = 0$ and $x \vee x' = 1$, where $0, 1 \in X$ are the elements of (BAi) above.

(BAiv) For any $x, y, z \in X$ we have $x \wedge (y \vee z) = (x \wedge y) \vee (x \wedge z)$.

Examples 1.3 (i) Let R be a commutative ring. Then, the ordered set $(\mathrm{Idem}(R), \le)$ defined above is a Boolean algebra (cf. Proposition 1.2).

(ii) For any set X the power set $\mathcal{P}(X)$, ordered by inclusion, is a Boolean algebra. Here, $0 = \emptyset$, $1 = X$, $A \wedge B = A \cap B$, $A \vee B = A \cup B$ and $A' = X \setminus A$ for any two subsets $A, B \subseteq X$.

(iii) If X is a topological space, then the set $L(X)$ of all clopen (closed and open) subsets $A \subseteq X$, ordered by inclusion, is a Boolean algebra (with formulae for 0, 1, $\wedge$, $\vee$ and complements as in (ii) above).

Lemma 1.4 *Let $(X, \le)$ be a Boolean algebra.*

(i) For any element $x \in X$ the element x' in condition (BAiii) above is unique. We refer to it as the complement of x. In particular, $x'' = x$.

(ii) For any two elements $x, y \in X$ with complements x' and y' respectively, we have $x \le y$ if and only if $y' \le x'$.

(iii) For any two elements $x, y \in X$ with complements x' and y' respectively, we have the following rules (de Morgan laws)

$$x' \wedge y' = (x \vee y)' \quad and \quad x' \vee y' = (x \wedge y)'.$$

Here, we denote by $(x \vee y)'$, $(x \wedge y)'$ the complements of $x \vee y$ and $x \wedge y$ respectively.

Proof. (i) If $x_0 \in X$ is such that $x \wedge x_0 = 0$ and $x \vee x_0 = 1$, then we have

$$x' = x' \wedge 1 = x' \wedge (x \vee x_0) = (x' \wedge x) \vee (x' \wedge x_0) = 0 \vee (x' \wedge x_0) = x' \wedge x_0$$

and hence $x' \le x_0$. Working in the same way, with the roles of x' and x_0 reversed, it follows that $x_0 \le x'$. Therefore, we must have $x' = x_0$.

(ii) If $x \le y$ then $y' \wedge x \le y' \wedge y = 0$, i.e. $y' \wedge x = 0$. It follows that

$$y' = y' \wedge 1 = y' \wedge (x \vee x') = (y' \wedge x) \vee (y' \wedge x') = 0 \vee (y' \wedge x') = y' \wedge x'$$

and hence $y' \le x'$. Conversely, if $y' \le x'$ then $x = x'' \le y'' = y$.

(iii) This is an immediate consequence of (ii) above. $\qquad\square$

Given two Boolean algebras $(X, \le)$ and $(Y, \le)$, a morphism

$$u : (X, \le) \longrightarrow (Y, \le)$$

is a map $u : X \longrightarrow Y$, such that $u(0_X) = 0_Y$, $u(1_X) = 1_Y$, $u(x \wedge y) = u(x) \wedge u(y)$ and $u(x') = u(x)'$ for all $x, y \in X$. Then, we also have $u(x \vee y) = u(x) \vee u(y)$

for all $x, y \in X$; this follows from Lemma 1.4(i),(iii). If u is bijective, then the inverse map u^{-1} is also a morphism of Boolean algebras. In that case, u is called an isomorphism of Boolean algebras.

Examples 1.5 (i) Let $\varphi : R \longrightarrow S$ be a homomorphism of commutative rings. Then, the induced map $\mathrm{Idem}(\varphi) : \mathrm{Idem}(R) \longrightarrow \mathrm{Idem}(S)$ is a morphism of Boolean algebras.

(ii) Let $f : X \longrightarrow Y$ be a map of sets. Then, the map $A \mapsto f^{-1}(A)$, $A \subseteq Y$, is a morphism of Boolean algebras $f^* : \mathcal{P}(Y) \longrightarrow \mathcal{P}(X)$.

(iii) Let $f : X \longrightarrow Y$ be a continuous map of topological spaces. Then, the map $A \mapsto f^{-1}(A)$, $A \in L(Y)$, is a morphism of Boolean algebras $L(f) : L(Y) \longrightarrow L(X)$.

III. GROUP ALGEBRAS. The basic type of rings that will be examined in this book is that of group algebras. If k is a commutative ring and G a group, then the group algebra kG is defined as follows: As a k-module, kG is free with basis the set G, whereas the multiplication of kG is the unique k-bilinear extension of the multiplication of G. Then, kG is a unital k-algebra, which is commutative if and only if the group G is abelian. We note that any element $a \in kG$ can be written uniquely as a sum $\sum_{g \in G} a_g g$, where $a_g \in k$ for all $g \in G$ and $a_g = 0$ for all but finitely many g's. For any $g \in G$ we consider the map

$$p_g : kG \longrightarrow k ,$$

which is defined by letting $a \mapsto a_g$, $a \in kG$ (where a can be written as a sum $\sum_{x \in G} a_x x$ as above). Then, p_g is k-linear, whereas

$$p_g(ab) = \sum \{ p_x(a) p_y(b) : x, y \in G \text{ and } xy = g \}$$

for all $a, b \in kG$ and all $g \in G$.

The group algebra kG associated with a pair (k, G) as above is functorial in both k and G: If $\sigma : k \longrightarrow k'$ is a homomorphism of commutative rings and G a group, there is a unique ring homomorphism $\widetilde{\sigma} : kG \longrightarrow k'G$ with $\widetilde{\sigma}(r1) = \sigma(r)1$ for all $r \in k$ and $\widetilde{\sigma}(g) = g$ for all $g \in G$. On the other hand, if k is a fixed commutative ring and $f : G \longrightarrow G'$ a group homomorphism, there is a unique k-algebra homomorphism $\widetilde{f} : kG \longrightarrow kG'$ with $\widetilde{f}(g) = f(g)$ for all $g \in G$.

Let k be a commutative ring and G a group. We note that the group algebra associated with k and the trivial group $\{1\}$ is naturally identified with k. Hence, considering the (unique) group homomorphism $G \longrightarrow \{1\}$, we obtain a k-algebra homomorphism

$$\varepsilon : kG \longrightarrow k ,$$

which is defined by letting $\varepsilon(a) = \sum_{g \in G} a_g$ for any $a = \sum_{g \in G} a_g g \in kG$. We shall refer to ε as the augmentation homomorphism. The kernel $I_G(k)$ of ε is the augmentation ideal of kG; it is easily seen that $I_G(k)$ is a free k-module with basis consisting of the elements $g - 1$, $g \in G \setminus \{1\}$.

Remarks 1.6 Let k be a commutative ring, G a group and $H \leq G$ a subgroup. Then, the inclusion $H \hookrightarrow G$ induces an inclusion of k-algebras $kH \hookrightarrow kG$. In particular, we may regard kG as a (left or right) kH-module.

(i) Let S be a set of representatives of the left cosets of H in G. Then, the left kH-module kG is free with basis S. Indeed, G is the disjoint union of the cosets Hs, $s \in S$, and hence $kG = \bigoplus_{s \in S} kHs$ as left kH-modules. The claim follows since $kHs \simeq kH$ (as left kH-modules) for all $s \in S$. In the same way, the right kH-module kG is free with basis any set S' of representatives of the right cosets of H in G.

(ii) Assume that H is a normal subgroup of G and consider the quotient group $\overline{G} = G/H$. We denote by $\overline{g}$ the canonical image in $\overline{G}$ of any element $g \in G$. Then, there is an isomorphism of k-algebras

$$k \otimes_{kH} kG \simeq k\overline{G} \, ,$$

which identifies $1 \otimes g \in k \otimes_{kH} kG$ with $\overline{g} \in k\overline{G}$ for any $g \in G$. Here, we regard kG as a left kH-module as above and k as a right kH-module by means of the associated augmentation homomorphism.

Let k be a commutative ring, G a group and R a k-algebra. It is easily seen that any k-algebra homomorphism from kG to R restricts to a group homomorphism from G to the group of units $U(R)$. Conversely, any group homomorphism as above can be uniquely extended to a k-algebra homomorphism from kG to R. Therefore, there is a natural identification

$$\mathrm{Hom}_{k-Alg}(kG, R) \simeq \mathrm{Hom}_{Grp}(G, U(R)) \, .$$

In particular, let us consider a k-module M. Then, a kG-module structure on M, which extends the given k-module structure, is determined by a group homomorphism from G to the group $\mathrm{Aut}_k M$ of k-linear automorphisms of M, i.e. by a k-linear action of G on M. A special case where that situation occurs is when the k-module M is free on a G-set X; the resulting kG-module is referred to as a permutation module. We recall that an action of a group is called free if all stabilizers are trivial.

Lemma 1.7 *Let k be a commutative ring, G a group and X a free G-set. Then, the associated permutation kG-module $M = \bigoplus_{x \in X} k \cdot x$ is free.*

Proof. The decomposition $X = \bigcup_i X_i$ of X into the disjoint union of G-orbits induces a decomposition of kG-modules $M = \bigoplus_i M_i$, where $M_i = \bigoplus_{x \in X_i} k \cdot x$ for all i. Since $X_i \simeq G$ as G-sets, it follows that $M_i \simeq kG$ as kG-modules for all i. $\qquad\square$

The theory of modules over group algebras possesses two important special features. First of all, the map $g \mapsto g^{-1}$, $g \in G$, extends to an involution (anti-automorphism of order 2)

$$\tau : kG \longrightarrow kG \ .$$

We may regard τ as an isomorphism between the group algebra kG and its opposite algebra $(kG)^{op}$. Consequently, there is no need to distinguish between left and right kG-modules: If M is a right kG-module, then we may define a left kG-module structure on M by letting $g \cdot m = mg^{-1}$ for all $g \in G$ and all $m \in M$. On the other hand, let us consider the group homomorphism

$$D : G \longrightarrow G \times G \ ,$$

which is given by $g \mapsto (g,g)$, $g \in G$. Then, if M, N are two kG-modules, we may endow the tensor product $M \otimes_k N$ with the structure of a kG-module by means of the composition

$$G \xrightarrow{D} G \times G \longrightarrow \operatorname{Aut}_k M \times \operatorname{Aut}_k N \longrightarrow \operatorname{Aut}_k (M \otimes_k N) \ .$$

Here, the second arrow is defined by using the given kG-module structures on M and N, whereas the third one is defined by letting $(\alpha, \beta) \mapsto \alpha \otimes \beta$ for all $\alpha \in \operatorname{Aut}_k M$ and $\beta \in \operatorname{Aut}_k N$. We shall refer to the resulting G-action on $M \otimes_k N$ as the diagonal action.

Lemma 1.8 *Let k be a commutative ring and G a group.*
   (i) The kG-module $kG \otimes_k kG$ (with diagonal G-action) is free.
   (ii) Let M, N be two projective kG-modules. Then, the diagonal kG-module $M \otimes_k N$ is also projective.

Proof. (i) The diagonal kG-module $kG \otimes_k kG$ is precisely the permutation module associated with the action of G by left translations on the set $G \times G$. Since the latter action is free, the result follows from Lemma 1.7.

   (ii) Since both M and N are direct summands of free kG-modules, the tensor product $M \otimes_k N$ is a direct summand of a direct sum of copies of $kG \otimes_k kG$. Hence, the result follows invoking (i) above.   □

We conclude this subsection with a basic result in the representation theory of finite groups.

Theorem 1.9 *(Maschke) Let k be a field and G a finite group whose order is invertible in k. Then, any kG-module is projective.*

Proof. Let V be a kG-module and V_0 the k-vector space obtained from V by restriction of scalars. We consider the kG-module $kG \otimes_k V_0$ obtained from V_0 by extension of scalars and the k-linear maps $i : V \longrightarrow kG \otimes_k V_0$ and $\pi : kG \otimes_k V_0 \longrightarrow V$, which are defined by letting

$$i(v) = \frac{1}{|G|} \sum_{g \in G} g \otimes g^{-1}v \quad \text{and} \quad \pi(g \otimes v) = gv$$

for all $v \in V$ and $g \in G$. It is easily seen that both i and π are kG-linear, whereas $\pi \circ i$ is the identity map on V. Therefore, i identifies the kG-module V with a direct summand of the free kG-module $kG \otimes_k V_0$.   □

1.1.2 Basic Notions from Analysis

In this subsection, we discuss about bounded linear operators on a Hilbert space $\mathcal{H}$ and develop some properties of the strong and weak operator topologies on $\mathcal{B}(\mathcal{H})$. We also define the reduced group C^*-algebra and the von Neumann algebra associated with a discrete group.

I. LINEAR OPERATORS ON HILBERT SPACES. Let $\mathcal{H}$ be a Hilbert space and $\mathcal{B}(\mathcal{H})$ the corresponding algebra of bounded linear operators. For any $a \in \mathcal{B}(\mathcal{H})$ its adjoint a^* is characterized by the equalities

$$<a(\xi), \eta> = <\xi, a^*(\eta)> ,$$

which are valid for all vectors $\xi, \eta \in \mathcal{H}$. Moreover, we have

$$\overline{\operatorname{im} a^*} = (\ker a)^\perp \quad \text{and} \quad \ker a^* = (\overline{\operatorname{im} a})^\perp = (\operatorname{im} a)^\perp .$$

A subset of $\mathcal{B}(\mathcal{H})$ is called self-adjoint if it contains the adjoints of its elements. For any $a \in \mathcal{B}(\mathcal{H})$ the operator norm $\|a\|$ is the supremum of the set $\{\|a(\xi)\| : \|\xi\| \leq 1\} \subseteq [0, \infty)$. The algebra $\mathcal{B}(\mathcal{H})$ is complete under that norm. A C^*-algebra $\mathcal{A}$ of operators acting on $\mathcal{H}$ is a closed self-adjoint subalgebra of $\mathcal{B}(\mathcal{H})$; in contrast to our general algebraic convention, we will not always assume that $\mathcal{A}$ is unital. As an example, the (non-unital) C^*-algebra $\mathcal{K}(\mathcal{H})$ of compact operators is the closure of the (non-unital) algebra $\mathcal{F}(\mathcal{H})$ of finite rank operators in $\mathcal{B}(\mathcal{H})$.

For any positive integer n the algebra $\mathbf{M}_n(\mathcal{B}(\mathcal{H}))$ acts on the n-fold direct sum $\mathcal{H}^n$ (whose elements are viewed as column-vectors) by left multiplication. In this way, we identify $\mathbf{M}_n(\mathcal{B}(\mathcal{H}))$ with the algebra $\mathcal{B}(\mathcal{H}^n)$. The following assertions are easily verified (cf. Exercise 1.3.2):

(i) If $A = (a_{ij})_{i,j} \in \mathbf{M}_n(\mathcal{B}(\mathcal{H}))$ then the adjoint $A^* = (b_{ij})_{i,j}$ is given by letting $b_{ij} = a_{ji}^*$ for all i, j.

(ii) If $(A_m)_m$ is a sequence in $\mathbf{M}_n(\mathcal{B}(\mathcal{H}))$ and $A_m = (a_{ij,m})_{i,j}$ for all m, then $\lim_m A_m = 0$ (in the operator norm topology of $\mathbf{M}_n(\mathcal{B}(\mathcal{H})) \simeq \mathcal{B}(\mathcal{H}^n)$) if and only if $\lim_m a_{ij,m} = 0$ for all i, j.

In particular, if $\mathcal{A}$ is a C^*-algebra of operators acting on $\mathcal{H}$, then $\mathbf{M}_n(\mathcal{A})$ is a C^*-algebra of operators acting on $\mathcal{H}^n$.

A linear operator $u \in \mathcal{B}(\mathcal{H})$ is an isometry if $u^*u = uu^* = 1$. The operator $v \in \mathcal{B}(\mathcal{H})$ is called a partial isometry if there are closed linear subspaces $V, V' \subseteq \mathcal{H}$, such that v maps V isometrically onto V' and vanishes on the orthogonal complement $V^\perp$. An operator $s \in \mathcal{B}(\mathcal{H})$ is called positive if the complex number $<s(\xi), \xi>$ is real and non-negative for all vectors $\xi \in \mathcal{H}$. For any $a \in \mathcal{B}(\mathcal{H})$ there is a unique positive operator $|a| \in \mathcal{B}(\mathcal{H})$ with $|a|^2 = a^* a$; the operator $|a|$ is called the absolute value of a (cf. [60, Theorem 12.33]). In fact, if $\mathcal{A} \subseteq \mathcal{B}(\mathcal{H})$ is any C^*-algebra and $a \in \mathcal{A}$, then $|a| \in \mathcal{A}$ as well [loc.cit.].

Lemma 1.10 *Let $a \in \mathcal{B}(\mathcal{H})$ be an operator and $s = |a|$ its absolute value. Then, $\ker s = \ker a$ and $\overline{\operatorname{im} s} = \overline{\operatorname{im} a^*}$.*

Proof. First of all, we note that $\ker a = \ker a^* a$. Indeed, the inclusion $\ker a \subseteq \ker a^* a$ is obvious, whereas for any $\xi \in \ker a^* a$ we have

$$\| a(\xi) \|^2 = <a(\xi), a(\xi)> = <a^* a(\xi), \xi> = 0$$

and hence $a(\xi) = 0$. Applying the above conclusion to the self-adjoint operator s, it follows that $\ker s = \ker s^2$. Since $s^2 = a^* a$, it follows that $\ker s = \ker a$. The operator s being self-adjoint, we have $\overline{\operatorname{im} s} = (\ker s)^\perp = (\ker a)^\perp = \overline{\operatorname{im} a^*}$, as needed. □

Proposition 1.11 *(polar decomposition) Let $a \in \mathcal{B}(\mathcal{H})$ be an operator and $s = |a|$ its absolute value. Then, there is a partial isometry v, which maps $\overline{\operatorname{im} a^*}$ isometrically onto $\overline{\operatorname{im} a}$ and vanishes on the orthogonal complement $\left(\overline{\operatorname{im} a^*}\right)^\perp$, such that $a = vs$. Moreover, the partial isometry v is uniquely determined by these properties.*

Proof. For any vector $\xi \in \mathcal{H}$ we have

$$<a(\xi), a(\xi)> = <a^* a(\xi), \xi> = <s^2(\xi), \xi> = <s(\xi), s(\xi)>$$

and hence $\| a(\xi) \| = \| s(\xi) \|$. It follows that the map $s(\xi) \mapsto a(\xi)$, $s(\xi) \in \operatorname{im} s$, is well-defined and extends to an isometry

$$v_0 : \overline{\operatorname{im} s} \longrightarrow \overline{\operatorname{im} a} .$$

We extend v_0 to a partial isometry $v \in \mathcal{B}(\mathcal{H})$, by letting v vanish on $\left(\overline{\operatorname{im} s}\right)^\perp$. It is clear that $a = vs$, whereas $\overline{\operatorname{im} s} = \overline{\operatorname{im} a^*}$ (cf. Lemma 1.10).

In order to prove the uniqueness of v, assume that v' is another partial isometry, which maps $\overline{\operatorname{im} a^*}$ isometrically onto $\overline{\operatorname{im} a}$ and vanishes on the orthogonal complement $\left(\overline{\operatorname{im} a^*}\right)^\perp$, such that $a = v's$. Then, $vs = v's$ and hence v and v' agree on the image of s. Since $\overline{\operatorname{im} s} = \overline{\operatorname{im} a^*}$, the (continuous) operators v and v' agree on $\overline{\operatorname{im} a^*}$. On the other hand, both operators vanish on the orthogonal complement $\left(\overline{\operatorname{im} a^*}\right)^\perp$. Hence, it follows that $v = v'$. □

Let $(e_i)_i$ be an orthonormal basis of $\mathcal{H}$. An operator $a \in \mathcal{B}(\mathcal{H})$ is said to be of trace class if the family $(<|a|(e_i), e_i>)_i$ is summable. We denote by $\mathcal{L}^1(\mathcal{H})$ the set of trace class operators and define

$$\| \cdot \|_1 : \mathcal{L}^1(\mathcal{H}) \longrightarrow [0, \infty)$$

by letting $\| a \|_1 = \sum_i <|a|(e_i), e_i>$ for all $a \in \mathcal{L}^1(\mathcal{H})$. The following result describes a few properties of $\mathcal{L}^1(\mathcal{H})$.

Theorem 1.12 *(cf. [56, §3.4]) Let $(e_i)_i$ be an orthonormal basis of $\mathcal{H}$.*

(i) The set $\mathcal{L}^1(\mathcal{H})$ is a subspace of $\mathcal{B}(\mathcal{H})$, which does not depend upon the choice of the orthonormal basis $(e_i)_i$. Moreover, we have $\mathcal{F}(\mathcal{H}) \subseteq \mathcal{L}^1(\mathcal{H}) \subseteq \mathcal{K}(\mathcal{H})$.

(ii) The map $\|\cdot\|_1$ defined above is a norm (that will be referred to as the Schatten 1-norm), which does not depend upon the choice of the orthonormal basis $(e_i)_i$. That norm endows $\mathcal{L}^1(\mathcal{H})$ with the structure of a Banach space.

(iii) The subspace $\mathcal{L}^1(\mathcal{H})$ is a self-adjoint ideal of $\mathcal{B}(\mathcal{H})$. For all $a \in \mathcal{L}^1(\mathcal{H})$ and $b \in \mathcal{B}(\mathcal{H})$ we have $\|ab\|_1 \leq \|a\|_1 \cdot \|b\|$ and $\|ba\|_1 \leq \|b\| \cdot \|a\|_1$.

(iv) For any operator $a \in \mathcal{L}^1(\mathcal{H})$ the family $(<a(e_i), e_i>)_i$ is absolutely summable and the sum $\sum_i <a(e_i), e_i>$ does not depend upon the choice of the orthonormal basis $(e_i)_i$. The linear map

$$Tr : \mathcal{L}^1(\mathcal{H}) \longrightarrow \mathbf{C} \ ,$$

which is defined by letting $Tr(a) = \sum_i <a(e_i), e_i>$ for all $a \in \mathcal{L}^1(\mathcal{H})$, is such that $Tr(ab) = Tr(ba)$ for all $a \in \mathcal{L}^1(\mathcal{H})$ and $b \in \mathcal{B}(\mathcal{H})$. ☐

II. THE STRONG AND WEAK OPERATOR TOPOLOGIES ON $\mathcal{B}(\mathcal{H})$. Let $\mathcal{H}$ be a Hilbert space. Besides the operator norm topology, the algebra $\mathcal{B}(\mathcal{H})$ can be also endowed with the strong operator topology (SOT). The latter is the locally convex topology which is induced by the family of semi-norms $(Q_\xi)_{\xi \in \mathcal{H}}$, where

$$Q_\xi(a) = \|a(\xi)\|$$

for all $\xi \in \mathcal{H}$ and $a \in \mathcal{B}(\mathcal{H})$. Hence, a net of operators $(a_\lambda)_\lambda$ in $\mathcal{B}(\mathcal{H})$ is SOT-convergent to 0 if and only if $\lim_\lambda a_\lambda(\xi) = 0$ for all $\xi \in \mathcal{H}$. The weak operator topology (WOT) on $\mathcal{B}(\mathcal{H})$ is the locally convex topology which is induced by the family of semi-norms $(P_{\xi,\eta})_{\xi,\eta \in \mathcal{H}}$, where

$$P_{\xi,\eta}(a) = |< a(\xi), \eta >|$$

for all $\xi, \eta \in \mathcal{H}$ and $a \in \mathcal{B}(\mathcal{H})$. In other words, a net of operators $(a_\lambda)_\lambda$ in $\mathcal{B}(\mathcal{H})$ is WOT-convergent to $0 \in \mathcal{B}(\mathcal{H})$ if and only if $\lim_\lambda <a_\lambda(\xi), \eta > = 0$ for all $\xi, \eta \in \mathcal{H}$.

Remarks 1.13 (i) Let $(a_\lambda)_\lambda$ be a net of operators on $\mathcal{H}$. Then, we have

$$\|\cdot\| - \lim_\lambda a_\lambda = 0 \Longrightarrow \text{SOT} - \lim_\lambda a_\lambda = 0 \Longrightarrow \text{WOT} - \lim_\lambda a_\lambda = 0 \ .$$

If the Hilbert space $\mathcal{H}$ is not finite dimensional, none of the implications above can be reversed (cf. Exercise 1.3.3).

(ii) For any $a \in \mathcal{B}(\mathcal{H})$ we consider the left (resp. right) multiplication operator

$$L_a : \mathcal{B}(\mathcal{H}) \longrightarrow \mathcal{B}(\mathcal{H}) \ (\text{resp. } R_a : \mathcal{B}(\mathcal{H}) \longrightarrow \mathcal{B}(\mathcal{H})) \ ,$$

which is defined by letting $L_a(b) = ab$ (resp. $R_a(b) = ba$) for all $b \in \mathcal{B}(\mathcal{H})$. It is easily seen that the operators L_a and R_a are WOT-continuous. On the other hand, if the Hilbert space $\mathcal{H}$ is not finite dimensional, then the multiplication in $\mathcal{B}(\mathcal{H})$ is not (jointly) WOT-continuous (cf. Exercise 1.3.3).

Proposition 1.14 *Let $(a_\lambda)_\lambda$ be a bounded net of operators on $\mathcal{H}$. Then, the following conditions are equivalent:*

(i) $WOT - \lim_\lambda a_\lambda = 0$.

(ii) There is an orthonormal basis $(e_i)_i$ of the Hilbert space $\mathcal{H}$, such that $\lim_\lambda <a_\lambda(e_i), e_j> = 0$ for all i, j.

(iii) There is a subset $B \subseteq \mathcal{H}$, whose closed linear span is $\mathcal{H}$, such that $\lim_\lambda <a_\lambda(\xi), \eta> = 0$ for all $\xi, \eta \in B$.

(iv) There is a dense subset $X \subseteq \mathcal{H}$, such that $\lim_\lambda <a_\lambda(\xi), \eta> = 0$ for all $\xi, \eta \in X$.

Proof. It is clear that (i)→(ii)→(iii), whereas the implication (iii)→(iv) follows by letting X be the (algebraic) linear span of B. It only remains to show that (iv)→(i). To that end, assume that $M > 0$ is such that $\| a_\lambda \| \leq M$ for all λ and consider two vectors $\xi, \eta \in \mathcal{H}$. For any positive ϵ we may choose two vectors $\xi', \eta' \in X$, such that $\| \xi - \xi' \| < \epsilon$ and $\| \eta - \eta' \| < \epsilon$. Since

$$<a_\lambda(\xi), \eta> - <a_\lambda(\xi'), \eta'> = <a_\lambda(\xi - \xi'), \eta> + <a_\lambda(\xi'), \eta - \eta'>,$$

it follows that

$$
\begin{aligned}
|<a_\lambda(\xi), \eta> - <a_\lambda(\xi'), \eta'>| &\leq |<a_\lambda(\xi - \xi'), \eta>| + |<a_\lambda(\xi'), \eta - \eta'>| \\
&\leq \| a_\lambda \| \cdot \| \xi - \xi' \| \cdot \| \eta \| + \\
&\quad \| a_\lambda \| \cdot \| \xi' \| \cdot \| \eta - \eta' \| \\
&\leq M\epsilon(\| \xi \| + \| \eta \| + \epsilon).
\end{aligned}
$$

Since $\lim_\lambda <a_\lambda(\xi'), \eta'> = 0$, we may choose λ_0 such that $|<a_\lambda(\xi'), \eta'>| < \epsilon$ for all $\lambda \geq \lambda_0$. It follows that $|<a_\lambda(\xi), \eta>| < \epsilon(1 + M(\| \xi \| + \| \eta \| + \epsilon))$ for all $\lambda \geq \lambda_0$ and hence $\lim_\lambda <a_\lambda(\xi), \eta> = 0$, as needed. $\square$

Theorem 1.15 *Let $\mathcal{H}$ be a separable Hilbert space, r a positive real number and $\mathcal{B}(\mathcal{H})_r = \{a \in \mathcal{B}(\mathcal{H}) : \| a \| \leq r\}$ the closed r-ball of $\mathcal{B}(\mathcal{H})$. Then, the topological space $(\mathcal{B}(\mathcal{H})_r, WOT)$ is compact and metrizable.*

Proof. In order to prove compactness, we consider for any two vectors $\xi, \eta \in \mathcal{H}$ the closed disc

$$D_{\xi,\eta} = \{z \in \mathbf{C} : |z| \leq r\|\xi\| \cdot \|\eta\|\} \subseteq \mathbf{C}$$

and the product space $\prod_{\xi,\eta \in \mathcal{H}} D_{\xi,\eta}$. In view of Tychonoff's theorem, the latter space is compact. We now define the map

$$f : \mathcal{B}(\mathcal{H})_r \longrightarrow \prod_{\xi,\eta \in \mathcal{H}} D_{\xi,\eta},$$

by letting $f(a) = (<a(\xi), \eta>)_{\xi,\eta}$ for all $a \in \mathcal{B}(\mathcal{H})_r$. It is clear that f is a homeomorphism of $(\mathcal{B}(\mathcal{H})_r, WOT)$ onto its image. Therefore, the compactness of $(\mathcal{B}(\mathcal{H})_r, \mathrm{WOT})$ will follow, as soon as we prove that the image $\mathrm{im}\, f$ of f is closed in $\prod_{\xi,\eta\in\mathcal{H}} D_{\xi,\eta}$. To that end, let $(z_{\xi,\eta})_{\xi,\eta}$ be an element in the closure of $\mathrm{im}\, f$. Then, the family $(z_{\xi,\eta})_{\xi,\eta}$ is easily seen to be linear in ξ and quasi-linear in η, whereas $|z_{\xi,\eta}| \le r \|\xi\| \cdot \|\eta\|$ for all ξ, η. Hence, there is a vector $a_\xi \in \mathcal{H}$ with $\| a_\xi \| \le r \|\xi\|$, such that $z_{\xi,\eta} = <a_\xi, \eta>$ for all $\xi, \eta \in \mathcal{H}$. Using the linearity of the family $(z_{\xi,\eta})_{\xi,\eta}$ in the first variable, it follows that there is an operator $a \in \mathcal{B}(\mathcal{H})_r$, such that $a_\xi = a(\xi)$ for all $\xi \in \mathcal{H}$. Then, $(z_{\xi,\eta})_{\xi,\eta} = f(a) \in \mathrm{im}\, f$, as needed.

In order to prove metrizability, we fix an orthonormal basis $(e_n)_{n=0}^\infty$ of the separable Hilbert space $\mathcal{H}$ and define for any $a, b \in \mathcal{B}(\mathcal{H})_r$

$$d_r(a,b) = \sum_{n,m} \frac{1}{2^{n+m}} |<(b-a)(e_n), e_m>| \ .$$

It is easily seen that d_r is a metric on $\mathcal{B}(\mathcal{H})_r$, which induces, in view of Proposition 1.14, the weak operator topology on $\mathcal{B}(\mathcal{H})_r$. $\square$

Our next goal is to prove a result of J. von Neumann, describing the closure of unital self-adjoint subalgebras of $\mathcal{B}(\mathcal{H})$ in the weak and strong operator topologies in purely algebraic terms. To that end, we consider for any subset $X \subseteq \mathcal{B}(\mathcal{H})$ the commutant

$$X' = \{a \in \mathcal{B}(\mathcal{H}) : ax = xa \ \text{for all } x \in X\} \ .$$

The bicommutant X'' of X is the commutant of X'. It is clear that $X \subseteq X''$.

Lemma 1.16 *For any $X \subseteq \mathcal{B}(\mathcal{H})$ the commutant X' is WOT-closed.*

Proof. For any operator $x \in \mathcal{B}(\mathcal{H})$ we consider the linear endomorphisms L_x and R_x of $\mathcal{B}(\mathcal{H})$, which are given by left and right multiplication with x respectively. Then, $X' = \bigcap_{x\in X} \ker(L_x - R_x)$ and hence the result follows from Remark 1.13(ii). $\square$

If n is a positive integer and $X \subseteq \mathcal{B}(\mathcal{H})$ a set of operators, we consider the set $X \cdot I_n = \{xI_n : x \in X\} \subseteq \mathbf{M}_n(\mathcal{B}(\mathcal{H})) \simeq \mathcal{B}(\mathcal{H}^n)$. Then, the following two properties are easily verified (cf. Exercise 1.3.5):

(i) The commutant $(X \cdot I_n)'$ of $X \cdot I_n$ in $\mathbf{M}_n(\mathcal{B}(\mathcal{H})) \simeq \mathcal{B}(\mathcal{H}^n)$ is the set $\mathbf{M}_n(X')$ of matrices with entries in the commutant X' of X in $\mathcal{B}(\mathcal{H})$.

(ii) The bicommutant $(X \cdot I_n)''$ of $X \cdot I_n$ in $\mathbf{M}_n(\mathcal{B}(\mathcal{H})) \simeq \mathcal{B}(\mathcal{H}^n)$ is the set $X'' \cdot I_n$, where X'' is the bicommutant of X in $\mathcal{B}(\mathcal{H})$.

Lemma 1.17 *Let $\mathcal{A}$ be a self-adjoint subalgebra of $\mathcal{B}(\mathcal{H})$ and $V \subseteq \mathcal{H}$ a closed $\mathcal{A}$-invariant subspace. Then:*

(i) The orthogonal complement $V^\perp$ is $\mathcal{A}$-invariant.

(ii) If p is the orthogonal projection onto V, then $p \in \mathcal{A}'$.

(iii) The subspace V is $\mathcal{A}''$-invariant.

Proof. (i) Let $\xi \in V^\perp$ and $a \in \mathcal{A}$. Then, for any vector $\eta \in V$ we have $a^*(\eta) \in \mathcal{A}V \subseteq V$ and hence $< a(\xi), \eta > = < \xi, a^*(\eta) > = 0$. Therefore, it follows that $a(\xi) \in V^\perp$.

(ii) We fix an operator $a \in \mathcal{A}$ and note that the subspaces V and $V^\perp$ are a-invariant, in view of our assumption and (i) above. It follows easily from this that the operators ap and pa coincide on both V and $V^\perp$. Hence, $ap = pa$.

(iii) Let $\xi \in V$, $a'' \in \mathcal{A}''$ and consider the orthogonal projection p onto V. In view of (ii) above, we have $a''p = pa''$ and hence $a''(\xi) = a''p(\xi) = pa''(\xi) \in V$, as needed.   □

We are now ready to state and prove von Neumann's theorem.

Theorem 1.18 (*von Neumann bicommutant theorem*) *Let $\mathcal{A} \subseteq \mathcal{B}(\mathcal{H})$ be a self-adjoint subalgebra containing the identity operator. Then, $\overline{\mathcal{A}}^{SOT} = \overline{\mathcal{A}}^{WOT} = \mathcal{A}''$, where we denote by $\overline{\mathcal{A}}^{SOT}$ (resp. $\overline{\mathcal{A}}^{WOT}$) the SOT-closure (resp. WOT-closure) of $\mathcal{A}$ in $\mathcal{B}(\mathcal{H})$.*

Proof. It is clear that $\overline{\mathcal{A}}^{SOT} \subseteq \overline{\mathcal{A}}^{WOT}$. Since $\mathcal{A} \subseteq \mathcal{A}''$, it follows from Lemma 1.16 that $\overline{\mathcal{A}}^{WOT} \subseteq \mathcal{A}''$. Hence, it only remains to show that $\mathcal{A}'' \subseteq \overline{\mathcal{A}}^{SOT}$. In order to verify this, we consider an operator $a'' \in \mathcal{A}''$, a positive real number ϵ, a positive integer n and vectors $\xi_1, \ldots, \xi_n \in \mathcal{H}$. We have to show that the SOT-neighborhood

$$\mathcal{N}_{\epsilon, \xi_1, \ldots, \xi_n}(a'') = \{a \in \mathcal{B}(\mathcal{H}) : \|(a - a'')\xi_i\| < \epsilon \text{ for all } i = 1, \ldots, n\}$$

of a'' intersects $\mathcal{A}$ non-trivially. To that end, we consider the self-adjoint subalgebra $\mathcal{A} \cdot I_n \subseteq \mathbf{M}_n(\mathcal{B}(\mathcal{H}))$ acting on the Hilbert space $\mathcal{H}^n$ by left multiplication and the closed subspace

$$V = \overline{\{(a(\xi_1), \ldots, a(\xi_n)) : a \in \mathcal{A}\}} \subseteq \mathcal{H}^n .$$

It is clear that V is $\mathcal{A} \cdot I_n$-invariant. Invoking Lemma 1.17(iii) and the discussion preceding it, we conclude that the subspace V is left invariant under the action of the operator $a''I_n \in \mathbf{M}_n(\mathcal{B}(\mathcal{H}))$. Since $1 \in \mathcal{A}$, we have $(\xi_1, \ldots, \xi_n) \in V$ and hence $(a''(\xi_1), \ldots, a''(\xi_n)) \in V$. Therefore, there is an operator $a \in \mathcal{A}$, such that

$$\|(a''(\xi_1), \ldots, a''(\xi_n)) - (a(\xi_1), \ldots, a(\xi_n))\| < \epsilon .$$

Then, $\|a''(\xi_i) - a(\xi_i)\| < \epsilon$ for all $i = 1, \ldots, n$ and hence $a \in \mathcal{N}_{\epsilon, \xi_1, \ldots, \xi_n}(a'')$, as needed.   □

A von Neumann algebra of operators acting on $\mathcal{H}$ is a self-adjoint subalgebra $\mathcal{N} \subseteq \mathcal{B}(\mathcal{H})$, which is WOT-closed and contains the identity 1. Equivalently, in view of von Neumann's bicommutant theorem, a von Neumann algebra $\mathcal{N}$ is a self-adjoint subalgebra of $\mathcal{B}(\mathcal{H})$, such that $\mathcal{N} = \mathcal{N}''$. It is clear that any von

Neumann algebra $\mathcal{N}$ as above is closed under the operator norm topology of $\mathcal{B}(\mathcal{H})$; in particular, $\mathcal{N}$ is a unital C^*-algebra.

For any positive integer n we identify $\mathbf{M}_n(\mathcal{B}(\mathcal{H}))$ with the algebra of bounded linear operators on $\mathcal{H}^n$. Then, a net $(A_\lambda)_\lambda$ in $\mathbf{M}_n(\mathcal{B}(\mathcal{H}))$ with $A_\lambda = (a_{ij,\lambda})_{i,j}$ converges to the zero matrix in the weak operator topology of $\mathbf{M}_n(\mathcal{B}(\mathcal{H}))$ if and only if the nets $(a_{ij,\lambda})_\lambda$ converge to zero in the weak operator topology of $\mathcal{B}(\mathcal{H})$ for all i,j (cf. Exercise 1.3.2). In particular, if $\mathcal{N}$ is a von Neumann algebra of operators acting on $\mathcal{H}$, then $\mathbf{M}_n(\mathcal{N})$ is a von Neumann algebra of operators acting on $\mathcal{H}^n$. An alternative proof of that assertion is provided in Exercise 1.3.5.

We have noted that for any C^*-algebra $\mathcal{A} \subseteq \mathcal{B}(\mathcal{H})$ and any operator $a \in \mathcal{A}$ the absolute value $s = |a|$ is contained in $\mathcal{A}$ as well. Nevertheless, if $a = vs$ is the polar decomposition of a (cf. Proposition 1.11), then the partial isometry v is not always contained in $\mathcal{A}$ (cf. Exercise 1.3.6). We now prove that $v \in \mathcal{A}$ if $\mathcal{A}$ is a von Neumann algebra.

Proposition 1.19 *Let $\mathcal{N}$ be a von Neumann algebra of operators acting on the Hilbert space $\mathcal{H}$. We consider an operator $a \in \mathcal{N}$, its absolute value $s = |a|$ and the polar decomposition $a = vs$. Then, $v \in \mathcal{N}$.*

Proof. Since $\mathcal{N} = \mathcal{N}''$, it suffices to show that v commutes with any element of the commutant $\mathcal{N}'$. To that end, consider an operator $b \in \mathcal{N}'$. Then, b commutes with both a and s (since $s \in \mathcal{N}$). Therefore, $bvs = ba = ab = vsb = vbs$ and hence the operators bv and vb agree on the closed subspace $\overline{\operatorname{im} s} = \overline{\operatorname{im} a^*}$ (cf. Lemma 1.10). In order to show that these operators are equal, it suffices to show that they also agree on the orthogonal complement $\left(\overline{\operatorname{im} a^*}\right)^\perp = \ker a$. Since $ab = ba$, the subspace $\ker a$ is easily seen to be b-invariant and hence both operators bv and vb vanish therein. $\qquad\square$

Let $\mathcal{A} \subseteq \mathcal{B}(\mathcal{H})$ be a self-adjoint algebra of operators containing 1 and $\mathcal{N}$ its WOT-closure. Then, any operator $a \in \mathcal{N}$ can be approximated (in the weak operator topology) by a net $(a_\lambda)_\lambda$ of operators from $\mathcal{A}$. The following result implies that the net $(a_\lambda)_\lambda$ can be chosen to be bounded.

Theorem 1.20 *(Kaplansky density theorem; cf. [36, Theorem 5.3.5]) Let $\mathcal{A}$ be a self-adjoint subalgebra of $\mathcal{B}(\mathcal{H})$ containing 1 and $\mathcal{N}$ its WOT-closure. Then, for any positive real number r the r-ball $\mathcal{A}_r = \mathcal{A} \cap \mathcal{B}(\mathcal{H})_r$ of $\mathcal{A}$ is WOT-dense in the r-ball $\mathcal{N}_r = \mathcal{N} \cap \mathcal{B}(\mathcal{H})_r$ of $\mathcal{N}$.* $\qquad\square$

III. OPERATOR ALGEBRAS ASSOCIATED WITH A DISCRETE GROUP. In this book, we are primarily interested in those operator algebras that are associated with (discrete) groups. Given such a group G, we consider the Hilbert space $\ell^2 G$ of square summable complex-valued functions on G with canonical orthonormal basis $(\delta_g)_{g \in G}$. In other words, $\ell^2 G$ consists of vectors of the form $\sum_{g \in G} r_g \delta_g$, where the r_g's are complex numbers such that $\sum_{g \in G} |r_g|^2 < \infty$. The inner product of two vectors $\xi = \sum_{g \in G} r_g \delta_g$ and $\xi' = \sum_{g \in G} r'_g \delta_g$ is given by

$$<\xi, \xi'> = \sum_{g \in G} r_g \overline{r'_g} \, .$$

For any element $g \in G$ we consider the linear endomorphism L_g of $\ell^2 G$, which is defined by letting

$$L_g \left(\sum_{x \in G} r_x \delta_x \right) = \sum_{x \in G} r_x \delta_{gx}$$

for any vector $\sum_{x \in G} r_x \delta_x \in \ell^2 G$. It is easily seen that $L_1 = 1$ and $L_{gh} = L_g L_h$ for all $g, h \in G$. Moreover, L_g is an isometry and hence $L_g^* = L_g^{-1} = L_{g^{-1}}$ for all $g \in G$. We consider the $\mathbf{C}$-linear map

$$L : \mathbf{C}G \longrightarrow \mathcal{B}(\ell^2 G) \, ,$$

which extends the map $g \mapsto L_g$, $g \in G$. For any element $a \in \mathbf{C}G$ we denote its image in $\mathcal{B}(\ell^2 G)$ by L_a.

Lemma 1.21 *Let G be a group and $L : \mathbf{C}G \longrightarrow \mathcal{B}(\ell^2 G)$ the linear map defined above. Then:*
(i) L is an injective algebra homomorphism.
(ii) The subalgebra $L(\mathbf{C}G) \subseteq \mathcal{B}(\ell^2 G)$ is self-adjoint.

Proof. (i) It is clear that L is an algebra homomorphism. For any $a = \sum_{g \in G} a_g g \in \mathbf{C}G$, where $a_g \in \mathbf{C}$ for all $g \in G$, we have $L_a = \sum_{g \in G} a_g L_g$ and hence $L_a(\delta_1) = \sum_{g \in G} a_g \delta_g \in \ell^2 G$. It follows readily from this that L is injective.

(ii) Let $a = \sum_{g \in G} a_g g \in \mathbf{C}G$, where $a_g \in \mathbf{C}$ for all $g \in G$, and consider the associated operator $L_a = \sum_{g \in G} a_g L_g \in L(\mathbf{C}G)$. Since $L_g^* = L_{g^{-1}}$ for all $g \in G$, it follows that $L_a^* = \sum_{g \in G} \overline{a_g} L_{g^{-1}} \in L(\mathbf{C}G)$. $\square$

We define the reduced C^*-algebra $C_r^* G$ of G to be the closure in the operator norm topology of $L(\mathbf{C}G)$ in $\mathcal{B}(\ell^2 G)$; then, $C_r^* G$ is a unital C^*-algebra. We also define the group von Neumann algebra $\mathcal{N}G$ as the WOT-closure of $L(\mathbf{C}G)$ in $\mathcal{B}(\ell^2 G)$. Since $\mathcal{N}G$ is closed under the operator norm topology, it contains $C_r^* G$; hence, there are inclusions $L(\mathbf{C}G) \subseteq C_r^* G \subseteq \mathcal{N}G \subseteq \mathcal{B}(\ell^2 G)$.

Remark 1.22 Assume that the group G is finite of order n. Then, the Hilbert space $\ell^2 G$ is identified with $\mathbf{C}^n$ and hence $\mathcal{B}(\ell^2 G) \simeq \mathbf{M}_n(\mathbf{C})$. Moreover, all three topologies defined above on $\mathcal{B}(\ell^2 G)$ (i.e. operator norm topology, SOT and WOT) coincide with the standard Cartesian product topology on $\mathbf{M}_n(\mathbf{C}) \simeq \mathbf{C}^{n^2}$. Since any linear subspace is closed therein, it follows that $L(\mathbf{C}G) = C_r^* G = \mathcal{N}G$.

1.1.3 The K_0-group of a Ring

In this subsection, we define the K_0-group of a ring by means of finitely generated projective modules and idempotent matrices and establish the equivalence of the two approaches. We then extend the definition of K_0 to the case of non-unital rings.

I. FINITELY GENERATED PROJECTIVE MODULES AND K_0. We consider a (unital) ring R and recall that an R-module P is called projective if it satisfies anyone of the following equivalent conditions:

(i) Any epimorphism of R-modules $f : M \longrightarrow M'$ induces an epimorphism of abelian groups $f_* : \mathrm{Hom}_R(P, M) \longrightarrow \mathrm{Hom}_R(P, M')$.

(ii) Any R-module epimorphism $M \longrightarrow P$ splits.

(iii) P is a direct summand of a free R-module.

Moreover, P is a finitely generated projective R-module if and only if it is a direct summand of R^n, for some $n \in \mathbf{N}$. It follows that the direct sum $P \oplus Q$ of two finitely generated projective R-modules P and Q is finitely generated and projective as well. Let $\mathrm{Proj}(R)$ be the set of isomorphism classes of finitely generated projective R-modules. If P is a finitely generated projective R-module we denote by $[P]$ its class in $\mathrm{Proj}(R)$. The operation $([P], [Q]) \mapsto [P \oplus Q]$, $[P], [Q] \in \mathrm{Proj}(R)$, endows $\mathrm{Proj}(R)$ with the structure of an abelian monoid with zero element the class of the zero module.

Remarks 1.23 (i) Let $\varphi : R \longrightarrow S$ be a ring homomorphism. Then, the map $[P] \mapsto [S \otimes_R P]$, $[P] \in \mathrm{Proj}(R)$, defines a morphism of abelian monoids

$$\mathrm{Proj}(\varphi) : \mathrm{Proj}(R) \longrightarrow \mathrm{Proj}(S) .$$

In this way, $R \mapsto \mathrm{Proj}(R)$ becomes a functor from the category of rings to that of abelian monoids.

(ii) Assume that R is a commutative ring. In this case, the tensor product of two finitely generated projective R-modules is also a finitely generated projective R-module. Therefore, the abelian monoid $\mathrm{Proj}(R)$ can be endowed with a multiplicative structure, by letting $[P] \cdot [Q] = [P \otimes_R Q]$ for any two finitely generated projective R-modules P, Q. This multiplication law is distributive over addition and has a unit element $[R]$; we shall refer to such a structure as a semiring. This semiring is commutative, i.e. $[P] \cdot [Q] = [Q] \cdot [P]$ for any two finitely generated projective R-modules P, Q. Furthermore, if S is another commutative ring and $\varphi : R \longrightarrow S$ a ring homomorphism, then the map $\mathrm{Proj}(\varphi)$ defined in (i) above is a morphism of semirings (i.e. besides being additive, it is also multiplicative and unit-preserving). In this way, $R \mapsto \mathrm{Proj}(R)$ becomes a functor from the category of commutative rings to that of commutative semirings.

(iii) For any ring R there is a unique morphism of abelian monoids $\iota_R : \mathbf{N} \longrightarrow \mathrm{Proj}(R)$ with $1 \mapsto [R]$. If $\varphi : R \longrightarrow S$ is a ring homomorphism, then $\iota_S = \mathrm{Proj}(\varphi) \circ \iota_R$. In the special case where the ring R is commutative, ι_R is a morphism of semirings (for the semiring structure on $\mathrm{Proj}(R)$ defined in (ii) above).

We now define the Grothendieck group of an abelian semigroup. This construction generalizes the passage from the (additive) semigroup $\mathbf{N}$ of natural numbers to the group $\mathbf{Z}$ of integers.

Proposition 1.24 *Let $(S, +)$ be an abelian semigroup. Then, there exists a unique (up to isomorphism) abelian group $(G(S), +)$ and a morphism $\mu : S \longrightarrow G(S)$ of abelian semigroups, such that for any abelian group G and any morphism of abelian semigroups $\lambda : S \longrightarrow G$ there exists a unique abelian group homomorphism $\Lambda : G(S) \longrightarrow G$ with $\Lambda \circ \mu = \lambda$. The abelian group $G(S)$ is called the Grothendieck group of S.*

Proof. We define an equivalence relation $\sim$ on the Cartesian product $S \times S$, by letting $(s, t) \sim (s', t')$ if and only if there exists an element $s_0 \in S$, such that $s + t' + s_0 = s' + t + s_0 \in S$. Then, the quotient $G(S) = S \times S/\sim$ becomes an abelian group by defining $\overline{(s, t)} + \overline{(s', t')} = \overline{(s + s', t + t')}$ for any two elements $\overline{(s, t)}, \overline{(s', t')} \in G(S)$. Here, the identity element is the class of (s, s) for any $s \in S$, whereas the opposite of the class of (s, t) is the class of (t, s). Moreover, the map

$$\mu : S \longrightarrow G(S) \, ,$$

which is defined by letting $\mu(s) = \overline{(s + s, s)}$ for all $s \in S$, is additive and has the universal property in the statement. Indeed, let G be an abelian group and $\lambda : S \longrightarrow G$ a morphism of abelian semigroups. Then, the map $\Lambda : G(S) \longrightarrow G$, which is given by $\Lambda\left(\overline{(s, t)}\right) = \lambda(s) - \lambda(t)$ for all $s, t \in S$, is well-defined; in fact, Λ is the unique group homomorphism with $\Lambda \circ \mu = \lambda$. The uniqueness (up to isomorphism) of the pair $(G(S), \mu)$ is an immediate consequence of the universal property. $\square$

Remarks 1.25 Let $(S, +)$ be an abelian semigroup and $(G(S), \mu)$ the pair defined in Proposition 1.24.

(i) Any element of the group $G(S)$ can be expressed as the difference between two elements of $\mu(S) \subseteq G(S)$. In particular, $\mu(S)$ generates $G(S)$.

(ii) Assume that S is an abelian monoid with zero element 0. Then, $\mu(0)$ is the zero element of $G(S)$.

(iii) For any $s, s' \in S$ we have $\mu(s) = \mu(s') \in G(S)$ if and only if there exists $s'' \in S$ with $s + s'' = s' + s'' \in S$. In particular, the morphism μ is injective if and only if the semigroup S has the cancellation property.

(iv) Assume that S is a semiring; by this, we mean that S has an associative multiplication law, which is distributive over addition and possesses a unit element 1. Then, there is an induced multiplicative law on $G(S)$, which is defined by letting $\overline{(s, t)} \cdot \overline{(s', t')} = \overline{(ss' + tt', st' + s't)}$ for any two elements $\overline{(s, t)}, \overline{(s', t')} \in G(S)$. The triple $(G(S), +, \cdot)$ is a ring with unit $\mu(1)$, whereas μ is a homomorphism of semirings. The ring $G(S)$ is commutative if this is the case for the semiring S.

(v) Let S' be another abelian semigroup and $\varphi : S \longrightarrow S'$ a morphism of semigroups. Then, there exists a unique abelian group homomorphism $G(\varphi) :$

$G(S) \longrightarrow G(S')$ with $G(\varphi) \circ \mu = \mu' \circ \varphi$ (here, μ' is the morphism corresponding to S'). In this way, $S \mapsto G(S)$ becomes a functor from the category of abelian semigroups to that of abelian groups. If both S and S' are semirings and φ is, in addition, multiplicative and unit-preserving, then $G(\varphi)$ is a homomorphism of rings (for the ring structures defined in (iv) above).

Lemma 1.26 *Let G be an abelian group and $S \subseteq G$ a semigroup, which generates G (as a group). Then, the Grothendieck group $G(S)$ is isomorphic with G.*

Proof. It is easily seen that the pair (G, i), where $i : S \longrightarrow G$ is the inclusion map, has the universal property of Proposition 1.24. □

If R is a ring then the K-theory group $K_0(R)$ is the Grothendieck group associated with the abelian semigroup $\mathrm{Proj}(R)$. For any finitely generated projective R-module P we denote its class in $K_0(R)$ by $[P]$ as well. The zero element of $K_0(R)$ is the class $[0]$ of the zero module. A typical element of $K_0(R)$ is a difference $[P] - [P']$, for suitable finitely generated projective R-modules P, P'; in fact, P' can be chosen to be of the form R^n for some $n \in \mathbf{N}$. If P, P' are two finitely generated projective R-modules, then $[P] = [P'] \in K_0(R)$ if and only if there exists a finitely generated projective R-module P'' such that the R-modules $P \oplus P''$ and $P' \oplus P''$ are isomorphic. In this case too, we may assume that $P'' = R^n$ for some $n \in \mathbf{N}$. Being a composition of functors, the map $R \mapsto K_0(R)$ is a functor from the category of rings to that of abelian groups. In particular, a ring homomorphism $\varphi : R \longrightarrow S$ induces a group homomorphism $K_0(\varphi) : K_0(R) \longrightarrow K_0(S)$.

Remarks 1.27 (i) Assume that the ring R is commutative. Then, $K_0(R)$ is a commutative ring as well, with multiplication given by letting $[P] \cdot [Q] = [P \otimes_R Q]$ for any two finitely generated projective R-modules P, Q. The unit element is the class of the free module R. If S is another commutative ring and $\varphi : R \longrightarrow S$ a ring homomorphism, then the induced additive map $K_0(\varphi) : K_0(R) \longrightarrow K_0(S)$ is a ring homomorphism as well. In this way, $R \mapsto K_0(R)$ is a functor from the category of commutative rings to itself.

(ii) For any ring R there is a group homomorphism $\iota_R : \mathbf{Z} \longrightarrow K_0(R)$, which maps 1 onto $[R]$. The reduced K-theory group $\widetilde{K}_0(R)$ is defined as the cokernel of ι_R, i.e. we let $\widetilde{K}_0(R) = \mathrm{coker}\, \iota_R$. If $\varphi : R \longrightarrow S$ is a ring homomorphism, then $\iota_S = K_0(\varphi) \circ \iota_R$. If the ring R is commutative then ι_R is a homomorphism of rings.

If R is a ring and $e \in R$ an idempotent, then the left ideal Re is a finitely generated projective R-module. Indeed, if $e' = 1 - e$ is the complementary idempotent of e, then $Re \oplus Re' = R$. Moreover, if R is commutative, then $Re \otimes_R Re \simeq Re$ and hence the class $[Re]$ of Re is an idempotent of the commutative ring $K_0(R)$ (cf. Remark 1.27(i) above).

Proposition 1.28 *Let R be a commutative ring and consider the map*

$$\sigma = \sigma_R : Idem(R) \longrightarrow Idem(K_0(R)) ,$$

which is defined by letting $\sigma(e) = [Re]$ for any idempotent $e \in R$. Then:

(i) σ is a morphism of Boolean algebras (cf. Example 1.3(i)).

(ii) If S is another commutative ring and $\varphi : R \longrightarrow S$ a ring homomorphism, then the following diagram is commutative

$$\begin{array}{ccc} Idem(R) & \xrightarrow{\sigma_R} & Idem(K_0(R)) \\ {\scriptstyle Idem(\varphi)} \downarrow & & \downarrow {\scriptstyle Idem(K_0(\varphi))} \\ Idem(S) & \xrightarrow{\sigma_S} & Idem(K_0(S)) \end{array}$$

Proof. (i) It is clear that σ maps 0, 1 onto [0] and [R] respectively. In order to show that σ preserves $\wedge$, we note that for any two idempotents $e, f \in R$ there is an isomorphism $Re \otimes_R Rf \simeq Ref$, which shows that

$$\sigma(e \wedge f) = \sigma(ef) = [Ref] = [Re] \cdot [Rf] = [Re] \wedge [Rf] = \sigma(e) \wedge \sigma(f) .$$

Finally, for any idempotent $e \in R$ with complement $e' = 1 - e$ we have $Re \oplus Re' = R$ and hence $\sigma(e') = [Re'] = [R] - [Re] = [Re]'$, i.e. σ preserves complements.

(ii) For any idempotent $e \in R$ we have $S \otimes_R Re \simeq S\varphi(e)$ and hence $K_0(\varphi)[Re] = [S\varphi(e)]$, as needed. $\qquad\square$

II. K_0 AND IDEMPOTENT MATRICES. If R is a ring then another (equivalent) approach to the definition of $\mathrm{Proj}(R)$, and hence to that of $K_0(R)$, is via idempotent matrices with entries in R. The embedding $A \mapsto \begin{pmatrix} A & 0 \\ 0 & 0 \end{pmatrix}$ of $\mathbf{M}_n(R)$ into $\mathbf{M}_{n+1}(R)$, for $n \geq 1$, defines the inductive system $(\mathbf{M}_n(R))_n$; let $\mathbf{M}(R)$ be the corresponding limit. In a similar way, the group $GL_n(R)$ embeds in $GL_{n+1}(R)$ by the map $G \mapsto \begin{pmatrix} G & 0 \\ 0 & 1 \end{pmatrix}$. The limit of the corresponding inductive system $(GL_n(R))_n$ is the group $GL(R)$. The conjugation actions of the groups $GL_n(R)$ on $\mathbf{M}_n(R)$ for all $n \geq 1$ induce an action of $GL(R)$ on $\mathbf{M}(R)$; by an obvious abuse of language, we refer to the latter action as action by conjugation. We note that the set of idempotent matrices $\mathrm{Idem}(\mathbf{M}(R)) = \varinjlim_n \mathrm{Idem}(\mathbf{M}_n(R))$ is $GL(R)$-invariant.

For all n we identify the ring $\mathbf{M}_n(R)$ with the opposite of the endomorphism ring of the (left) R-module R^n and regard any matrix $A \in \mathbf{M}_n(R)$ as the R-linear map $\widetilde{A} : R^n \longrightarrow R^n$, which is given by right multiplication with A on the row vectors of R^n. For any idempotent matrix $E \in \mathbf{M}_n(R)$ we have $\mathrm{im}\,\widetilde{E} \oplus \mathrm{im}\,\widetilde{E'} = R^n$, where $E' = I_n - E$; in particular, the R-module $P(E) = \mathrm{im}\,\widetilde{E}$ is finitely generated projective. We note that the isomorphism class of $P(E)$ depends only upon the class $[E]$ of the idempotent matrix E in the limit $\mathrm{Idem}(\mathbf{M}(R))$. In this way, we obtain a map

$$\theta : \mathrm{Idem}(\mathbf{M}(R)) \longrightarrow \mathrm{Proj}(R) , \qquad (1.1)$$

which is given by letting $\theta[E] = [P(E)]$ for any $[E] \in \mathrm{Idem}(\mathbf{M}(R))$. Since any finitely generated projective R-module is isomorphic with a direct summand of R^n for some n, it follows that θ is surjective.

Proposition 1.29 *Let R be a ring and E, E' two idempotent matrices with entries in R. Then, the finitely generated projective R-modules $P(E)$ and $P(E')$ defined above are isomorphic if and only if the classes $[E], [E'] \in \mathrm{Idem}(R)$ are conjugate under the action of $GL(R)$.*

Proof. Suppose that the classes $[E], [E']$ are conjugate under the action of $GL(R)$. Then, we may assume that $E, E' \in \mathbf{M}_n(R)$ for some $n \gg 0$, whereas there exists $G \in GL_n(R)$, such that $E' = GEG^{-1}$. It follows that $GE = E'G \in \mathbf{M}_n(R)$ and hence $\widetilde{E}\widetilde{G} = \widetilde{G}\widetilde{E'} \in \mathrm{End}_R R^n$. It follows readily from this that $\widetilde{G}$ restricts to an isomorphism between the images $P(E')$ and $P(E)$ of $\widetilde{E'}$ and $\widetilde{E}$ respectively.

Now assume that $E, E' \in \mathrm{Idem}(\mathbf{M}_n(R))$ are such that the R-modules $P(E)$ and $P(E')$ are isomorphic. Let $a : P(E) \longrightarrow P(E')$ be an R-linear isomorphism and $b : P(E') \longrightarrow P(E)$ its inverse. We define $\alpha \in \mathrm{End}_R R^n$ as the composition

$$R^n = P(E) \oplus P(I_n - E) \to P(E) \overset{a}{\to} P(E') \to P(E') \oplus P(I_n - E') = R^n ,$$

where the first (resp. the last) unlabelled arrow is the natural projection (resp. the natural inclusion). Then, there exists a matrix $A \in \mathbf{M}_n(R)$, such that $\alpha = \widetilde{A}$. In the same way, using b, we define $\beta \in \mathrm{End}_R R^n$ and let $B \in \mathbf{M}_n(R)$ be the matrix with $\beta = \widetilde{B}$. Since $\alpha\beta$ is the identity on $P(E')$, we have $E'BA = E'$. On the other hand, β vanishes on $P(I_n - E')$ and hence $(I_n - E')B = 0$. It follows that

$$E'B = B \quad \text{and} \quad BA = E' .$$

In the same way, we conclude that

$$EA = A \quad \text{and} \quad AB = E .$$

The equalities above imply that

$$BE = B(AB) = (BA)B = E'B = B$$

and

$$AE' = A(BA) = (AB)A = EA = A .$$

We now define the matrices $U, V \in \mathbf{M}_{2n}(R)$, by letting

$$U = \begin{pmatrix} A & I_n - E \\ I_n - E' & B \end{pmatrix} \quad \text{and} \quad V = \begin{pmatrix} B & I_n - E' \\ I_n - E & A \end{pmatrix} .$$

Using the equalities established above, it is easily verified that $UV = VU = I_{2n}$, whereas $V \begin{pmatrix} E & 0 \\ 0 & 0 \end{pmatrix} U = \begin{pmatrix} E' & 0 \\ 0 & 0 \end{pmatrix}$. Therefore, the classes of the idempotent matrices E and E' in $\mathbf{M}(R)$ are conjugate under the action of $GL(R)$, as needed. $\qquad\square$

We now consider the quotient set $V(R) = \mathrm{Idem}(\mathbf{M}(R))/GL(R)$ of $GL(R)$-orbits in $\mathrm{Idem}(\mathbf{M}(R))$. For any idempotent matrix $E \in \mathbf{M}_n(R)$ we denote by ρ_E the $GL(R)$-orbit of $[E] \in \mathrm{Idem}(\mathbf{M}(R))$. Then, Proposition 1.29 implies that the map θ of (1.1) induces, by passage to the quotient, a bijective map

$$\overline{\theta} : V(R) \longrightarrow \mathrm{Proj}(R) . \tag{1.2}$$

For all $n, n' \in \mathbf{N}$ we consider the map

$$\mathrm{Idem}(\mathbf{M}_n(R)) \times \mathrm{Idem}(\mathbf{M}_{n'}(R)) \longrightarrow V(R) ,$$

which maps a pair (E, E') onto the $GL(R)$-orbit $\rho_{E \oplus E'}$ of the idempotent matrix

$$E \oplus E' = \begin{pmatrix} E & 0 \\ 0 & E' \end{pmatrix} \in \mathbf{M}_{n+n'}(R) .$$

It is easily seen that $\rho_{E \oplus E'} = \rho_{E' \oplus E}$ for any idempotent matrices E, E', whereas the maps defined above induce, by passage to the direct limit, a well-defined map

$$\mathrm{Idem}(\mathbf{M}(R)) \times \mathrm{Idem}(\mathbf{M}(R)) \longrightarrow V(R) ,$$

which maps any pair $([E], [E'])$ onto the $GL(R)$-orbit $\rho_{E \oplus E'}$. Moreover, by passage to the quotients, we obtain a well-defined map

$$V(R) \times V(R) \longrightarrow V(R) ,$$

which maps any pair $(\rho_E, \rho_{E'})$ onto $\rho_{E \oplus E'}$. For any idempotent matrices E, E' as above, we denote, by an obvious abuse of notation, the element $\rho_{E \oplus E'}$ by $\rho_E \oplus \rho_{E'}$. In this way, the pair $(V(R), \oplus)$ is an abelian monoid with zero element ρ_0.

Proposition 1.30 *The map $\overline{\theta} : V(R) \longrightarrow Proj(R)$ of (1.2) is an isomorphism of abelian monoids.*

Proof. We already know that $\overline{\theta}$ is bijective, whereas $\overline{\theta}(\rho_0) = \theta[0] = [P(0)] = [0]$. In order to show that $\overline{\theta}$ is additive, we consider two idempotent matrices E, E' with entries in R and compute

$$\begin{aligned}
\overline{\theta}(\rho_E \oplus \rho_{E'}) &= \theta([E \oplus E']) \\
&= [P(E \oplus E')] \\
&= [P(E) \oplus P(E')] \\
&= [P(E)] + [P(E')] \\
&= \theta[E] + \theta[E'] \\
&= \overline{\theta}(\rho_E) + \overline{\theta}(\rho_{E'}) .
\end{aligned}$$

In the above chain of equalities, the third one follows since $P(E \oplus E') = \mathrm{im}\,(E \oplus E')\widetilde{\ } = \mathrm{im}\,\widetilde{E} \oplus \mathrm{im}\,\widetilde{E'} = P(E) \oplus P(E')$. $\hspace{1cm}\square$

Corollary 1.31 *Let R be a ring. Then, the map $\overline{\theta} : V(R) \longrightarrow Proj(R)$ of (1.2) induces a canonical identification between the Grothendieck group of the abelian monoid $V(R)$ and the abelian group $K_0(R)$.* $\hspace{1cm}\square$

Remark 1.32 Let R be a ring and V an abelian group. We assume that for any positive integer n we are given a map $\lambda_n : \mathrm{Idem}(\mathbf{M}_n(R)) \longrightarrow V$, in such a way that the following conditions are satisfied:

(i) For any idempotent matrix $E \in \mathbf{M}_n(R)$ we have $\lambda_n(E) = \lambda_{n+1}(E')$, where $E' = \begin{pmatrix} E & 0 \\ 0 & 0 \end{pmatrix} \in \mathbf{M}_{n+1}(R)$.

(ii) The λ_n's are invariant under conjugation, i.e. for any idempotent matrix $E \in \mathbf{M}_n(R)$ and any $G \in GL_n(R)$ we have $\lambda_n(E) = \lambda_n(GEG^{-1})$.

(iii) For any idempotent matrices $E_1 \in \mathbf{M}_{n_1}(R)$ and $E_2 \in \mathbf{M}_{n_2}(R)$ we have $\lambda_{n_1+n_2}(E_1 \oplus E_2) = \lambda_{n_1}(E_1) + \lambda_{n_2}(E_2)$.

Then, the λ_n's induce (in view of (i)) a well-defined map from the limit $\mathrm{Idem}(\mathbf{M}(R))$ to V, which is invariant under the action of $GL(R)$ (in view of (ii)). Since the resulting map $\lambda_* : V(A) \longrightarrow V$ is additive (in view of (iii)), the universal property of the Grothendieck group and Corollary 1.31 imply the existence of a group homomorphism $\lambda_* : K_0(R) \longrightarrow V$, which maps the class of any idempotent matrix $E \in \mathbf{M}_n(R)$ onto $\lambda_n(E)$.

Taking into account the description of the K_0-group of a ring in terms of idempotent matrices, the following density result should not be very surprising. Its proof uses the functional calculus for elements of Banach algebras.

Theorem 1.33 *(Karoubi density theorem; cf. [39, II.6.15]) We consider two unital Banach algebras $\mathcal{A}, \mathcal{B}$ and let $\iota : \mathcal{A} \longrightarrow \mathcal{B}$ be a continuous and injective homomorphism with dense image. We assume that for any positive integer n a matrix $(a_{ij})_{i,j} \in \mathbf{M}_n(\mathcal{A})$ is invertible (in $\mathbf{M}_n(\mathcal{A})$) if and only if the matrix $(\iota(a_{ij}))_{i,j} \in \mathbf{M}_n(\mathcal{B})$ is invertible (in $\mathbf{M}_n(\mathcal{B})$). Then, the induced map $K_0(\iota) : K_0(\mathcal{A}) \longrightarrow K_0(\mathcal{B})$ is an isomorphism.* $\hspace{1cm}\square$

III. NON-UNITAL RINGS AND K_0. We now extend the definition of the K_0-group to non-unital rings. To that end, we consider a non-unital ring I and let I^+ be the associated unital ring. Here, $I^+ = I \oplus \mathbf{Z}$ as an abelian group, whereas the product of any two elements $(x, n), (y, m) \in I^+$ is equal to $(xy + ny + mx, nm) \in I^+$. We consider the split extension

$$0 \longrightarrow I \longrightarrow I^+ \stackrel{\pi}{\longrightarrow} \mathbf{Z} \longrightarrow 0 \,,$$

where π is the projection $(x, n) \mapsto n$, $(x, n) \in I^+$, and define $K_0(I)$ as the kernel of the induced additive map $K_0(\pi) : K_0(I^+) \longrightarrow K_0(\mathbf{Z})$.

Remarks 1.34 (i) Any morphism $\varphi : I \longrightarrow J$ of non-unital rings extends uniquely to a ring homomorphism $\varphi^+ : I^+ \longrightarrow J^+$. Moreover, the additive map $K_0(\varphi^+) : K_0(I^+) \longrightarrow K_0(J^+)$ restricts to an additive map, denoted by $K_0(\varphi)$, from $K_0(I)$ to $K_0(J)$. In this way, $I \mapsto K_0(I)$ becomes a functor from the category of non-unital rings to that of abelian groups.

(ii) If I is a unital ring then I^+ decomposes into the direct product $I \times \mathbf{Z}$ and the definition of the group $K_0(I)$ given above coincides with that given previously for unital rings (cf. Exercise 1.3.7(ii)). Moreover, if $\varphi : I \longrightarrow J$ is a ring homomorphism then the additive map $K_0(\varphi)$ defined in (i) above coincides with that defined earlier.

Let I be an ideal in a unital ring R and define the double $D(R, I)$ of R along I by letting
$$D(R, I) = \{(x, y) \in R \times R : y - x \in I\} \,.$$
The ring $D(R, I)$ comes equipped with the two coordinate projections π_1 and π_2 onto R. We define the relative group $K_0(R, I)$ as the kernel of the additive map $K_0(\pi_1) : K_0(D(R, I)) \longrightarrow K_0(R)$. This group is identified with the K_0-group of I by the following basic result.

Theorem 1.35 (*excision in* K_0; *cf.* [58, §1.5]) *Let R be a unital ring, $I \subseteq R$ an ideal and $D(R, I)$ the double of R along I. We also consider the ring homomorphism $\varrho : I^+ \longrightarrow D(R, I)$, which is defined by $(x, n) \mapsto (n \cdot 1, x + n \cdot 1)$, $(x, n) \in I^+$. Then, the map $K_0(\varrho) : K_0(I^+) \longrightarrow K_0(D(R, I))$ restricts to an isomorphism $\rho : K_0(I) \longrightarrow K_0(R, I)$.* $\qquad\qquad\square$

1.1.4 Traces and the K_0-group

In order to study the K_0-group of a ring R, one has to examine the additive maps from $K_0(R)$ to various abelian groups.[2] In this subsection, we show that one method of constructing such additive maps is via traces. We study several properties of the additive map on $K_0(R)$ induced by a trace on R, in the unital as well as the non-unital case, and construct the universal trace defined by A. Hattori and J. Stallings.

I. TRACES AND THE HATTORI-STALLINGS RANK. Let R be a ring and V an abelian group. Then, a trace on R with values in V is an additive map $\tau : R \longrightarrow V$, such that $\tau(rr') = \tau(r'r)$ for all $r, r' \in R$. The set of traces on R with values in V is easily seen to be a subgroup of the group of all additive maps from R to V. If V' is another abelian group and $f : V \longrightarrow V'$ an additive map, then for any trace $\tau : R \longrightarrow V$ the composition $f \circ \tau$ is a trace on R with values in V'. On the other hand, if R' is another ring and $\varphi : R' \longrightarrow R$ a ring homomorphism, then for any trace $\tau : R \longrightarrow V$ the composition $\tau \circ \varphi$ is a V-valued trace on R'.

We now consider the subgroup $[R, R] \subseteq R$, which is generated by the commutators $rr' - r'r$, $r, r' \in R$, and let $T(R) = R/[R, R]$ be the corresponding

---

[2] In general, any abelian group A is determined by the functor $\operatorname{Hom}(A, \_)$ from the category of abelian groups to itself.

quotient. We note that any ring homomorphism $\varphi : R \longrightarrow S$ maps $[R, R]$ into $[S, S]$; hence, φ induces by passage to the quotients an additive map $T(\varphi) : T(R) \longrightarrow T(S)$. It is clear that any trace $\tau : R \longrightarrow V$ as above vanishes on $[R, R]$ and hence induces an additive map $\overline{\tau} : T(R) \longrightarrow V$. Conversely, any additive map from $T(R)$ to an abelian group can be pulled back to a trace on R. In this way, the group of traces on R with values in V is identified with the group $\mathrm{Hom}(T(R), V)$ of all homomorphisms from $T(R)$ to V. In particular, the identity of $T(R)$ corresponds to the $T(R)$-valued trace on R, which is given by the projection map $p_R : R \longrightarrow T(R)$. We may rephrase the identification of the group of V-valued traces on R with the group $\mathrm{Hom}(T(R), V)$ as a universal property of p_R, as follows:

Lemma 1.36 *Let R be a ring and $p_R : R \longrightarrow T(R)$ the trace defined above.*

(i) Consider an abelian group V and a trace $\tau : R \longrightarrow V$. Then, there is a unique group homomorphism $\overline{\tau} : T(R) \longrightarrow V$, such that $\tau = \overline{\tau} \circ p_R$.

(ii) If S is another ring and $\varphi : R \longrightarrow S$ a ring homomorphism, then the following diagram is commutative

$$
\begin{array}{ccc}
R & \xrightarrow{\; p_R \;} & T(R) \\
\varphi \downarrow & & \downarrow T(\varphi) \\
S & \xrightarrow{\; p_S \;} & T(S)
\end{array}
$$

Here, $p_S : S \longrightarrow T(S)$ is the universal trace defined on S.

Proof. Assertion (i) was established in the preceding discussion, whereas the proof of (ii) is an immediate consequence of the definitions.   □

Our next goal is to relate the traces defined on a ring R to traces defined on the matrix rings $\mathbf{M}_n(R)$, $n \geq 1$. To that end, we fix n and consider the additive map

$$\iota : R \longrightarrow \mathbf{M}_n(R) \,,$$

which maps any element $x \in R$ onto the matrix xE_{11}; here, we denote by E_{ij}, $i, j = 1, \ldots, n$, the matrix units of $\mathbf{M}_n(R)$. Since $E_{11} \in \mathbf{M}_n(R)$ is an idempotent, it follows that ι maps $[R, R]$ into the commutator subgroup $[\mathbf{M}_n(R), \mathbf{M}_n(R)]$ of $\mathbf{M}_n(R)$. Therefore, there is an induced additive map

$$\overline{\iota} : T(R) \longrightarrow T(\mathbf{M}_n(R)) \,.$$

On the other hand, the ordinary trace of matrices

$$\mathrm{tr} : \mathbf{M}_n(R) \longrightarrow R$$

is additive and maps $[\mathbf{M}_n(R), \mathbf{M}_n(R)]$ into $[R, R]$. Indeed, if $A = (a_{ij})_{i,j}$ and $B = (b_{ij})_{i,j}$ are two $n \times n$ matrices with entries in R, then an easy computation shows that $\mathrm{tr}(AB - BA) = \sum_{i,j}(a_{ij}b_{ji} - b_{ji}a_{ij}) \in [R, R]$. It follows that there is an induced additive map

$$\overline{\mathrm{tr}} : T(\mathbf{M}_n(R)) \longrightarrow T(R) \,.$$

Lemma 1.37 *Let R be a ring, n a positive integer and $\bar{\imath}$, $\overline{\mathrm{tr}}$ the additive maps defined above.*

(i) The maps $\bar{\imath}$ and $\overline{\mathrm{tr}}$ are inverses of each other; in particular, they are both bijective.

(ii) If S is another ring and $\varphi : R \longrightarrow S$ a ring homomorphism, then the following diagrams are commutative

$$
\begin{array}{ccc}
T(R) & \xrightarrow{\;\bar{\imath}\;} & T(\mathbf{M}_n(R)) \\
{\scriptstyle T(\varphi)}\downarrow & & \downarrow{\scriptstyle T(\varphi_n)} \\
T(S) & \xrightarrow{\;\bar{\imath}\;} & T(\mathbf{M}_n(S))
\end{array}
\qquad
\begin{array}{ccc}
T(\mathbf{M}_n(R)) & \xrightarrow{\;\overline{\mathrm{tr}}\;} & T(R) \\
{\scriptstyle T(\varphi_n)}\downarrow & & \downarrow{\scriptstyle T(\varphi)} \\
T(\mathbf{M}_n(S)) & \xrightarrow{\;\overline{\mathrm{tr}}\;} & T(S)
\end{array}
$$

Here, φ_n denotes the ring homomorphism induced by φ between the corresponding matrix rings, whereas we have used the same symbols ($\bar{\imath}$ and $\overline{\mathrm{tr}}$) to denote the maps defined above corresponding to the ring S.

Proof. (i) Since $\mathrm{tr}(xE_{11}) = x$ for all $x \in R$, it follows that $\mathrm{tr} \circ \iota$ is the identity on R and hence $\overline{\mathrm{tr}} \circ \bar{\imath}$ is the identity on $T(R)$. The abelian group $\mathbf{M}_n(R)$ is generated by the matrices of the form xE_{ij}, where $x \in R$ and $i, j = 1, \ldots, n$; therefore, the abelian group $T(\mathbf{M}_n(R))$ is generated by the classes of these matrices modulo commutators. Hence, in order to show that the composition $\bar{\imath} \circ \overline{\mathrm{tr}}$ is the identity map on $T(\mathbf{M}_n(R))$, it suffices to prove that

$$(\iota \circ \mathrm{tr})(xE_{ij}) - xE_{ij} \in [\mathbf{M}_n(R), \mathbf{M}_n(R)]$$

for all $x \in R$ and all $i, j = 1, \ldots, n$. If $i = j$, this follows since

$$
\begin{aligned}
(\iota \circ \mathrm{tr})(xE_{ii}) - xE_{ii} &= \iota(x) - xE_{ii} \\
&= xE_{11} - xE_{ii} \\
&= xE_{1i} \cdot E_{i1} - E_{i1} \cdot xE_{1i} \,.
\end{aligned}
$$

In the case where $i \neq j$, we have $\mathrm{tr}(xE_{ij}) = 0$ and hence

$$(\iota \circ \mathrm{tr})(xE_{ij}) - xE_{ij} = -xE_{ij} = E_{ij} \cdot xE_{ii} - xE_{ii} \cdot E_{ij} \,.$$

(ii) The commutativity of both diagrams is an immediate consequence of the definitions. □

We now consider the commutative diagram

$$
\begin{array}{ccc}
\mathbf{M}_n(R) & \xrightarrow{\;\mathrm{tr}\;} & R \\
{\scriptstyle p}\downarrow & & \downarrow{\scriptstyle p} \\
T(\mathbf{M}_n(R)) & \xrightarrow{\;\overline{\mathrm{tr}}\;} & T(R)
\end{array}
$$

where we denote by p the universal traces defined on R and $\mathbf{M}_n(R)$. The additive map $r^R = p \circ \mathrm{tr} = \overline{\mathrm{tr}} \circ p$ is a trace on $\mathbf{M}_n(R)$ with values in $T(R)$; in fact, if we identify the abelian groups $T(\mathbf{M}_n(R))$ and $T(R)$ by means of $\overline{\mathrm{tr}}$, then r^R is identified with the universal trace on $\mathbf{M}_n(R)$. The trace r^R will be also denoted by r^R_{HS} (or simply by r_{HS}) and referred to as the Hattori-Stallings trace.

Lemma 1.38 *Let $\varphi : R \longrightarrow S$ be a ring homomorphism. Then, the following diagram is commutative for any positive integer n*

$$\begin{array}{ccc} \mathbf{M}_n(R) & \xrightarrow{r_{HS}^R} & T(R) \\ \varphi_n \downarrow & & \downarrow T(\varphi) \\ \mathbf{M}_n(S) & \xrightarrow{r_{HS}^S} & T(S) \end{array}$$

Here, we denote by φ_n the homomorphism induced by φ between the corresponding matrix rings.

Proof. This is an immediate consequence of the definition of the Hattori-Stallings trace, in view of Lemmas 1.36(ii) and 1.37(ii). □

Proposition 1.39 *Let R be a ring, V an abelian group and $\tau : R \longrightarrow V$ a trace.*
  (i) For any positive integer n the map

$$\tau_n : \mathbf{M}_n(R) \longrightarrow V ,$$

which is defined by letting $\tau_n(A) = \sum_i \tau(a_{ii})$ for any matrix $A = (a_{ij})_{i,j} \in \mathbf{M}_n(R)$, is a V-valued trace on $\mathbf{M}_n(R)$.
  (ii) There is an additive map $\tau_* : K_0(R) \longrightarrow V$, which maps the class of any idempotent matrix $E \in \mathbf{M}_n(R)$ onto $\tau_n(E)$.

Proof. (i) We note that τ_n is the composition

$$\mathbf{M}_n(R) \xrightarrow{r_{HS}} T(R) \xrightarrow{\overline{\tau}} V ,$$

where r_{HS} is the Hattori-Stallings trace and $\overline{\tau}$ the additive map induced by τ. It follows readily from this that τ_n is indeed a trace.
  (ii) The existence of τ_* follows invoking Remark 1.32, since the restrictions $\tau_n' : \mathrm{Idem}(\mathbf{M}_n(R)) \longrightarrow V$ of τ_n, $n \geq 1$, define a sequence of maps satisfying conditions (i), (ii) and (iii) therein. □

In particular, the universal trace $p : R \longrightarrow T(R)$ induces for any positive integer n the Hattori-Stallings trace $r_{HS} = p_n : \mathbf{M}_n(R) \longrightarrow T(R)$. Moreover, there is a group homomorphism

$$p_* : K_0(R) \longrightarrow T(R) ,$$

which maps the K-theory class of any idempotent matrix $E \in \mathbf{M}_n(R)$ onto the Hattori-Stallings trace $r_{HS}(E) = p_n(E)$, i.e. onto the residue class of the ordinary trace $\mathrm{tr}(E) \in R$ in the quotient group $T(R) = R/[R, R]$. By an obvious abuse of notation, we denote p_* by r_{HS} (or r_{HS}^R if the dependence upon the ring R is to be emphasized) and refer to it as the Hattori-Stalling rank map. If P is a finitely generated projective R-module and $E \in \mathbf{M}_n(R)$ an idempotent matrix with $[P] = \bar{\theta}(\rho_E)$ (cf. Proposition 1.30), then the Hattori-Stallings rank $r_{HS}(P)$ of P is defined to be the Hattori-Stallings trace (rank) $r_{HS}(E)$ of E. This definition is independent of the choice of E and depends only upon the isomorphism class of P.

Proposition 1.40 *(i) Let R be a ring and $f : V \longrightarrow V'$ a homomorphism of abelian groups. We consider a trace $\tau : R \longrightarrow V$ and the V'-valued trace $f \circ \tau$ on R. Then, the induced additive map $(f \circ \tau)_* : K_0(R) \longrightarrow V'$ is the composition*

$$K_0(R) \xrightarrow{\tau_*} V \xrightarrow{f} V' ,$$

where $\tau_* : K_0(R) \longrightarrow V$ is the additive map induced by the trace τ.

(ii) Let $\varphi : R \longrightarrow S$ be a ring homomorphism and V an abelian group. We consider a trace $\tau : S \longrightarrow V$ and the V-valued trace $\tau \circ \varphi$ on R. Then, the induced additive map $(\tau \circ \varphi)_* : K_0(R) \longrightarrow V$ is the composition

$$K_0(R) \xrightarrow{K_0(\varphi)} K_0(S) \xrightarrow{\tau_*} V ,$$

where $\tau_* : K_0(S) \longrightarrow V$ is the additive map induced by the trace τ.

Proof. (i) Let $E = (e_{ij})_{i,j} \in \mathbf{M}_n(R)$ be an idempotent matrix and compute

$$\begin{aligned}
(f \circ \tau_*)[E] &= f(\tau_*[E]) \\
&= f\left(\sum_i \tau(e_{ii})\right) \\
&= \sum_i f(\tau(e_{ii})) \\
&= \sum_i (f \circ \tau)(e_{ii}) \\
&= (f \circ \tau)_*[E] .
\end{aligned}$$

Since the abelian group $K_0(R)$ is generated by the $[E]$'s (Remark 1.25(i)), it follows that $f \circ \tau_* = (f \circ \tau)_*$.

(ii) Let $E = (e_{ij})_{i,j} \in \mathbf{M}_n(R)$ be an idempotent matrix and $E' = (\varphi(e_{ij}))_{i,j}$ the corresponding idempotent matrix with entries in S. Then,

$$\begin{aligned}
(\tau_* \circ K_0(\varphi))[E] &= \tau_*(K_0(\varphi)[E]) \\
&= \tau_*[E'] \\
&= \sum_i \tau(\varphi(e_{ii})) \\
&= \sum_i (\tau \circ \varphi)(e_{ii}) \\
&= (\tau \circ \varphi)_*[E]
\end{aligned}$$

and hence we conclude that $\tau_* \circ K_0(\varphi) = (\tau \circ \varphi)_*$. $\qquad\square$

Corollary 1.41 *Let $\varphi : R \longrightarrow S$ be a ring homomorphism. Then, the following diagram is commutative*

$$K_0(R) \xrightarrow{r_{HS}^R} T(R)$$
$$K_0(\varphi) \downarrow \qquad \downarrow T(\varphi)$$
$$K_0(S) \xrightarrow{r_{HS}^S} T(S)$$

Proof. This is an immediate consequence of Proposition 1.40, in view of Lemma 1.36(ii). □

Remark 1.42 Let $\mathcal{H}$ be a Hilbert space, $\mathcal{A}$ a self-adjoint subalgebra of the algebra $\mathcal{B}(\mathcal{H})$ of all bounded operators on $\mathcal{H}$ and $\tau : \mathcal{A} \longrightarrow \mathbf{C}$ a trace. We fix a positive integer n and consider the induced $\mathbf{C}$-valued trace τ_n on $\mathbf{M}_n(\mathcal{A})$ (cf. Proposition 1.39(i)). The matrix algebra $\mathbf{M}_n(\mathcal{A}) \subseteq \mathbf{M}_n(\mathcal{B}(\mathcal{H}))$ is viewed as a self-adjoint subalgebra of $\mathcal{B}(\mathcal{H}^n)$.

(i) We endow $\mathcal{A} \subseteq \mathcal{B}(\mathcal{H})$ with the operator norm topology (resp. weak operator topology) and assume that τ is continuous. Then, the trace τ_n is continuous as well, where $\mathbf{M}_n(\mathcal{A}) \subseteq \mathcal{B}(\mathcal{H}^n)$ is endowed with the corresponding operator norm topology (resp. weak operator topology).

(ii) Assume that the trace τ is positive; this means that $\tau(a^*a)$ is a non-negative real number for all $a \in \mathcal{A}$. Then, τ_n is positive as well. Indeed, let us consider a matrix $A = (a_{ij})_{i,j} \in \mathbf{M}_n(\mathcal{A})$ and the adjoint matrix $A^* = (b_{ij})_{i,j}$, where $b_{ij} = a_{ji}^*$ for all i, j. Then, the j-th entry along the diagonal of the product $A^* A$ is equal to $\sum_i b_{ji} a_{ij}$. Hence,

$$\tau_n(A^*A) = \sum_j \tau \Big(\sum_i b_{ji} a_{ij} \Big)$$
$$= \sum_j \tau \Big(\sum_i a_{ij}^* a_{ij} \Big)$$
$$= \sum_{i,j} \tau(a_{ij}^* a_{ij})$$

In view of the positivity of τ, $\tau_n(A^*A)$ is a sum of non-negative real numbers and hence $\tau_n(A^*A) \geq 0$, as needed.

(iii) Assume that the trace τ is positive and faithful; faithfulness means that the complex number $\tau(a^*a)$ vanishes for some element $a \in \mathcal{A}$ only if $a = 0$. In that case, the positive trace τ_n is faithful as well. Indeed, if $A = (a_{ij})_{i,j} \in \mathbf{M}_n(\mathcal{A})$ is a matrix with $\tau_n(A^*A) = 0$, then the computation in (ii) above shows that $\tau(a_{ij}^* a_{ij}) = 0$ and hence $a_{ij} = 0$ for all i, j.

II. THE NON-UNITAL CASE. We now consider a non-unital ring I and let τ be a trace on I with values in an abelian group V; by this, we mean that τ is an additive map which vanishes on the commutators $xy - yx$, $x, y \in I$. Then, τ extends uniquely to an additive map τ^+ on the associated unital ring $I^+ = I \oplus \mathbf{Z}$ that satisfies $\tau^+(0, 1) = 0$; in fact, τ^+ is a trace. We define the map $\tau_* : K_0(I) \longrightarrow V$ as the restriction of $(\tau^+)_* : K_0(I^+) \longrightarrow V$ to the subgroup $K_0(I) \subseteq K_0(I^+)$. If the ring I is unital, the additive map τ_* just defined is easily seen to coincide with that defined in Proposition 1.39(ii) (cf. Exercise 1.3.7(iii)).

Example 1.43 Let $\mathcal{H}$ be a Hilbert space, $\mathcal{L}^1(\mathcal{H})$ the ideal of trace-class operators and $\mathrm{Tr} : \mathcal{L}^1(\mathcal{H}) \longrightarrow \mathbf{C}$ the trace defined in Theorem 1.12(iv). Then, the induced additive map $\mathrm{Tr}_* : K_0(\mathcal{L}^1(\mathcal{H})) \longrightarrow \mathbf{C}$ is injective with image the subgroup $\mathbf{Z} \subseteq \mathbf{C}$ (cf. [58, Lemma 6.3.14]).

The following result will be needed in Chap. 5.

Proposition 1.44 *Let $\varphi, \psi : \mathcal{A} \longrightarrow \mathcal{B}$ be two homomorphisms of non-unital rings and $I \subseteq \mathcal{B}$ an ideal, such that $\psi(a) - \varphi(a) \in I$ for all $a \in \mathcal{A}$. We consider an abelian group V and an additive map $\tau : I \longrightarrow V$ that vanishes on elements of the form $xy - yx$ for all $x \in I$ and $y \in \mathcal{B}$; in particular, τ is a trace on I. Let $\tau' : \mathcal{A} \longrightarrow V$ be the additive map, which is defined by letting $\tau'(a) = \tau(\psi(a) - \varphi(a))$ for all $a \in \mathcal{A}$. Then:*

(i) The map τ' is a trace on $\mathcal{A}$.

(ii) The image of the additive map $\tau'_* : K_0(\mathcal{A}) \longrightarrow V$ is contained in the image of the additive map $\tau_* : K_0(I) \longrightarrow V$.

Proof. (i) For any two elements $a, a' \in \mathcal{A}$ we compute

$$
\begin{aligned}
\tau'(aa') &= \tau[\psi(aa') - \varphi(aa')] \\
&= \tau[\psi(a)\psi(a') - \varphi(a)\varphi(a')] \\
&= \tau[\psi(a)(\psi(a') - \varphi(a'))] + \tau[(\psi(a) - \varphi(a))\varphi(a')] \\
&= \tau[(\psi(a') - \varphi(a'))\psi(a)] + \tau[\varphi(a')(\psi(a) - \varphi(a))] \\
&= \tau[\psi(a')\psi(a) - \varphi(a')\varphi(a)] \\
&= \tau[\psi(a'a) - \varphi(a'a)] \\
&= \tau'(a'a) \, .
\end{aligned}
$$

(ii) Without any loss of generality, we may assume that the rings $\mathcal{A}$ and $\mathcal{B}$ are unital and the homomorphisms φ and ψ unit preserving. Let

$$
\mathcal{D} = D(\mathcal{B}, I) = \{(x, x') \in \mathcal{B} \times \mathcal{B} : x' - x \in I\}
$$

be the double of $\mathcal{B}$ along I (cf. the discussion following Remarks 1.34) and consider the ring homomorphism $\theta : \mathcal{A} \longrightarrow \mathcal{D}$, which is defined by letting $\theta(a) = (\varphi(a), \psi(a))$ for all $a \in \mathcal{A}$. The additive map $t : \mathcal{D} \longrightarrow V$, which is given by $t(x, x') = \tau(x' - x)$ for all $(x, x') \in \mathcal{D}$, is easily seen to be a trace.[3] Since the trace τ' is the composition

$$
\mathcal{A} \xrightarrow{\theta} \mathcal{D} \xrightarrow{t} V \, ,
$$

Proposition 1.40(ii) implies that $\tau'_* : K_0(\mathcal{A}) \longrightarrow V$ is the composition

$$
K_0(\mathcal{A}) \xrightarrow{K_0(\theta)} K_0(\mathcal{D}) \xrightarrow{t_*} V \, .
$$

Therefore, it suffices to show that the image of $t_* : K_0(\mathcal{D}) \longrightarrow V$ is contained in the image of $\tau_* : K_0(I) \longrightarrow V$.

---

[3] In fact, this assertion is really the special case of (i) above, where $\mathcal{A} = \mathcal{D}$ and φ, ψ are the two coordinate projection maps to $\mathcal{B}$.

The first coordinate projection map $\pi_1 : \mathcal{D} \longrightarrow \mathcal{B}$ splits by the diagonal map $\Delta : \mathcal{B} \longrightarrow \mathcal{D}$ and hence the additive map $K_0(\pi_1) : K_0(\mathcal{D}) \longrightarrow K_0(\mathcal{B})$ splits by the additive map $K_0(\Delta) : K_0(\mathcal{B}) \longrightarrow K_0(\mathcal{D})$. It follows that $K_0(\mathcal{D}) = K_0(\mathcal{B}, I) \oplus \operatorname{im} K_0(\Delta)$, where $K_0(\mathcal{B}, I) = \ker K_0(\pi_1)$ is the relative K_0-group. We claim that the additive map t_* vanishes on the subgroup $\operatorname{im} K_0(\Delta) \subseteq K_0(\mathcal{D})$. In order to verify this, we note that the composition

$$ K_0(\mathcal{B}) \xrightarrow{K_0(\Delta)} K_0(\mathcal{D}) \xrightarrow{t_*} V $$

is the additive map induced by the V-valued trace $t \circ \Delta$ on $\mathcal{B}$ (loc.cit.). The claim is proved, since $t \circ \Delta$ is the zero map. In particular, it follows that the image of $t_* : K_0(\mathcal{D}) \longrightarrow V$ is equal to the image of its restriction to the subgroup $K_0(\mathcal{B}, I) \subseteq K_0(\mathcal{D})$. We now consider the commutative diagram

$$
\begin{array}{ccc}
K_0(I) & \xrightarrow{\rho} & K_0(\mathcal{B}, I) \\
\downarrow & & \downarrow \\
K_0(I^+) & \xrightarrow{K_0(\varrho)} K_0(\mathcal{D}) \xrightarrow{t_*} V
\end{array}
$$

Here, I^+ is the unital ring associated with I, $\varrho : I^+ \longrightarrow \mathcal{D}$ the ring homomorphism which is defined by letting $\varrho(x, n) = (n \cdot 1, x + n \cdot 1)$ for all $(x, n) \in I^+$ and ρ the excision isomorphism of Theorem 1.35. It follows that the subgroup $t_*(K_0(\mathcal{B}, I)) \subseteq V$ is the image of the composition

$$ K_0(I) \hookrightarrow K_0(I^+) \xrightarrow{K_0(\varrho)} K_0(\mathcal{D}) \xrightarrow{t_*} V . \tag{1.3} $$

On the other hand, Proposition 1.40(ii) implies that $t_* \circ K_0(\varrho)$ is the additive map which is induced by the V-valued trace $t \circ \varrho$ on I^+. Since $t \circ \varrho$ is easily seen to coincide with the trace τ^+ on I^+, which is associated with the trace τ on I, the composition (1.3) is precisely the additive map $\tau_* : K_0(I) \longrightarrow V$. Therefore, we conclude that $t_*(K_0(\mathcal{B}, I)) = \tau_*(K_0(I))$. $\square$

1.2 The Idempotent Conjectures

In this section, we focus our attention to the group algebra kG of a group G with coefficients in a commutative ring k. We give an explicit description of the Hattori-Stallings rank on the K_0-group of kG and state Bass' conjecture in §1.2.1. In the following subsection, we examine the relation between torsion elements in G and idempotent elements in $\mathbf{C}G$ and state the idempotent conjecture for torsion-free groups. Finally, in §1.2.3, we obtain some interesting classes of groups that satisfy the idempotent conjecture, by using only elementary considerations.

1.2.1 The Hattori-Stallings Rank on $K_0(kG)$

Let k be a commutative ring, G a group and kG the associated group algebra. Then, the subgroup $[kG, kG] \subseteq kG$ is the k-linear span of the set $\{gh - hg :$

$g, h \in G\}$. In particular, $[kG, kG]$ is a k-submodule of kG and hence the quotient group $T(kG) = kG/[kG, kG]$ has the structure of a k-module. In order to give an explicit description of that k-module, we consider the set $\mathcal{C}(G)$ of conjugacy classes of elements of G and denote by $[g]$ the conjugacy class of any $g \in G$. The natural projection $G \longrightarrow \mathcal{C}(G)$ extends uniquely to a k-linear map between the respective free k-modules

$$p : kG \longrightarrow \bigoplus_{[g] \in \mathcal{C}(G)} k \cdot [g] \, .$$

In other words, we define $p(\sum_g a_g g) = \sum_g a_g [g] = \sum_{[g]} \alpha_{[g]} [g]$, where $\alpha_{[g]} = \sum_{x \in [g]} a_x$ for all $[g] \in \mathcal{C}(G)$. It is clear that p is surjective.

Lemma 1.45 *Let k be a commutative ring, G a group and p the k-linear map defined above. Then, $\ker p = [kG, kG]$.*

Proof. For all $g, h \in G$ the elements gh and hg are conjugate and hence $p(gh - hg) = [gh] - [hg] = 0$. It follows that $[kG, kG] \subseteq \ker p$. Conversely, assume that $a = \sum_g a_g g \in \ker p$ and fix a conjugacy class $[g] \in \mathcal{C}(G)$. Then, $\sum_{x \in [g]} a_x = 0$ and hence $\sum_{x \in [g]} a_x x = \sum_{x \in [g]} a_x (x - g) \in kG$. Since $x - g \in [kG, kG]$ for all $x \in [g]$, it follows that $\sum_{x \in [g]} a_x x \in [kG, kG]$. This is the case for any $[g] \in \mathcal{C}(G)$ and hence $a = \sum_{[g]} \left(\sum_{x \in [g]} a_x x \right) \in [kG, kG]$, as needed. $\square$

In view of Lemma 1.45, the map p induces an isomorphism $\bar{p}$ between the k-modules $T(kG) = kG/[kG, kG]$ and $\bigoplus_{[g] \in \mathcal{C}(G)} k \cdot [g]$. In the sequel, we identify these k-modules by means of $\bar{p}$ and write simply $T(kG) = \bigoplus_{[g] \in \mathcal{C}(G)} k \cdot [g]$. Then, the universal trace $kG \longrightarrow T(kG)$ (cf. Lemma 1.36) is identified with the map p defined above. For any conjugacy class $[g]$ we consider the coordinate projection

$$\pi_{[g]} : T(kG) = \bigoplus_{[x] \in \mathcal{C}(G)} k \cdot [x] \longrightarrow k \cdot [g] \simeq k$$

and write any element $r \in T(kG)$ as a sum $\sum_{[g]} \pi_{[g]}(r)[g]$. We now define for any $g \in G$ the additive map

$$r_g : K_0(kG) \longrightarrow k$$

as the composition

$$K_0(kG) \xrightarrow{r_{HS}} T(kG) \xrightarrow{\pi_{[g]}} k \, ,$$

where r_{HS} is the Hattori-Stallings rank associated with kG. It is clear that $r_g = r_h$ if the elements $g, h \in G$ are conjugate. By an obvious abuse of notation, we also denote by r_g the composition

$$\mathbf{M}_n(kG) \xrightarrow{r_{HS}} T(kG) \xrightarrow{\pi_{[g]}} k$$

for all $n \geq 1$. With this notation, for any idempotent matrix $E \in \mathbf{M}_n(kG)$ the Hattori-Stallings rank $r_{HS}(E)$, i.e. the residue class of the trace $\mathrm{tr}(E)$ in the quotient $T(kG)$, can be expressed as the sum $\sum_{[g]} r_g(E)[g]$.

In the next two results, we describe the functorial behavior of the r_g's with respect to coefficient ring and group homomorphisms.

Proposition 1.46 *Let* $\varphi : k \longrightarrow k'$ *be a homomorphism of commutative rings,* G *a group and* $\widetilde{\varphi} : kG \longrightarrow k'G$ *the extension of* φ *with* $\widetilde{\varphi}(g) = g$ *for all* $g \in G$. *We consider a positive integer* n, *an idempotent matrix* $E = (e_{ij})_{i,j} \in \mathbf{M}_n(kG)$ *with Hattori-Stallings rank* $r_{HS}(E) = \sum_{[g]} r_g(E)[g]$ *and the induced idempotent matrix* $E' = (\widetilde{\varphi}(e_{ij}))_{i,j} \in \mathbf{M}_n(k'G)$. *Then,* $r_g(E') = \varphi(r_g(E))$ *for all* $g \in G$.

Proof. It suffices to verify that the following diagram is commutative for all $g \in G$

$$
\begin{array}{ccccc}
\mathbf{M}_n(kG) & \xrightarrow{r_{HS}} & T(kG) & \xrightarrow{\pi_{[g]}} & k \\
\widetilde{\varphi}_n \downarrow & & T(\widetilde{\varphi}) \downarrow & & \downarrow \varphi \\
\mathbf{M}_n(k'G) & \xrightarrow{r_{HS}} & T(k'G) & \xrightarrow{\pi_{[g]}} & k'
\end{array}
$$

Here, we denote by $\widetilde{\varphi}_n$ the map induced by $\widetilde{\varphi}$ between the corresponding matrix rings, whereas $T(\widetilde{\varphi})$ is obtained from $\widetilde{\varphi}$ by passage to the quotients. The commutativity of the right-hand square is an immediate consequence of the definitions and hence the result follows from Lemma 1.38. $\qquad \square$

Proposition 1.47 *Let* k *be a commutative ring,* $f : G \longrightarrow G'$ *a homomorphism of groups and* $\widetilde{f} : kG \longrightarrow kG'$ *the* k-*algebra homomorphism extending* f. *We consider a positive integer* n, *an idempotent matrix* $E = (e_{ij})_{i,j} \in \mathbf{M}_n(kG)$ *with Hattori-Stallings rank* $r_{HS}(E) = \sum_{[g]} r_g(E)[g]$ *and the induced idempotent matrix* $E' = \left(\widetilde{f}(e_{ij})\right)_{i,j} \in \mathbf{M}_n(kG')$. *Then,*

$$
r_{g'}(E') = \sum \{r_g(E) : [g] \in \mathcal{C}(G) \ and \ [f(g)] = [g'] \in \mathcal{C}(G')\}
$$

for all $g' \in G'$.

Proof. We fix an element $g' \in G'$ and consider the k-linear map

$$
\phi_{[g']} : T(kG) \longrightarrow k \ ,
$$

which maps any conjugacy class $[g] \in T(kG)$ onto 1 if $[f(g)] = [g']$ and 0 if $[f(g)] \neq [g']$. With this notation, it suffices to verify that the following diagram is commutative

$$
\begin{array}{ccccc}
\mathbf{M}_n(kG) & \xrightarrow{r_{HS}} & T(kG) & \xrightarrow{\phi_{[g']}} & k \\
\widetilde{f}_n \downarrow & & T(\widetilde{f}) \downarrow & & \| \\
\mathbf{M}_n(kG') & \xrightarrow{r_{HS}} & T(kG') & \xrightarrow{\pi_{[g']}} & k
\end{array}
$$

Here, we denote by $\widetilde{f}_n$ the map induced by $\widetilde{f}$ between the corresponding matrix rings, whereas $T\left(\widetilde{f}\right)$ is the map obtained from $\widetilde{f}$ by passage to the quotients. As in the proof of Proposition 1.46, the commutativity of the right-hand square is an immediate consequence of the definitions; hence, the result follows from Lemma 1.38.   □

Corollary 1.48 *Let k be a field, G a group and E an idempotent matrix with entries in the group algebra kG and Hattori-Stallings rank $r_{HS}(E) = \sum_{[g]} r_g(E)[g] \in T(kG)$. Then, there is a non-negative integer n, such that $\sum_{[g]} r_g(E) = n \cdot 1 \in k$.*

Proof. We assume that $E \in \mathbf{M}_t(kG)$ for some $t \geq 1$ and consider the augmentation $\varepsilon : kG \longrightarrow k$ and the induced idempotent matrix $E' = \varepsilon_t(E)$ with entries in k. Then, $r_{HS}(E') = n \cdot 1 \in k$ for some non-negative integer n (cf. Exercise 1.3.8). The result follows from Proposition 1.47 by letting G' be the trivial group therein.   □

In the sequel, we shall be interested in the Hattori-Stallings rank

$$r_{HS} : K_0(kG) \longrightarrow T(kG) ,$$

in the special case where the coefficient ring is a subring of the field $\mathbf{C}$ of complex numbers. A pair (k, G), where $k \subseteq \mathbf{C}$ is a subring and G a group, will be said to satisfy Bass' conjecture if the map

$$r_g : K_0(kG) \longrightarrow k$$

is identically zero for all elements $g \in G$ with $g \neq 1$. In that case, the Hattori-Stallings rank of an idempotent matrix E with entries in kG will be equal to $n[1] \in T(kG)$ for a suitable non-negative integer $n = n(E)$ (cf. Corollary 1.48). We note that a pair (k, G) as above satisfies Bass' conjecture if this is the case for all pairs of the form (k_0, G_0), where k_0 is a finitely generated subring of k and G_0 a finitely generated subgroup of G (cf. Exercise 1.3.9).

Remark 1.49 Let k be a subring of the field $\mathbf{C}$ of complex numbers and G a group having an element g of finite order $n > 1$. If n is invertible in k, then the map r_g above is not identically zero; in particular, the pair (k, G) does not satisfy Bass' conjecture. Indeed, it is easily seen that the element $e = \frac{1}{n} \sum_{i=0}^{n-1} g^i \in kG$ is an idempotent and $r_g(e) = \frac{\alpha}{n}$, where $\alpha \in \{1, \ldots, n\}$ is the number of elements of the cyclic group $<g>$ which are conjugate to g (i.e. α is the cardinality of the intersection $<g> \cap [g]$).

It follows from Remark 1.49 that certain arithmetic restrictions are necessary conditions for a pair (k, G) as above to satisfy Bass' conjecture. In particular, the orders of the non-identity torsion elements of G can't be invertible in k. In that direction, we note the following simple result.

Lemma 1.50 *Let k be a subring of the field $\mathbf{C}$ of complex numbers. Then, the following conditions are equivalent:*

(i) $k \cap \mathbf{Q} = \mathbf{Z}$.

(ii) No integer $n > 1$ is invertible in k, i.e. $\mathbf{Z} \cap U(k) = \{\pm 1\}$.

(iii) No prime number p is invertible in k.

Proof. (i)→(ii): If a non-zero integer n is invertible in k, then $\frac{1}{n} \in k \cap \mathbf{Q} = \mathbf{Z}$ and hence $n = \pm 1$.

(ii)→(iii): This is clear.

(iii)→(i): First of all, we note that we always have $\mathbf{Z} \subseteq k \cap \mathbf{Q}$. In order to show the reverse inclusion, assume that there exist two relatively prime integers a and b with $b \neq 0$, such that $\frac{a}{b} \in k \setminus \mathbf{Z}$. In that case, there is a prime number p which divides b but not a. Hence, there are $x, y, b' \in \mathbf{Z}$, such that $b = pb'$ and $ax + py = 1$. Then,

$$\frac{1}{p} = \frac{ax}{p} + y = \frac{axb'}{b} + y = xb' \cdot \frac{a}{b} + y \cdot 1 \in \mathbf{Z} \cdot \frac{a}{b} + \mathbf{Z} \cdot 1 \subseteq k \,,$$

a contradiction.   □

A group G will be said to satisfy Bass' conjecture if the pair (k, G) satisfies Bass' conjecture for any subring $k \subseteq \mathbf{C}$ with $k \cap \mathbf{Q} = \mathbf{Z}$. In other words, G satisfies Bass' conjecture if the map

$$r_g : K_0(kG) \longrightarrow k$$

is identically zero for any subring $k \subseteq \mathbf{C}$ with $k \cap \mathbf{Q} = \mathbf{Z}$ and any group element $g \in G$ with $g \neq 1$. We note that a group satisfies Bass' conjecture if this is the case for all finitely generated subgroups of it (cf. Exercise 1.3.9). In the following chapters, we study Bass' conjecture for certain classes of groups, obtaining thereby some insight about its geometric significance.

1.2.2 Idempotents in $\mathbf{C}G$

Let G be a group and $\mathbf{C}G$ the corresponding complex group algebra. Besides the trivial idempotents 0 and 1, the algebra $\mathbf{C}G$ may contain non-trivial idempotents as well. For example, if G is a group having non-trivial torsion elements, then $\mathbf{C}G$ has non-trivial idempotents, in view of Remark 1.49. In fact, there is a general method for constructing non-trivial idempotents in $\mathbf{C}G$, by considering finite subgroups of G. More precisely, if $H \leq G$ is a finite subgroup, then $\mathbf{C}H$ is a subalgebra of $\mathbf{C}G$ and hence $\mathrm{Idem}(\mathbf{C}H) \subseteq \mathrm{Idem}(\mathbf{C}G)$. The Wedderburn-Artin theory (cf. [41, Chap. 1]) implies, in view of Maschke's theorem (Theorem 1.9), that the group algebra $\mathbf{C}H$ can be identified with a direct product of complex matrix algebras. In this way, idempotent matrices with complex entries provide us with idempotent elements in $\mathbf{C}H$ and hence in $\mathbf{C}G$ as well. Furthermore, identifying a matrix algebra with entries in $\mathbf{C}H$

with the appropriate direct product of complex matrix algebras, we obtain examples of idempotent matrices with entries in $\mathbf{C}H$ as well. Of course, in order to obtain explicit formulae for these matrices, one has to explicit the Wedderburn decomposition of $\mathbf{C}H$.

This technique for constructing idempotents and idempotent matrices with entries in group algebras is illustrated in the following example (where the verification of many details is left as an exercise to the reader).

Example 1.51 Let $G = S_3$ be the group of permutations on three letters a, b and c. Then, there is an isomorphism of complex algebras

$$\mathbf{C}G \simeq \mathbf{C} \times \mathbf{C} \times \mathbf{M}_2(\mathbf{C}) \tag{1.4}$$

(cf. [41, p.129]). The first (resp. second) copy of $\mathbf{C}$ corresponds to the trivial (resp. the sign) representation of G on $\mathbf{C}$, whereas the 2×2 matrix algebra comes from the unique (up to isomorphism) irreducible representation of G on $\mathbf{C}^2$, which is given by

$$(ab) \mapsto \begin{pmatrix} 0 & 1 \\ 1 & 0 \end{pmatrix} \quad \text{and} \quad (abc) \mapsto \begin{pmatrix} 0 & -1 \\ 1 & -1 \end{pmatrix} .$$

The central idempotents of $\mathbf{C}G$ that correspond to the idempotents $(1, 0, 0)$, $(0, 1, 0)$ and $(0, 0, I_2)$ of the Wedderburn decomposition (1.4) are the elements

$$e_1 = \frac{1}{6}(1 + (ab) + (ac) + (bc) + (abc) + (acb)) ,$$

$$e_2 = \frac{1}{6}(1 - (ab) - (ac) - (bc) + (abc) + (acb))$$

and

$$e_3 = 1 - e_1 - e_2 = \frac{1}{3}(2 - (abc) - (acb))$$

respectively. Moreover, the element of $\mathbf{C}G$ that corresponds to the idempotent $(0, 0, E)$, where $E = \begin{pmatrix} 1 & 1 \\ 0 & 0 \end{pmatrix} \in \mathbf{M}_2(\mathbf{C})$, is the idempotent

$$e = \frac{1}{3}(1 + (ab) - 2(ac) + (bc) - 2(abc) + (acb)) \in \mathbf{C}G .$$

In the same way, the element of $\mathbf{C}G$ that corresponds to $(0, 0, A)$, where $A = \begin{pmatrix} 0 & 1 \\ 0 & 0 \end{pmatrix} \in \mathbf{M}_2(\mathbf{C})$, is the element

$$a = \frac{1}{3}((ab) - (ac) - (abc) + (acb)) \in \mathbf{C}G.$$

The Wedderburn decomposition (1.4) of $\mathbf{C}G$ induces an algebra isomorphism between $\mathbf{M}_2(\mathbf{C}G)$ and the direct product

$$\mathbf{M}_2(\mathbf{C}) \times \mathbf{M}_2(\mathbf{C}) \times \mathbf{M}_2(\mathbf{M}_2(\mathbf{C})) = \mathbf{M}_2(\mathbf{C}) \times \mathbf{M}_2(\mathbf{C}) \times \mathbf{M}_4(\mathbf{C}) .$$

In this way, the idempotent

$$\left(\begin{pmatrix} 2 & -1 \\ 2 & -1 \end{pmatrix}, \begin{pmatrix} 0 & -3 \\ 0 & 1 \end{pmatrix}, \begin{pmatrix} I_2 & A \\ 0 & E \end{pmatrix} \right) \in \mathbf{M}_2(\mathbf{C}) \times \mathbf{M}_2(\mathbf{C}) \times \mathbf{M}_2(\mathbf{M}_2(\mathbf{C})) ,$$

where the matrices $A, E \in \mathbf{M}_2(\mathbf{C})$ are defined as above, corresponds to the idempotent matrix

$$\begin{pmatrix} 2e_1 + e_3 & -e_1 - 3e_2 + a \\ 2e_1 & -e_1 + e_2 + e \end{pmatrix} \in \mathbf{M}_2(\mathbf{C}G) ,$$

with the elements $e_1, e_2, e_3, e, a \in \mathbf{C}G$ defined as above.

A natural question that one may ask is whether the existence of non-trivial idempotents in a complex group algebra is due solely to the existence of non-trivial torsion elements in the group. In particular, the idempotent conjecture for a torsion-free group G asserts that the complex group algebra $\mathbf{C}G$ has no non-trivial idempotents. Concerning the role of the field of complex numbers in that conjecture, we can state the following result (see also Exercises 1.3.13(iv) and 1.3.14).

Proposition 1.52 *The following conditions are equivalent for a group G:*
(i) The complex group algebra $\mathbf{C}G$ has no non-trivial idempotents.
(ii) For any field F of characteristic 0 the group algebra FG has no non-trivial idempotents.
(iii) For any commutative $\mathbf{Q}$-algebra k having no non-trivial idempotents, the group algebra kG has no non-trivial idempotents.

Proof. The implications (iii)→(ii)→(i) are clear. We complete the proof by showing that (i)→(ii) and (ii)→(iii).

(i)→(ii): Let F be a field of characteristic 0 and $e \in FG$ an idempotent. Since e involves only finitely many elements of F, there exists a finitely generated subfield F_0 of F, such that $e \in F_0G$. Being a finitely generated extension of $\mathbf{Q}$, the field F_0 admits an embedding into $\mathbf{C}$ and hence we are reduced to the case where e is an idempotent in $\mathbf{C}G$.

(ii)→(iii): Let k be a commutative $\mathbf{Q}$-algebra having no non-trivial idempotents and consider an idempotent $e \in kG$. In order to show that e is trivial, we can assume (by passing, if necessary, to a finitely generated $\mathbf{Q}$-subalgebra of k as above) that k is Noetherian; cf. the proof of Corollary A.32 of Appendix A. Then, the nil radical $\operatorname{nil} k$ of k is nilpotent and hence we can further assume (by passing, if necessary, to $k/\operatorname{nil} k$) that k is reduced; cf. Exercise 1.3.11. In that case, Exercise 1.3.12 implies that k is a subring of a finite direct product of fields of characteristic 0, say $k \subseteq F_1 \times \cdots \times F_n$. Then, $kG \subseteq F_1G \times \cdots \times F_nG$ and we may consider the coordinates $e_i \in F_iG$ of e. Since $e_i \in \operatorname{Idem}(F_iG)$, our assumption implies that $e_i \in F_i$ for all i. On the other hand, the intersection

$$kG \cap (F_1 \times \cdots \times F_n) \subseteq F_1 G \times \cdots \times F_n G$$

is equal to k and hence $e \in k$. It follows that e is trivial.   □

Remarks 1.53 (i) Let k be a subring of the field $\mathbf{C}$ of complex numbers, such that the group algebra kG has no non-trivial idempotents for all groups G. Then, $k \cap \mathbf{Q} = \mathbf{Z}$.[4] Indeed, if $k \cap \mathbf{Q} \neq \mathbf{Z}$ then there is a prime number p which is invertible in k (cf. Lemma 1.50). In that case, Remark 1.49 shows that the group algebra of the finite cyclic group of order p with coefficients in k has non-trivial idempotents.

(ii) Even though it is not completely apparent at this point, the idempotent conjecture turns out to be closely related to Bass' conjecture. In fact, let k be a subring of the field $\mathbf{C}$ of complex numbers and G a group. Then, there is a generalized form of Bass' conjecture on the Hattori-Stallings rank of idempotent $n \times n$ matrices with entries in kG, which reduces to

- Bass' conjecture if $k \cap \mathbf{Q} = \mathbf{Z}$ and
- the idempotent conjecture if $k = \mathbf{C}$, G is torsion-free and $n = 1$.

For more details on this, see Exercise 3.3.7.

1.2.3 Some First Examples of Groups that Satisfy the Idempotent Conjecture

In this final subsection, we use some elementary considerations in order to identify certain classes of torsion-free groups G, for which the complex group algebra $\mathbf{C}G$ has no non-trivial idempotents.

I. ORDERED GROUPS. An ordered group G is a group which is endowed with a total order, such that for any elements $x, y, z \in G$ we have

$$x < y \implies xz < yz \text{ and } zx < zy . \tag{1.5}$$

We note that any ordered group G is torsion-free. Indeed, if $x \in G$ is an element with $x > 1$, then a simple inductive argument shows that $x^n > 1$ for all $n \geq 1$; in particular, x has infinite order. We conclude that any element $g \in G$ with $g \neq 1$ has infinite order, since either g or g^{-1} is > 1.

Remarks 1.54 (i) Let G be an ordered group and consider a subgroup $H \subseteq G$. Then, the restriction of the order relation of G endows H with the structure of an ordered group.

(ii) Let $(G_i)_{i \in I}$ be a family of ordered groups and assume that the index set I is well-ordered. Then, the lexicographic order endows the direct product

---

[4] Conversely, we shall prove that the group algebra of any group with coefficients in a subring k of $\mathbf{C}$ has no non-trivial idempotents if $k \cap \mathbf{Q} = \mathbf{Z}$; cf. Corollary 3.21.

$G = \prod_i G_i$ with the structure of an ordered group. More precisely, for any two elements $x = (x_i)_i$ and $y = (y_i)_i$ of G with $x \neq y$, we consider the index $i_0 = \min\{i \in I : x_i \neq y_i\}$ and define $x < y$ (in the group G) if $x_{i_0} < y_{i_0}$ (in the group G_{i_0}).

(iii) There are torsion-free groups that can't be ordered. For example, consider the group $G = \langle a, b \mid b^{-1}ab = a^{-1} \rangle$. Since G is the semi-direct product of $\langle b \rangle \simeq \mathbf{Z}$ by $\langle a \rangle \simeq \mathbf{Z}$, it is torsion-free. Assume that there is a total order on G that satisfies (1.5). If $a > 1$ then $ba^{-1} = ab > b$ and hence $a^{-1} > 1$, i.e. $1 > a$, a contradiction. A similar argument shows that the inequality $a < 1$ is impossible; therefore, the group G can't be ordered.

Proposition 1.55 *An abelian group G can be ordered if and only if G is torsion-free.*

Proof. We already know that any ordered group must be torsion-free. Conversely, assume that the abelian group G is torsion-free. Then, G can be viewed as a subgroup of the $\mathbf{Q}$-vector space $V = G \otimes \mathbf{Q}$. We choose a basis $(e_i)_{i \in I}$ of V and fix a well-ordering on the index set I. Then,

$$G \subseteq V = \bigoplus_i \mathbf{Q}e_i \subseteq \prod_i \mathbf{Q}e_i$$

and hence it suffices to show that the group $\prod_i \mathbf{Q}e_i$ can be ordered (cf. Remark 1.54(i)). Since the additive group $\mathbf{Q}$ of rational numbers can be ordered, the proof is finished by invoking Remark 1.54(ii). $\square$

The above result provides us with many examples of groups that can be ordered. It is known that the free product of ordered groups can be endowed with the structure of an ordered group as well (Vinogradov's theorem; cf. [55, Chap. 13, Theorem 2.7]). In particular, free groups can be ordered. The relevance of the class of ordered groups in the study of the idempotent conjecture stems from the following result.

Theorem 1.56 *If k is an integral domain and G an ordered group, then any idempotent of kG is contained in k. In particular, the group G satisfies the idempotent conjecture.*

Proof. For any non-zero element $a = \sum_g a_g g \in kG$, where $a_g \in k$ for all $g \in G$, we consider the finite subset $\Lambda_a = \{g \in G : a_g \neq 0\} \subseteq G$ and define the elements $\max(a), \min(a) \in G$ as the maximum and minimum elements of Λ_a respectively. Since k is assumed to be an integral domain, it follows easily that $\max(ab) = \max(a)\max(b)$ and $\min(ab) = \min(a)\min(b)$ for all $a, b \in kG \setminus \{0\}$. In particular, if $e \in kG$ is a non-zero idempotent then $\max(e) = \max(e)^2$ and $\min(e) = \min(e)^2$, i.e. $\max(e) = \min(e) = 1$. We conclude that $e \in k$, as needed. $\square$

II. REDUCTION MODULO THE AUGMENTATION IDEAL. In order to obtain more examples of torsion-free groups that satisfy the idempotent conjecture, we consider for any group G the augmentation homomorphism

$$\varepsilon : \mathbf{C}G \longrightarrow \mathbf{C}$$

and the augmentation ideal $I_G = \ker \varepsilon$. The resulting extension of scalars functor from the category of left $\mathbf{C}G$-modules to that of $\mathbf{C}$-vector spaces maps any left $\mathbf{C}G$-module M onto

$$M_G = \mathbf{C} \otimes_{\mathbf{C}G} M = M/I_G M = M/< (g-1)x; g \in G, x \in M > .$$

We are interested in projective $\mathbf{C}G$-modules P for which P_G vanishes and examine whether this vanishing implies that P itself is the zero module.

Example 1.57 Let G be a group and $g \in G$ a torsion element of order $n > 1$. Then, the idempotent $e = \frac{1}{n}(1 + g + \cdots + g^{n-1}) \in \mathbf{C}G$ is non-trivial and hence the projective $\mathbf{C}G$-module $P = \mathbf{C}G(1-e)$ is non-zero. Since $\varepsilon(e) = 1$, we have $1 - e \in I_G$. It follows that $P = \mathbf{C}G(1-e)^2 \subseteq I_G(1-e) = I_G P$ and hence $P_G = P/I_G P = 0$.

We consider the class $\mathcal{S}$, which consists of those groups G that satisfy the following condition: *For any non-zero projective $\mathbf{C}G$-module P, we have $P_G \neq 0$.* In view of Example 1.57, any group contained in $\mathcal{S}$ must be torsion-free. The importance of class $\mathcal{S}$ in the study of the idempotent conjecture is manifested by the following result.

Theorem 1.58 *Groups in class $\mathcal{S}$ satisfy the idempotent conjecture.*

Proof. Let G be a group in $\mathcal{S}$ and $e = \sum_g e_g g \in \mathbf{C}G$ an idempotent, where $e_g \in \mathbf{C}$ for all $g \in G$. Since the augmentation $\varepsilon(e) \in \mathbf{C}$ is an idempotent, we have $\varepsilon(e) = 0$ or 1. In order to show that e is trivial, we may assume, considering if necessary the idempotent $1 - e$ instead of e, that $\varepsilon(e) = 0$, i.e. that $e \in I_G$. We now consider the projective $\mathbf{C}G$-module $P = \mathbf{C}Ge$ and note that $P = \mathbf{C}Ge^2 \subseteq I_G e = I_G P$. It follows that $P_G = P/I_G P = 0$ and hence our assumption on G implies that $P = 0$. Therefore, $e = 0$ is a trivial idempotent, as needed. $\square$

For any ideal I of a ring R we define the ideal $I^\infty = \bigcap_{n=1}^\infty I^n$. Using this notation, we now give a criterion for a group G to be contained in $\mathcal{S}$.

Proposition 1.59 *Let G be a group and assume that the augmentation ideal I_G of the group algebra $\mathbf{C}G$ is such that $I_G^\infty = 0$. Then, $G \in \mathcal{S}$.*

Proof. Let P be a projective $\mathbf{C}G$-module. We choose a free $\mathbf{C}G$-module F containing P as a direct summand and note that $P \cap JF = JP$ for any ideal $J \subseteq \mathbf{C}G$. We claim that if $I \subseteq \mathbf{C}G$ is an ideal such that $P = IP$, then $P = I^\infty P$. Indeed, we have $P = I^n P \subseteq I^n F$ for all $n \in \mathbf{N}$ and hence $P \subseteq \bigcap_n (I^n F) = (\bigcap_n I^n) F = I^\infty F$. Therefore, $P = P \cap I^\infty F = I^\infty P$.

We now assume that $P_G = P / I_G P$ vanishes. Then, $P = I_G P$ and hence $P = I_G^\infty P = 0$, in view of our hypothesis. $\qquad\square$

Example 1.60 The additive group $\mathbf{Q}$ of rational numbers is an $\mathcal{S}$-group. Indeed, letting $G = \mathbf{Q}$, the group algebra $\mathbf{C}G$ is identified with the algebra $\mathbf{C}[t^a : a \in \mathbf{Q}]$ of functions on the positive real line, and the augmentation ideal I_G with the ideal $(t^a - 1 : a \in \mathbf{Q})$. We shall prove that $G \in \mathcal{S}$ by showing that $I_G^\infty = 0$ (cf. Proposition 1.59). To that end, we note that for any $x \in I_G^n$ the i-th derivative $\frac{d^i x}{dt^i}$ vanishes at $t = 1$ for all $i < n$. Let $x = \sum_{j=1}^m x_j t^{k_j} \in I_G^\infty$, where the x_j's are complex numbers and the k_j's distinct rational numbers. Then, the derivative $\frac{d^i x}{dt^i}\big|_{t=1}$ vanishes for all $i \geq 0$ and hence the x_j's satisfy the system of linear equations

$$\sum_{j=1}^m k_j (k_j - 1) \cdots (k_j - i + 1) x_j = 0, \quad i = 0, 1, \ldots, m - 1 \ .$$

It is an easy exercise, which is left to the reader, to verify that the determinant of this linear system is equal to the product $\prod_{j<j'} (k_{j'} - k_j)$, which is non-zero since the k_j's are distinct. Hence, we must have $x_j = 0$ for all $j = 1, \ldots, m$, i.e. $x = 0$. It follows that the ideal I_G^∞ is trivial, as needed.

We now prove that $\mathcal{S}$ is closed under subgroups, extensions, direct products and free products. In this way, invoking Theorem 1.58, we obtain many examples of groups that satisfy the idempotent conjecture.

Proposition 1.61 *The class $\mathcal{S}$ is closed under subgroups and extensions.*

Proof. In order to show that $\mathcal{S}$ is closed under subgroups, let G be an $\mathcal{S}$-group and consider a subgroup $H \leq G$. If P is a projective $\mathbf{C}H$-module with $\mathbf{C} \otimes_{\mathbf{C}H} P = 0$, then $P' = \mathbf{C}G \otimes_{\mathbf{C}H} P$ is a projective $\mathbf{C}G$-module with

$$\mathbf{C} \otimes_{\mathbf{C}G} P' = \mathbf{C} \otimes_{\mathbf{C}G} (\mathbf{C}G \otimes_{\mathbf{C}H} P) = \mathbf{C} \otimes_{\mathbf{C}H} P = 0 \ .$$

Since $G \in \mathcal{S}$, we have $P' = 0$. On the other hand, $\mathbf{C}H$ is a direct summand of $\mathbf{C}G$ as a right $\mathbf{C}H$-module and hence $P = \mathbf{C}H \otimes_{\mathbf{C}H} P$ is a direct summand of P' as an abelian group. It follows that P must be zero as well and hence $H \in \mathcal{S}$.

In order to show the closure of $\mathcal{S}$ under extensions, let N be a normal subgroup of a group G and assume that both N and the quotient group $\overline{G} = G/N$ are contained in $\mathcal{S}$. If P is a projective $\mathbf{C}G$-module with $\mathbf{C} \otimes_{\mathbf{C}G} P = 0$, then $\overline{P} = \mathbf{C}\overline{G} \otimes_{\mathbf{C}G} P$ is a projective $\mathbf{C}\overline{G}$-module with

$$\mathbf{C} \otimes_{\mathbf{C}\overline{G}} \overline{P} = \mathbf{C} \otimes_{\mathbf{C}\overline{G}} (\mathbf{C}\overline{G} \otimes_{\mathbf{C}G} P) = \mathbf{C} \otimes_{\mathbf{C}G} P = 0 \ .$$

Since $\overline{G} \in \mathcal{S}$, we have $\overline{P} = 0$. On the other hand, as a $\mathbf{C}$-vector space, $\overline{P}$ is isomorphic to $\mathbf{C} \otimes_{\mathbf{C}N} P = P_N$, where P is now viewed as a $\mathbf{C}N$-module by restriction of scalars (cf. Remark 1.6(ii)). Since the $\mathbf{C}N$-module P is projective (Lemma 1.1(iii)) and $N \in \mathcal{S}$, we conclude that $P = 0$. Hence, it follows that $G \in \mathcal{S}$. $\qquad\square$

In order to obtain another interesting criterion for a group G to be contained in $\mathcal{S}$, we need the following technical result.

Lemma 1.62 *Let G be a group, $(H_\lambda)_\lambda$ a chain of subgroups of G and H the intersection of the H_λ's.*
(i) The intersection of the family $(I_{H_\lambda}\mathbf{C}G)_\lambda$ of left $\mathbf{C}H$-submodules of $\mathbf{C}G$ is equal to $I_H\mathbf{C}G$.
(ii) If P is a projective $\mathbf{C}G$-module and $\mathbf{C} \otimes_{\mathbf{C}H_\lambda} P = 0$ for all λ, then $\mathbf{C} \otimes_{\mathbf{C}H} P = 0$.

Proof. (i) It is clear that $I_H\mathbf{C}G$ is contained in the intersection of the $I_{H_\lambda}\mathbf{C}G$'s. In order to show the reverse inclusion, let $x \in \mathbf{C}G$ be an element with $x \notin I_H\mathbf{C}G$. We consider a set T of left H-coset representatives in G and note that there is a decomposition of left $\mathbf{C}H$-modules

$$\mathbf{C}G = \bigoplus_{t \in T} \mathbf{C}H \cdot t.$$

Then, we can write $x = x_1 t_1 + \cdots + x_n t_n$, where $x_i \in \mathbf{C}H$ and $t_i \in T$ for all $i = 1, \ldots, n$. Since $x \notin I_H\mathbf{C}G$, there exists i_0 with $x_{i_0} \notin I_H$. Assuming that the t_i's are distinct, we have $t_i t_j^{-1} \notin H$ for all $i \neq j$; hence, there exists λ_{ij} with $t_i t_j^{-1} \notin H_{\lambda_{ij}}$ for all $i \neq j$. We can now find an index λ such that $H_\lambda \subseteq H_{\lambda_{ij}}$ for all $i \neq j$. Then, $t_i t_j^{-1} \notin H_\lambda$ and hence the t_i's form part of a system T_λ of left H_λ-coset representatives in G. Since $x_i \in \mathbf{C}H \subseteq \mathbf{C}H_\lambda$ for all $i = 1, \ldots, n$, it follows that $x = x_1 t_1 + \cdots + x_n t_n$ is the expression of x following the left $\mathbf{C}H_\lambda$-module decomposition

$$\mathbf{C}G = \bigoplus_{t \in T_\lambda} \mathbf{C}H_\lambda \cdot t.$$

Since $\mathbf{C}H \cap I_{H_\lambda} = I_H$, we have $x_{i_0} \notin I_{H_\lambda}$; it follows that $x \notin I_{H_\lambda}\mathbf{C}G$.

(ii) Let F be a free $\mathbf{C}G$-module containing P as a direct summand. Since $P/I_{H_\lambda}P = \mathbf{C} \otimes_{\mathbf{C}H_\lambda} P = 0$, we have $P = I_{H_\lambda}P$ and hence $P \subseteq I_{H_\lambda}F$ for all λ. It follows from (i) above that the intersection of the family $(I_{H_\lambda}F)_\lambda$ is equal to $I_H F$; therefore, $P \subseteq I_H F$. Since P is a direct summand of F as a $\mathbf{C}G$-module and hence as a $\mathbf{C}H$-module as well, the intersection $P \cap I_H F$ is easily seen to coincide with $I_H P$. It follows that $P = P \cap I_H F = I_H P$, i.e. $\mathbf{C} \otimes_{\mathbf{C}H} P = P/I_H P = 0$, as needed. $\qquad\square$

Proposition 1.63 *Let G be a group and assume that any non-trivial subgroup $H \leq G$ admits a non-trivial homomorphism into a group which is contained in $\mathcal{S}$. Then, $G \in \mathcal{S}$.*

Proof. We note that, in view of the closure of $\mathcal{S}$ under subgroups (cf. Proposition 1.61), our assumption can be restated as follows: Any non-trivial subgroup $H \leq G$ has a non-trivial quotient group which is contained in $\mathcal{S}$. In order to show that $G \in \mathcal{S}$, let us consider a projective $\mathbf{C}G$-module P with $\mathbf{C} \otimes_{\mathbf{C}G} P = 0$. Then, the family $\mathcal{F}$ which consists of those subgroups H of G for which $\mathbf{C} \otimes_{\mathbf{C}H} P = 0$ is non-empty, since $G \in \mathcal{F}$. We order $\mathcal{F}$ by the relation opposite to inclusion and note that Lemma 1.62(ii) implies that the hypothesis of Zorn's lemma is satisfied. Hence, there is a subgroup H of G which is minimal with respect to the property that $\mathbf{C} \otimes_{\mathbf{C}H} P = 0$. If H is trivial then $\mathbf{C} \otimes_{\mathbf{C}H} P = \mathbf{C} \otimes_{\mathbf{C}} P = P$ and hence $P = 0$. If H is non-trivial there exists (by assumption) a proper normal subgroup $N \trianglelefteq H$ such that $\overline{H} = H/N \in \mathcal{S}$. The $\mathbf{C}\overline{H}$-module $\overline{P} = \mathbf{C}\overline{H} \otimes_{\mathbf{C}H} P$ is projective and

$$\mathbf{C} \otimes_{\mathbf{C}\overline{H}} \overline{P} = \mathbf{C} \otimes_{\mathbf{C}\overline{H}} (\mathbf{C}\overline{H} \otimes_{\mathbf{C}H} P) = \mathbf{C} \otimes_{\mathbf{C}H} P = 0 \, .$$

The group $\overline{H}$ being an $\mathcal{S}$-group, it follows that $\overline{P} = 0$. Since $\overline{P} \simeq \mathbf{C} \otimes_{\mathbf{C}N} P$ as $\mathbf{C}$-vector spaces (cf. Remark 1.6(ii)), we conclude that $N \in \mathcal{F}$, contradicting the minimality of H. $\square$

Proposition 1.64 *The class $\mathcal{S}$ is closed under direct products and free products.*

Proof. Let $(G_i)_i$ be a family of $\mathcal{S}$-groups. We shall prove that both groups $\prod_i G_i$ and $*_i G_i$ are contained in $\mathcal{S}$, by using the criterion established in Proposition 1.63.

To that end, let H be a non-trivial subgroup of the direct product $\prod_i G_i$. Then, H maps by the restriction of a suitable coordinate projection onto a non-trivial subgroup of one of the G_i's. Hence, the criterion of Proposition 1.63 can be applied.

Now let K be a non-trivial subgroup of the free product $*_i G_i$. Then, a theorem of Kurosh (cf. [65, Chap. 1, Theorem 14] impies that K decomposes into the free product of a free group and a free product of certain subgroups of G, which are isomorphic with subgroups of the G_i's. It follows that K admits a non-trivial homomorphism into one of the G_i's or else into $\mathbf{Z} \subseteq \mathbf{Q}$. Since $\mathbf{Q} \in \mathcal{S}$ (cf. Example 1.60), the criterion of Proposition 1.63 can be applied in this case as well. $\square$

Proposition 1.65 *The class $\mathcal{S}$ contains all torsion-free abelian groups.*

Proof. A torsion-free abelian group G is contained in the $\mathbf{Q}$-vector space $V = \mathbf{Q} \otimes G$. Such a vector space V is a direct sum of copies of $\mathbf{Q}$ and hence contained in the corresponding direct product of these copies. The proof is finished, since $\mathbf{Q}$ is an $\mathcal{S}$-group (Example 1.60), whereas $\mathcal{S}$ is closed under subgroups (Proposition 1.61) and direct products (Proposition 1.64). $\square$

1.3 Exercises

1. Given two rings A and B, an (A, B)-bimodule is an abelian group M, which is a left A-module and a right B-module, in such a way that $a(mb) = (am)b$ for all $a \in A$, $b \in B$ and $m \in M$.

(i) Let S be a ring and L, N two left S-modules. Show that the composition of maps endows the abelian group $\mathrm{Hom}_S(L, N)$ with the structure of an $(\mathrm{End}_S N, \mathrm{End}_S L)$-bimodule. In particular, if R is another ring and L is an (S, R)-bimodule, then $\mathrm{Hom}_S(L, N)$ can be naturally viewed as a left R-module.

(ii) Let R be a ring, L a right R-module and M a left R-module. Show that the abelian group $L \otimes_R M$ can be endowed with the structure of a module over $\mathrm{End}_R L \otimes \mathrm{End}_R M$, by letting $(f \otimes g) \cdot (l \otimes m) = f(l) \otimes g(m)$ for all $f \in \mathrm{End}_R L$, $g \in \mathrm{End}_R M$, $l \in L$ and $m \in M$. In particular, if S is another ring and L is an (S, R)-bimodule, then $L \otimes_R M$ can be naturally viewed as a left S-module.

(iii) Let R, S be two rings, L an (S, R)-bimodule, M a left R-module and N a left S-module. Show that there is an isomorphism of abelian groups

$$\lambda : \mathrm{Hom}_S(L \otimes_R M, N) \overset{\sim}{\longrightarrow} \mathrm{Hom}_R(M, \mathrm{Hom}_S(L, N)) \, ,$$

which is defined by letting $\lambda(f) : M \longrightarrow \mathrm{Hom}_R(L, N)$ be the map with $\lambda(f)(m)(l) = f(l \otimes m)$ for all $f \in \mathrm{Hom}_S(L \otimes_R M, N)$, $m \in M$ and $l \in L$. Here, the S-module structure on $L \otimes_R M$ is that defined in (ii) and the R-module structure on $\mathrm{Hom}_S(L, N)$ that defined in (i) above.

(iv) Let $\varphi : R \longrightarrow S$ be a ring homomorphism, M a left R-module and N a left S-module. Letting L be the (S, R)-bimodule S (with right R-module structure induced by φ), show that the isomorphism of (iii) above reduces to the identification λ, which appears in the text preceding Lemma 1.1.

2. Let $\mathcal{H}$ be a Hilbert space, n a positive integer and consider the identification between $\mathbf{M}_n(\mathcal{B}(\mathcal{H}))$ and $\mathcal{B}(\mathcal{H}^n)$ considered in the beginning of §1.1.2.

(i) Let $a \in \mathcal{B}(\mathcal{H})$ and $\xi = (\xi_1, \ldots, \xi_n), \eta = (\eta_1, \ldots, \eta_n) \in \mathcal{H}^n$. Show that $< aE_{ij}(\xi), \eta > = < a(\xi_j), \eta_i >$ for all $i, j = 1, \ldots, n$. In particular, $(aE_{ij})^* = a^* E_{ji}$ for all $i, j = 1, \ldots, n$.

(ii) Show that the adjoint $A^* = (b_{ij})_{i,j}$ of a matrix $A = (a_{ij})_{i,j} \in \mathbf{M}_n(\mathcal{B}(\mathcal{H}))$ is such that $b_{ij} = a_{ji}^*$ for all $i, j = 1, \ldots, n$.

(iii) Let $(a_\lambda)_\lambda$ be a net in $\mathcal{B}(\mathcal{H})$ and fix two indices $i, j \in \{1, \ldots, n\}$. Show that the net $(a_\lambda E_{ij})_\lambda$ converges to the zero operator of $\mathcal{H}^n$ in the norm (resp. the strong, resp. the weak) operator topology if and only if the net $(a_\lambda)_\lambda$ converges to the zero operator of $\mathcal{H}$ in the norm (resp. the strong, resp. the weak) operator topology.

(iv) Let $(A_\lambda)_\lambda$ be a net of matrices in $\mathbf{M}_n(\mathcal{B}(\mathcal{H}))$ and write $A_\lambda = (a_{ij,\lambda})_{i,j}$ for all λ. Show that the net $(A_\lambda)_\lambda$ converges to the zero operator of $\mathcal{H}^n$ in the norm (resp. the strong, resp. the weak) operator topology if and only if the net $(a_{ij,\lambda})_\lambda$ converges to the zero operator of $\mathcal{H}$ in the norm (resp. the strong, resp. the weak) operator topology for all $i, j = 1, \ldots, n$.
(*Hint:* Use (iii) above and note that $A_\lambda = \sum_{i,j} a_{ij,\lambda} E_{ij}$ for all λ and $a_{ij,\lambda} E_{11} = E_{1i} A_\lambda E_{j1}$ for all i, j, λ.)

3. Let $\ell^2\mathbf{N}$ be the Hilbert space of all square summable sequences of complex numbers and consider the operators $a, b \in \mathcal{B}(\ell^2\mathbf{N})$, which are defined by letting $a(\xi_0, \xi_1, \xi_2, \ldots) = (\xi_1, \xi_2, \ldots)$ and $b(\xi_0, \xi_1, \xi_2, \ldots) = (0, \xi_0, \xi_1, \xi_2, \ldots)$ for all $(\xi_0, \xi_1, \xi_2, \ldots) \in \ell^2\mathbf{N}$.
   (i) Show that $\|a^n\| = \|b^n\| = 1$ for all $n \geq 1$.
   (ii) Show that the sequence $(a^n)_n$ is SOT-convergent to 0, but not norm-convergent to 0. In particular, the sequence $(a^n)_n$ is WOT-convergent to 0.
   (iii) Show that the sequence $(b^n)_n$ is WOT-convergent to 0, but not SOT-convergent to 0.
   (*Hint:* In order to show WOT-convergence to 0, use Proposition 1.14.)
   (iv) Show that the sequence $(a^n b^n)_n$ is not WOT-convergent to 0. In particular, multiplication in $\mathcal{B}(\ell^2\mathbf{N})$ is not jointly WOT-continuous.
4. Let R be a ring. For any subset $X \subseteq R$ we define the commutant X' of X in R, by letting $X' = \{r \in R : xr = rx \text{ for all } x \in X\}$. Show that:
   (i) X' is a subring of R for any subset $X \subseteq R$.
   (ii) If X, Y are subsets of R with $X \subseteq Y$, then $Y' \subseteq X'$.
   (iii) $X \subseteq X''$ for any subset $X \subseteq R$; here, X'' (the bicommutant of X in R) is defined as the commutant of $X' \subseteq R$.
   (iv) $X' = X'''$ for any subset $X \subseteq R$.
   (v) The commutant R' of R is the center $Z(R) \subseteq R$ and $R'' = R$.
5. Let R be a ring, n a positive integer and $\mathbf{M}_n(R)$ the corresponding matrix ring. For any subset $X \subseteq R$ we consider the subset $\mathbf{M}_n(X)$ (resp. $X \cdot I_n$) of $\mathbf{M}_n(R)$, which consists of all $n \times n$ matrices with entries in X (resp. of all matrices of the form xI_n, $x \in X$). Show that:
   (i) The commutant $(\mathbf{M}_n(X))'$ of $\mathbf{M}_n(X)$ in $\mathbf{M}_n(R)$ is equal to $X' \cdot I_n$, where X' is the commutant of X in R. In particular, the center $Z(\mathbf{M}_n(R))$ of $\mathbf{M}_n(R)$ is equal to $Z(R) \cdot I_n$, where $Z(R)$ is the center of R.
   (ii) The commutant $(X \cdot I_n)'$ of $X \cdot I_n$ in $\mathbf{M}_n(R)$ is equal to $\mathbf{M}_n(X')$.
   (iii) The bicommutant $(\mathbf{M}_n(X))''$ of $\mathbf{M}_n(X)$ in $\mathbf{M}_n(R)$ is equal to $\mathbf{M}_n(X'')$, where X'' is the bicommutant of X in R. In particular, if S is a subring of R, such that $S'' = S$, then $\mathbf{M}_n(S)$ is a subring of $\mathbf{M}_n(R)$, such that $(\mathbf{M}_n(S))'' = \mathbf{M}_n(S)$.
6. Let $\mathcal{A}$ be the C^*-algebra of continuous complex-valued functions on $[0, 1]$. We may regard $\mathcal{A}$ as a subalgebra of the algebra of bounded linear operators on the Hilbert space $L^2[0, 1]$ (under the Lebesgue measure), by identifying any continuous function on $[0, 1]$ with the corresponding multiplication operator on $L^2[0, 1]$. Let $f \in \mathcal{A}$ be the function which is defined by letting $f(x) = \max\{0, x - \frac{1}{2}\}$ for all $x \in [0, 1]$. Show that the partial isometry $v \in \mathcal{B}(L^2[0, 1])$ in the polar decomposition of $f \in \mathcal{A} \subseteq \mathcal{B}(L^2[0, 1])$ is not contained in $\mathcal{A}$.
   (*Hint:* Since the function f is positive, v is the orthogonal projection onto the closure of the range of the multiplication operator associated with f.)

7. Our goal in this Exercise is to show that the definitions of §1.1.3 and §1.1.4 given for a not necessarily unital ring I coincide with those given for unital rings if I is unital.

(i) Let R, S be two unital rings. Show that the coordinate projection maps from the product $R \times S$ to R and S induce an isomorphism of abelian groups $K_0(R \times S) \simeq K_0(R) \oplus K_0(S)$.

(ii) Let I be a unital ring and I^+ the ring defined in the beginning of §1.1.3.III. Show that there is an isomorphism of rings $I^+ \simeq I \times \mathbf{Z}$, which identifies the map $\pi : I^+ \longrightarrow \mathbf{Z}$ of loc.cit. with the second coordinate projection from $I \times \mathbf{Z}$ onto $\mathbf{Z}$. In particular, conclude that the K_0-group of the unital ring I can be identified with the kernel of the additive map $K_0(\pi) : K_0(I^+) \longrightarrow K_0(\mathbf{Z})$.

(iii) Let I be a unital ring, τ a trace on I with values in an abelian group V and $\tau_* : K_0(I) \longrightarrow V$ the induced additive map. Let τ^+ be the V-valued trace on I^+ defined in the beginning of §1.1.4.II. Show that, under the identification of $K_0(I^+)$ with the direct sum $K_0(I) \oplus K_0(\mathbf{Z})$ resulting from (i) and (ii) above, the restriction of the additive map $(\tau^+)_* : K_0(I^+) \longrightarrow V$ to $K_0(I)$ coincides with τ_*.

8. Let k be a field and E an idempotent $n \times n$ matrix with entries in k. Show that $r_{HS}(E) = r \cdot 1 \in k$, where $r \in \{0, 1, \ldots, n\}$ is the rank of the matrix E, as defined in Linear Algebra (i.e. r is the maximum number of linearly independent rows of E).

9. Let k be a subring of the field $\mathbf{C}$ of complex numbers and G a group.

(i) Show that the pair (k, G) satisfies Bass' conjecture if and only if the pairs (k', G) satisfy Bass' conjecture for any finitely generated subring $k' \subseteq k$.

(ii) Show that the pair (k, G) satisfies Bass' conjecture if the pairs (k, G') satisfy Bass' conjecture for any finitely generated subgroup $G' \subseteq G$.

10. Let G be a group and consider the set

$$\mathrm{Fin}(G) = \{o(g) : g \in G \text{ is an element of finite order}\} \subseteq \mathbf{N} \, .$$

We define Λ_G (resp. Λ_G^+) to be the subring (resp. the additive subgroup) of $\mathbf{Q}$ generated by the set $\{1/s : s \in \mathrm{Fin}(G)\}$. Show that the following conditions are equivalent for a subring k of the field $\mathbf{C}$ of complex numbers:

(i) $k \cap \Lambda_G = \mathbf{Z}$.

(ii) $k \cap \Lambda_G^+ = \mathbf{Z}$.

(iii) If $g \in G$ is a torsion element with $g \neq 1$, then its order $o(g)$ is not invertible in k.

Moreover, show that if the group G is finite of order n, then the above conditions are also equivalent to

(iv) $k \cap \mathbf{Z} \cdot \frac{1}{n} = \mathbf{Z}$.

11. Let R be a commutative ring having no non-trivial idempotents and consider its nil radical nil R. Show that the quotient $\overline{R} = R/\mathrm{nil}\,R$ has no non-trivial idempotents as well.[5]
    (*Hint:* If $r \in R$ is such that $\overline{r} \in \overline{R}$ is an idempotent, then $r^n(1-r)^n = 0 \in R$ for some $n \in \mathbf{N}$. Show that the elements $r^n x$ and $(1-r)^n y$ are complementary idempotents of R for suitable $x, y \in R$.)
12. The goal of this Exercise is to show that any commutative reduced Noetherian ring can be embedded into a finite direct product of fields. We shall assume some familiarity with the concept of localization at a prime ideal (cf. Appendix A). Let R be a Noetherian ring and $\operatorname{Spec} R$ its prime spectrum. For any ideal $I \subseteq R$ we denote by $V(I)$ the set of those prime ideals $\wp \in \operatorname{Spec} R$ that contain I; in particular, $\operatorname{Spec} R = V(0)$.
    (i) Show that there are finitely many prime ideals $\wp_1, \ldots, \wp_n \subseteq R$, such that $\operatorname{Spec} R = \bigcup_{i=1}^n V(\wp_i)$.
    (*Hint:* Argue by contradiction and assume that $I \subseteq R$ is an ideal maximal with respect to the property that $V(I)$ is not a finite union of subsets of the form $V(\wp)$, $\wp \in \operatorname{Spec} R$.)
    (ii) Let $\wp_1, \ldots, \wp_n \subseteq R$ be prime ideals, such that $\operatorname{Spec} R = \bigcup_i V(\wp_i)$ as in (i) above, and assume that $r \in R$ is an element whose image in $R_{\wp_i}$ vanishes for all $i = 1, \ldots, n$. Then, show that r is nilpotent.
    (iii) Assume that the ring R is reduced and let $\wp \subseteq R$ be a minimal prime ideal. Then, show that the localization $R_\wp$ is a field. Conclude that R embeds into a finite direct product of fields $F_1 \times \cdots \times F_n$.
13. Let G be a group. The goal of this Exercise is to complement Proposition 1.52 and provide yet another condition, which is equivalent to conditions (i), (ii) and (iii) therein. To that end, we consider a commutative ring k, the group algebra kG and an idempotent $e \in kG$.
    (i) Show that $(r-e)^n = r^n - (r^n - (r-1)^n)e \in kG$ for all $r \in k = k \cdot 1 \subseteq kG$ and $n \geq 0$.
    (ii) Assume that $r, x \in k$ are two elements, such that $r^n x = 0 = (1-r)^n x$ for some $n \gg 0$. Then, show that $x = 0$.
    (iii) Let $I \subseteq k$ be a nilpotent ideal, $\overline{k} = k/I$ the corresponding quotient ring and $\overline{e} \in \overline{k}G$ the image of e under the quotient homomorphism $kG \longrightarrow \overline{k}G$. If $\overline{e}$ is contained in $\overline{k} = \overline{k} \cdot 1 \subseteq \overline{k}G$, then show that e is contained in $k = k \cdot 1 \subseteq kG$.
    (iv) Assume that G satisfies the equivalent conditions of Proposition 1.52, whereas the commutative ring k is a $\mathbf{Q}$-algebra. Then, show that any idempotent of kG is contained in $k = k \cdot 1 \subseteq kG$.
14. (i) Let G be a group and assume that the complex group algebra $\mathbf{C}G$ has a non-trivial idempotent. Show that for all but finitely many prime numbers p there exists a finite field $\mathbf{F}$ of characteristic p, such that the group algebra $\mathbf{F}G$ has a non-trivial idempotent as well.
    (*Hint:* Let $e = \sum_{g \in G} e_g g \in \mathbf{C}G$ be a non-trivial idempotent and con-

sider suitable quotients of the commutative ring $k = \mathbf{Z}[e_g : g \in G]$; cf. Corollaries A.27 and A.22(ii) of Appendix A.)

(ii) Let p be a prime number and consider a cyclic group G of order p. Show that the complex group algebra $\mathbf{C}G$ has non-trivial idempotents, whereas for any field $\mathbf{F}$ of characteristic p the group algebra $\mathbf{F}G$ has no non-trivial idempotents.

(*Hint:* Consider the left regular representation $L : \mathbf{F}G \longrightarrow \mathbf{M}_p(\mathbf{F})$ and show that $\operatorname{tr}(L(e)) = 0 \in \mathbf{F}$ for any idempotent $e \in \mathbf{F}G$.)

15. Let G be a group, $N \trianglelefteq G$ a normal subgroup and $\overline{G} = G/N$ the corresponding quotient. Prove the following generalization of Theorem 1.58: If $N \in \mathcal{S}$ and $\overline{G}$ satisfies the idempotent conjecture, then G satisfies the idempotent conjecture as well.

Notes and Comments on Chap. 1. The basic results from Algebra and Analysis that are needed in the book can be found in graduate textbooks, such as [42, 60] and [61]. The standard reference for the algebraic properties of group rings is the encyclopedic treatise of D. Passman [55]. The K_0-group of a ring was defined by A. Grothendieck, in order to generalize the Riemann-Roch theorem in Algebraic Geometry [7]. For a more complete introduction to the K-theory of rings, the reader may consult J. Rosenberg's book [58]. The universal trace on a ring was introduced by J. Stallings in [66], in order to give an algebraic proof of D. Gottlieb's theorem on finite $K(G, 1)$-complexes, and, independently, by A. Hattori [32]. The universal trace for group algebras was used by H. Bass in [4, 5], who reformulated accordingly the theorem of R. Swan on integral representations of finite groups [70]; in that direction, see also §2.2. The idempotent conjecture for the complex group algebra of a torsion-free group is a classical problem in ring theory. The fact that ordered groups satisfy the conjecture is folklore; see, for example, [55]. The class $\mathcal{S}$ was introduced in [23], by relaxing a condition that was imposed on a group by R. Strebel [68], in his studies on the derived series, and A. Strojnowski [69].

2

Motivating Examples

2.1 The Case of Abelian Groups

Let k be a subring of $\mathbf{C}$ with $k \cap \mathbf{Q} = \mathbf{Z}$, G an abelian group, kG the corresponding group ring and $E \in \mathbf{M}_N(kG)$ an idempotent matrix. Since the group ring kG is commutative, we have $T(kG) = kG$ and the Hattori-Stallings rank $r_{HS}(E)$ of E is precisely its trace $\mathrm{tr}(E)$. Hence, Bass' conjecture for G asserts that $\mathrm{tr}(E) \in k \cdot 1 \subseteq kG$. Our goal in this section is to prove that this is indeed the case. We consider the commutative ring $R = kG$, its prime spectrum $\mathrm{Spec}\, R$ and the finitely generated projective R-module $P = \mathrm{im}\, \widetilde{E}$, where $\widetilde{E}$ is the endomorphism of R^N corresponding to E. We construct a certain map $r(P) : \mathrm{Spec}\, R \longrightarrow \mathbf{N}$, which is locally constant if $\mathrm{Spec}\, R$ is endowed with the Zariski topology. It will turn out that the space $\mathrm{Spec}\, R$ is connected and hence the map $r(P)$ must be constant. If n is the constant value of $r(P)$, then we prove that the trace of E is equal to $n \cdot 1 \in R$. Consequently, Bass' conjecture for G will follow.

Before specializing to the case of the group ring of an abelian group in §2.1.3, the discussion will concern a general commutative ring. For such a ring, we construct the geometric rank associated with finitely generated projective modules in §2.1.1 and study its relation to the Hattori-Stallings rank in §2.1.2. This approach will put Bass' conjecture in the case of abelian groups into a more geometric perspective and exhibit the equivalence of the conjecture to the assertion that the prime spectrum of the corresponding group ring is connected. Having that case in mind, one may regard the validity of Bass' conjecture for an arbitrary (not necessarily abelian) group as a generalized connectedness assertion. On the other hand, the geometric approach of this section is formally similar to the approach followed in §5.1.2, where we exhibit the equivalence of the idempotent conjecture for the reduced group C^*-algebra associated with a torsion-free abelian group to the connectedness of its dual group.

2.1.1 The Geometric Rank Function

We fix a commutative ring R and denote by Spec R the set of all prime ideals $\wp \subseteq R$; this is the prime spectrum of R. For any $\wp \in$ Spec R the localization $R_\wp$ is a local ring (cf. Appendix A). We also consider a finitely generated projective R-module P. Then, Proposition A.10 of Appendix A implies that for any $\wp \in$ Spec R the finitely generated projective $R_\wp$-module $P \otimes_R R_\wp$ is free; let $r_\wp(P)$ be its rank. In this way, we associate with any prime ideal of R a non-negative integer. Let

$$r(P) : \text{Spec } R \longrightarrow \mathbf{N}$$

be the map $\wp \mapsto r_\wp(P)$, $\wp \in$ Spec R. It is clear that $r(P)$ depends only upon the isomorphism class of P.

Definition 2.1 *Let P be a finitely generated projective R-module. The map $r(P)$ defined above is called the geometric rank of P.*

Remark 2.2 The geometric rank function $r(P)$ associated with a finitely generated projective R-module P is the algebraic analogue of the following geometric situation:[1] If $\pi : E \longrightarrow B$ is a vector bundle over a manifold B, then its rank $r(E) : B \longrightarrow \mathbf{N}$ is the function which assigns to any point $b \in B$ the dimension $r_b(E)$ of the fibre $\pi^{-1}(b)$, which is itself a finite dimensional vector space. The rank function $r(E)$ is locally constant. In particular, if the base B is connected then the function $r(E)$ is constant.

Our next goal is to define the Zariski topology on the prime spectrum Spec R of R. To that end, we consider for any ideal $I \subseteq R$ the set

$$V(I) = \{\wp \in \text{Spec } R : \wp \supseteq I\} .$$

The following lemma describes some basic properties of the $V(I)$'s.

Lemma 2.3 *(i) $\bigcap_a V(I_a) = V(\sum_a I_a)$ for any family $(I_a)_a$ of ideals in R.*
(ii) $V(I) \cup V(J) = V(IJ)$ for any ideals $I, J \subseteq R$.
(iii) $V(0) = $ Spec R and $V(R) = \emptyset$.
(iv) $V(I) = \emptyset$ only if $I = R$.
(v) $V(I) = $ Spec R if and only if $I \subseteq \text{nil } R$.
(vi) If $I, J \subseteq R$ are two ideals and $V(I) \subseteq V(J)$, then for any $x \in J$ there exists $n \in \mathbf{N}$ such that $x^n \in I$.

Proof. (i) This is clear, since a (prime) ideal $\wp$ contains all of the I_a's if and only if it contains their sum.

(ii) This property simply asserts that a prime ideal contains the product of two ideals I and J if and only if it contains one of them.

(iii) This is immediate.

---

[1] For details on this analogy, see [31].

(iv) This too is clear since any proper ideal I is contained in a maximal (and hence prime) ideal.

(v) We note that any nilpotent element is contained in any prime ideal. Therefore, if the ideal I consists of nilpotent elements then I is contained in all prime ideals. Conversely, if $s \in I$ is not nilpotent then the localization $R[S^{-1}]$ of R at the multiplicatively closed subset $S \subseteq R$ generated by s is a non-zero ring; as such, it possesses a maximal ideal $\mathbf{m}$. The inverse image $\wp = \{r \in R : r/1 \in \mathbf{m}\}$ of $\mathbf{m}$ under the natural ring homomorphism $R \longrightarrow R[S^{-1}]$ is a prime ideal of R with $s \notin \wp$. In particular, $I \not\subseteq \wp$ and hence $\wp \notin V(I)$.

(vi) Assume that $V(I) \subseteq V(J)$ and suppose that there is an element $x \in J$, such that no power of it lies in I. Then, the image $\overline{J}$ of J in the residue ring $\overline{R} = R/I$ is not contained in the nil radical $\mathrm{nil}\,\overline{R}$. Therefore, we may invoke (v) above and conclude that there is a prime ideal $\overline{\wp} = \wp/I \in \operatorname{Spec} \overline{R}$ with $\overline{J} \not\subseteq \overline{\wp}$. But then $\wp$ is a prime ideal of R containing I with $J \not\subseteq \wp$, contradicting our assumption that $V(I) \subseteq V(J)$. $\qquad\square$

It follows from assertions (i), (ii) and (iii) of Lemma 2.3 that the $V(I)$'s form the collection of closed sets for a certain topology on $\operatorname{Spec} R$.

Definition 2.4 *The topology on $\operatorname{Spec} R$ with closed sets those of the form $V(I)$, where $I \subseteq R$ is an ideal, is called Zariski topology.*

Lemma 2.5 *The prime spectrum $\operatorname{Spec} R$, endowed with the Zariski topology, is compact.*

Proof. Consider a family of closed subsets $(V(I_a))_a$ of $\operatorname{Spec} R$ corresponding to a family of ideals $(I_a)_a$ of R and suppose that the intersection $\bigcap_a V(I_a)$ is empty. If I is the sum of the I_a's then $V(I) = \bigcap_a V(I_a) = \emptyset$ (Lemma 2.3(i)) and hence $I = R$ (Lemma 2.3(iv)). Then, $1 \in R$ is contained in the sum of a finite subfamily of the I_a's; say $1 \in \sum_{i=1}^{n} I_{a_i}$. This means that $\sum_{i=1}^{n} I_{a_i} = R$ and hence we may reverse the arguments above, in order to conclude that $\bigcap_{i=1}^{n} V(I_{a_i}) = \emptyset$. We have therefore proved that the space $\operatorname{Spec} R$ has the finite intersection property; hence, it is compact. $\qquad\square$

Example 2.6 Let k be an algebraically closed field and $R = k[X,Y]$ the k-algebra of polynomials in two variables. In that case, it can be shown (cf. [3]) that there are three types of prime ideals in R:

(i) ideals of the form $(X - a, Y - b)$, where $a, b \in k$,
(ii) principal ideals generated by irreducible polynomials and
(iii) the zero ideal 0.

One may picture the maximal ideals of R as the points of the affine plane k^2, by letting a maximal ideal $(X - a, Y - b)$ correspond to the point with coordinates (a, b). Then, any prime ideal $\wp = (f)$, where $f \in R$ is an irreducible polynomial, corresponds to the plane curve with equation $f = 0$, in the sense that the maximal ideals contained in $V(\wp)$ are precisely those of the form $(X - a, Y - b)$, where $a, b \in k$ are such that $f(a,b) = 0$.

Having defined the Zariski topology on $\operatorname{Spec} R$ and regarding $\mathbf{N}$ as a discrete space, the geometric rank function $r(P)$ associated with a finitely generated projective R-module P is a map between two topological spaces.

Remark 2.7 Let X, Y be two topological spaces and $f : X \longrightarrow Y$ a map between them. The map f is locally constant if for any $x \in X$ there exists an open set $U \subseteq X$, such that $x \in U$ and the restriction $f \mid_U$ is constant. If f is locally constant then f is clearly continuous. Conversely, if the space Y is discrete and f is continuous then f is locally constant. Indeed, for any $x \in X$ the singleton $V = \{f(x)\}$ is open in Y and hence $U = f^{-1}(V)$ is an open neighborhood of x in X, on which f is constant.

Theorem 2.8 *Let P be a finitely generated projective R-module. Then, the geometric rank function $r(P)$ is continuous (i.e. locally constant).*

Proof. Let $\wp \subseteq R$ be a prime ideal and $n = r_\wp(P)$. Then, $P_\wp \simeq P \otimes_R R_\wp \simeq R_\wp^n$ as $R_\wp$-modules and hence there are elements $p_1, \ldots, p_n \in P$ which map onto an $R_\wp$-basis $p_1/1, \ldots, p_n/1$ of $P_\wp$. This means that the R-linear map $\varphi : R^n \longrightarrow P$, which maps the basis elements of R^n onto the p_i's, becomes an isomorphism when localized at $\wp$. Using Proposition A.7 of Appendix A, we may choose an element $u \in R \setminus \wp$, such that φ becomes an isomorphism after inverting u. We now invoke Corollary A.3(i) of Appendix A and conclude that φ becomes an isomorphism when localized at any multiplicatively closed subset containing u. In particular, the localization of φ at any prime ideal $\wp' \subseteq R$ with $u \notin \wp'$ is an isomorphism

$$\varphi \otimes 1 : R_{\wp'}^n \xrightarrow{\sim} P \otimes_R R_{\wp'} \,.$$

It follows that $r_{\wp'}(P) = n$ for all prime ideals $\wp' \subseteq R$ with $u \notin \wp'$ and hence $r(P)$ is constant on the open neighborhood $\operatorname{Spec} R \setminus V(u)$ of $\wp$. $\qquad\square$

2.1.2 K-theory and the Geometric Rank

In this subsection, we keep the commutative ring R fixed and consider its prime spectrum $\operatorname{Spec} R$. The set of continuous (i.e. locally constant) maps from $\operatorname{Spec} R$ to the discrete space $\mathbf{N}$ is a semi-ring with addition and multiplication defined pointwise; this semi-ring will be denoted by $[\operatorname{Spec} R, \mathbf{N}]$. We note that the constant function with value 0 (resp. 1) is the 0 (resp. 1) of the semi-ring $[\operatorname{Spec} R, \mathbf{N}]$. The geometric rank function associated with finitely generated projective R-modules defines a map

$$r : \operatorname{Proj}(R) \longrightarrow [\operatorname{Spec} R, \mathbf{N}] \,.$$

The next result shows that r is a morphism of semirings, for the semiring structure on $\operatorname{Proj}(R)$ defined in Remark 1.23(ii).

Lemma 2.9 *(i)* $r(P \oplus Q) = r(P) + r(Q)$ *and* $r(P \otimes_R Q) = r(P)r(Q)$ *in* $[\operatorname{Spec} R, \mathbf{N}]$ *for any finitely generated projective R-modules P and Q.*

(ii) $r(0)$ *(resp. $r(R)$) is the constant function with value 0 (resp. 1).*

Proof. Consider a prime ideal $\wp \in \operatorname{Spec} R$.

(i) Let $r_\wp(P) = n$ and $r_\wp(Q) = m$. The rank $r_\wp(P \oplus Q)$ of $P \oplus Q$ at $\wp$ is, by definition, the rank of the free $R_\wp$-module

$$(P \oplus Q) \otimes_R R_\wp \simeq (P \otimes_R R_\wp) \oplus (Q \otimes_R R_\wp) \simeq R_\wp^n \oplus R_\wp^m = R_\wp^{n+m}$$

and hence $r_\wp(P \oplus Q) = n + m$. Similarly, the rank $r_\wp(P \otimes_R Q)$ of $P \otimes_R Q$ at $\wp$ is the rank of the free $R_\wp$-module

$$(P \otimes_R Q) \otimes_R R_\wp \simeq (P \otimes_R R_\wp) \otimes_{R_\wp} (Q \otimes_R R_\wp) \simeq R_\wp^n \otimes_{R_\wp} R_\wp^m = R_\wp^{nm}$$

and hence $r_\wp(P \otimes Q) = nm$.

(ii) This is clear, since the localization of 0 (resp. R) at $\wp$ is the zero module (resp. the free $R_\wp$-module of rank one). □

The Grothendieck group of the semi-ring $[\operatorname{Spec} R, \mathbf{N}]$ is the commutative ring $[\operatorname{Spec} R, \mathbf{Z}]$ with addition and multiplication defined pointwise. This is a special case of the following lemma.

Lemma 2.10 *Let X be a topological space. Then, the Grothendieck group of the semi-ring $[X, \mathbf{N}]$ of locally constant $\mathbf{N}$-valued functions on X is the commutative ring $[X, \mathbf{Z}]$ of locally constant $\mathbf{Z}$-valued functions on X.*

Proof. For any $f \in [X, \mathbf{Z}]$ we consider the functions $f^+, f^- \in [X, \mathbf{N}]$, which are defined by $f^+ = \max\{f, 0\}$ and $f^- = -\min\{f, 0\}$. Then, $f = f^+ - f^-$ and hence $[X, \mathbf{N}]$ generates the group $[X, \mathbf{Z}]$. The result then follows from Lemma 1.26. □

Corollary 2.11 *There is a ring homomorphism $r : K_0(R) \longrightarrow [\operatorname{Spec} R, \mathbf{Z}]$, which is characterized by $[P] \mapsto r(P)$, whenever P is a finitely generated projective R-module.*

Proof. This is an immediate consequence of Lemmas 2.9 and 2.10, in view of Remark 1.25(v). □

We refer to the ring homomorphism of the corollary above as the geometric rank corresponding to R. In order to describe some of its properties and relate it to the Hattori-Stallings rank r_{HS}, we have to study the ring of locally constant integer-valued functions on $\operatorname{Spec} R$.

In view of the compactness of $\operatorname{Spec} R$ (Lemma 2.5), any locally constant function $f : \operatorname{Spec} R \longrightarrow \mathbf{Z}$ has a finite image, say $\{a_1, \ldots, a_n\}$. The inverse image of the singleton $\{a_i\}$ under f is a clopen subset $X_i \subseteq \operatorname{Spec} R$, $i = 1, \ldots, n$. The next result shows that the decompositions of $\operatorname{Spec} R$ into the disjoint union of clopen subsets $X_1, \ldots, X_n$ correspond bijectively to the decompositions of the ring R into the direct product of ideals $R_1, \ldots, R_n$.

Proposition 2.12 *(i) Assume that R admits a decomposition into a product of the form $\prod_{i=1}^{n} R_i$ and let $e_1, \ldots, e_n$ be the orthogonal idempotents corresponding to that decomposition. Then, $\operatorname{Spec} R$ admits a decomposition into the disjoint union of clopen subsets $X_1, \ldots, X_n$, where $X_i = \operatorname{Spec} R \setminus V(e_i)$ for all $i = 1, \ldots, n$.*

(ii) Any decomposition of $\operatorname{Spec} R$ into the disjoint union of clopen subsets $X_1, \ldots, X_n$ arises from a unique direct product decomposition of R as above.

Proof. (i) Since the e_i's sum up to 1, the ideal generated by them is R. Using the assertions established in Lemma 2.3, we conclude that $\bigcap_{i=1}^{n} V(e_i) = V(\sum_{i=1}^{n} Re_i) = V(R) = \emptyset$ and hence $\bigcup_{i=1}^{n} X_i = \operatorname{Spec} R$. Since the e_i's are orthogonal, we have $V(e_i) \cup V(e_j) = V(e_i e_j) = V(0) = \operatorname{Spec} R$ for $i \neq j$ and hence $X_i \cap X_j = \emptyset$ for $i \neq j$. Therefore, the open subspaces $X_1, \ldots, X_n$ are mutually disjoint and cover $\operatorname{Spec} R$. It follows that each X_i, being the complement of a union of open sets, is also closed.

(ii) Conversely, assume that $\operatorname{Spec} R$ decomposes into the disjoint union of open subsets $X_1, \ldots, X_n$. Then, there are ideals $I_1, \ldots, I_n$ of R such that $X_i = \operatorname{Spec} R \setminus V(I_i)$ for all $i = 1, \ldots, n$. Since the X_i's cover $\operatorname{Spec} R$, $V(\sum_{i=1}^{n} I_i) = \bigcap_{i=1}^{n} V(I_i) = \emptyset$ and hence we may invoke Lemma 2.3(iv) in order to conclude that $\sum_{i=1}^{n} I_i = R$. Therefore, there are elements $x_i \in I_i$, $i = 1, \ldots, n$, such that $\sum_{i=1}^{n} x_i = 1$. For any $i \neq j$ the intersection $X_i \cap X_j$ is empty and hence $V(I_i I_j) = V(I_i) \cup V(I_j) = \operatorname{Spec} R$. In view of Lemma 2.3(v), we conclude that $I_i I_j \subseteq \operatorname{nil} R$ whenever $i \neq j$. In particular, there exists $t \gg 0$ such that $(x_i x_j)^t = 0$ for all $i \neq j$. Since $\sum_{i=1}^{n} x_i = 1$, we have

$$1 = \left(\sum_{i=1}^{n} x_i \right)^{nt-n+1} = \sum \left\{ \prod_{i=1}^{n} x_i^{k_i} : \sum_{i=1}^{n} k_i = nt - n + 1 \right\}. \quad (2.1)$$

Let Λ be the set of n-tuples of non-negative integers $(k_1, \ldots, k_n)$ for which $\sum_{i=1}^{n} k_i = nt - n + 1$. We define a partition of Λ into the disjoint union of subsets $\Lambda_1, \ldots, \Lambda_n$ as follows: Λ_1 consists of those n-tuples $(k_1, \ldots, k_n) \in \Lambda$ for which $k_1 \geq t$, Λ_2 consists of those n-tuples $(k_1, \ldots, k_n) \in \Lambda$ for which $k_1 < t$ and $k_2 \geq t$ and, in general,

$$\Lambda_i = \{(k_1, \ldots, k_n) \in \Lambda : k_1 < t, \ldots, k_{i-1} < t \text{ and } k_i \geq t\}$$

for all $i = 1, \ldots, n$. It is clear that the Λ_i's are mutually disjoint. In order to show that they cover Λ, we consider an n-tuple $(k_1, \ldots, k_n) \in \Lambda$ and assume that $k_i < t$ for all i. Then, $k_i \leq t - 1$ for all i and hence $nt - n + 1 = k_1 + \cdots + k_n \leq n(t-1) = nt - n$, a contradiction. Therefore, there are i's for which $k_i \geq t$. If i_0 is the smallest such i, then $(k_1, \ldots, k_n) \in \Lambda_{i_0}$. We now define the elements $e_1, \ldots, e_n \in R$ by letting

$$e_i = \sum \left\{ \prod_{i=1}^{n} x_i^{k_i} : (k_1, \ldots, k_n) \in \Lambda_i \right\}$$

for all $i = 1, \ldots, n$. Since the Λ_i's form a partition of Λ, (2.1) implies that $1 = \sum_{i=1}^{n} e_i$. We note that for any n-tuple $(k_1, \ldots, k_n) \in \Lambda_i$ we have $k_i \geq t$;

therefore, it follows that e_i is a multiple of x_i^t for all i. In particular, $e_i \in I_i$ for all i and $e_i e_j = 0$ for all $i \neq j$. Hence, the e_i's are orthogonal idempotents summing up to $1 \in R$. Working as in (i) above, the e_i's define a decomposition of $\operatorname{Spec} R$ into the disjoint union of the clopen subsets $Y_1, \ldots, Y_n$, where $Y_i = \operatorname{Spec} R \backslash V(e_i)$ for all $i = 1, \ldots, n$. Since $e_i \in I_i$, we have $V(I_i) \subseteq V(e_i)$ and hence

$$Y_i = \operatorname{Spec} R \setminus V(e_i) \subseteq \operatorname{Spec} R \setminus V(I_i) = X_i$$

for all i. Using the fact that the Y_i's cover $\operatorname{Spec} R$, whereas the X_i's are mutually disjoint, it is easily seen that the above inclusions can't be proper; hence, we have $X_i = Y_i$ for all i.

It only remains to show that the idempotents $e_1, \ldots, e_n$ are uniquely determined by the X_i's. Indeed, suppose that $e'_1, \ldots, e'_n$ is another sequence of idempotents of R, such that $X_i = \operatorname{Spec} R \setminus V(e'_i)$ for all $i = 1, \ldots, n$; then, $V(e_i) = V(e'_i)$ for all i. The following Lemma shows that $e_i = e'_i$ for all i, thereby finishing the proof of the Proposition. $\square$

Lemma 2.13 *Let $e, e' \in R$ be two idempotents, such that $V(e) = V(e') \subseteq \operatorname{Spec} R$. Then, $e = e'$.*

Proof. Since $V(e) \subseteq V(e')$, Lemma 2.3(vi) implies that there exists $n \in \mathbf{N}$ such that $e' = e'^n \in Re$. But then $e' = xe$ for some $x \in R$ and hence $e' = xe = x(ee) = (xe)e = e'e$. Since we also have $V(e') \subseteq V(e)$, a symmetric argument shows that $e = ee' = e'e = e'$. $\square$

We have thus established a close relationship between the set $\operatorname{Idem}(R)$ of idempotents of R and the set $L(\operatorname{Spec} R)$ of clopen subsets of $\operatorname{Spec} R$. The correspondence established in Proposition 2.12 preserves the relevant order structures (cf. Examples 1.3), as shown in the following result.

Proposition 2.14 *The map $u : \operatorname{Idem}(R) \longrightarrow L(\operatorname{Spec} R)$, which is given by $e \mapsto \operatorname{Spec} R \setminus V(e)$, $e \in \operatorname{Idem}(R)$, is an isomorphism of Boolean algebras.*

Proof. The map u is bijective, in view of Proposition 2.12, and obviously preserves 0 and 1. The argument given in the proof of Proposition 2.12(i) shows that u preserves complements. Therefore, it suffices to show that u is $\wedge$-preserving. But this is clear, since

$$
\begin{aligned}
u(e \wedge e') &= u(ee') \\
&= \operatorname{Spec} R \setminus V(ee') \\
&= \operatorname{Spec} R \setminus (V(e) \cup V(e')) \\
&= (\operatorname{Spec} R \setminus V(e)) \cap (\operatorname{Spec} R \setminus V(e')) \\
&= u(e) \cap u(e') \\
&= u(e) \wedge u(e')
\end{aligned}
$$

for all $e, e' \in \operatorname{Idem}(R)$. $\square$

Corollary 2.15 *The commutative ring R has no non-trivial idempotents if and only if the topological space $\operatorname{Spec} R$ is connected.* $\square$

We now consider the Boolean algebra morphism

$$\nu : L(\operatorname{Spec} R) \longrightarrow \operatorname{Idem}(K_0(R)) \, ,$$

which is defined as the composition

$$L(\operatorname{Spec} R) \xrightarrow{u^{-1}} \operatorname{Idem}(R) \xrightarrow{\sigma} \operatorname{Idem}(K_0(R)) \, .$$

Here, σ is the morphism of Proposition 1.28. If $X \subseteq \operatorname{Spec} R$ is a clopen subset and $e \in R$ the corresponding idempotent (so that $X = \operatorname{Spec} R \setminus V(e)$), then $\nu(X)$ is the class of the finitely generated projective R-module Re in the K-theory group $K_0(R)$. The $K_0(R)$-valued discrete integral associated with ν (cf. Appendix B) is a homomorphism of commutative rings

$$\mathcal{I}_\nu : [\operatorname{Spec} R, \mathbf{Z}] \longrightarrow K_0(R) \, .$$

For any integer-valued locally constant function f on $\operatorname{Spec} R$, there exist integers $a_1, \ldots, a_n$ and a decomposition of $\operatorname{Spec} R$ into the disjoint union of clopen subsets $X_1, \ldots, X_n$, such that $f = \sum_{i=1}^n a_i \chi_{X_i}$. If $e_1, \ldots, e_n$ are the orthogonal idempotents of R corresponding to the above decomposition of $\operatorname{Spec} R$ and $R = \prod_{i=1}^n R_i$ the induced direct product decomposition (so that $R_i = Re_i$ for all i), then

$$\mathcal{I}_\nu(f) = \sum_{i=1}^n a_i [R_i] \in K_0(R) \, .$$

In particular, if f is $\mathbf{N}$-valued then the a_i's are non-negative and we may consider the finitely generated projective R-module $\bigoplus_{i=1}^n R_i^{a_i}$, which we denote by R^f. Then, the formula above becomes

$$\mathcal{I}_\nu(f) = \left[R^f \right] \in K_0(R) \, .$$

Theorem 2.16 *The geometric rank* $r : K_0(R) \longrightarrow [\operatorname{Spec} R, \mathbf{Z}]$ *admits the ring homomorphism* $\mathcal{I}_\nu$ *as a right inverse. In particular,* r *is surjective.*

Proof. We have to prove that the composition $r \circ \mathcal{I}_\nu$ is the identity map on $[\operatorname{Spec} R, \mathbf{Z}]$. Let us consider a locally constant $\mathbf{N}$-valued function f on $\operatorname{Spec} R$ and the finitely generated projective R-module $P = R^f$. We shall prove that the geometric rank $r(P)$ of P is equal to f; since the group $[\operatorname{Spec} R, \mathbf{Z}]$ is generated by these f's (cf. Lemma 2.10), this will finish the proof. We know that f determines a decomposition of $\operatorname{Spec} R$ into the disjoint union of clopen subsets $X_1, \ldots, X_n$, in such a way that its restriction on X_i is constant, say with value $a_i \in \mathbf{N}$, $i = 1, \ldots, n$. Let $e_1, \ldots, e_n$ be the orthogonal idempotents of R corresponding to this decomposition of $\operatorname{Spec} R$ and consider the ideals $R_i = Re_i$, $i = 1, \ldots, n$. Then, $P = \bigoplus_{i=1}^n R_i^{a_i}$. In order to show that $r(P) = f$, let us fix a prime $\wp \in \operatorname{Spec} R$. Of course, $\wp$ lies in one of the X_i's; without loss of generality, we assume that $\wp \in X_1$. Since f is constant on X_1 with value a_1, we have $f(\wp) = a_1$. Before computing the rank $r_\wp(P)$ of P at $\wp$, we note that $X_1 = \operatorname{Spec} R \setminus V(e_1)$ and hence $\wp \notin V(e_1)$, i.e. $e_1 \notin \wp$. We need the following lemma.

Lemma 2.17 *Let $\wp \subseteq R$ be a prime ideal. Consider a sequence $e_1, \ldots, e_n$ of orthogonal idempotents of R that sum up to 1 and the associated decomposition of R into the direct product of ideals $R_1, \ldots, R_n$ (so that $R_i = Re_i$ for all i). If $e_1 \notin \wp$ then:*

(i) $e_1/1 = 1/1 \in R_\wp$ and hence $(R_1)_\wp = R_\wp$,

(ii) $e_i/1 = 0/1 \in R_\wp$ and hence $(R_i)_\wp = 0$ for all $i > 1$.

Proof of Lemma 2.17. (i) Since $e_1 \notin \wp$, $e_1/1$ is a unit of the localization $R_\wp$. Being also an idempotent, it must be equal to $1/1 \in R_\wp$. Hence, $(R_1)_\wp = R_\wp \cdot (e_1/1) = R_\wp$.

(ii) Since $e_1 e_i = 0$ for all $i > 1$, it follows from (i) that $e_i/1 = 0/1 \in R_\wp$ for all $i > 1$. Therefore, $(R_i)_\wp = R_\wp \cdot (e_i/1) = 0$ for all $i > 1$. $\qquad\square$

Proof of Theorem 2.16 (cont.). In view of Lemma 2.17, we have

$$P_\wp = (\bigoplus_{i=1}^n R_i^{a_i})_\wp = \bigoplus_{i=1}^n (R_i^{a_i})_\wp = \bigoplus_{i=1}^n (R_i)_\wp^{a_i} = (R_1)_\wp^{a_1} = (R_\wp)^{a_1}.$$

Taking into account the definition of the geometric rank function $r(P)$, it follows that $r_\wp(P) = a_1 = f(\wp)$ and hence the proof is finished. $\qquad\square$

Remark 2.18 Theorem 2.16 implies that the K-theory ring $K_0(R)$ of the commutative ring R is the semi-direct product of $[\operatorname{Spec} R, \mathbf{Z}]$ by the ideal $\ker r$. Since the ring $[\operatorname{Spec} R, \mathbf{Z}]$ is reduced, the ideal $\ker r$ contains the nil radical $\operatorname{nil} K_0(R)$. In fact, one can show that $\ker r$ consists of nilpotent elements and hence $\ker r = \operatorname{nil} K_0(R)$ (cf. Exercise 2.3.5).

We now consider the isomorphism of Boolean algebras

$$u^{-1} : L(\operatorname{Spec} R) \longrightarrow \operatorname{Idem}(R),$$

defined in Proposition 2.14. Recall that if X is a clopen subset of $\operatorname{Spec} R$, then $u^{-1}(X)$ is the idempotent $e \in R$, which is such that $X = \operatorname{Spec} R \setminus V(e)$. The discrete R-valued integral associated with u^{-1} (cf. Appendix B) is a homomorphism of commutative rings

$$\mathcal{I}_{u^{-1}} : [\operatorname{Spec} R, \mathbf{Z}] \longrightarrow R.$$

If f is an integer-valued locally constant function on $\operatorname{Spec} R$, then there are suitable integers $a_1, \ldots, a_n$ and a decomposition of $\operatorname{Spec} R$ into the disjoint union of clopen subsets $X_1, \ldots, X_n$, such that $f = \sum_{i=1}^n a_i \chi_{X_i}$. If $e_1, \ldots, e_n$ are the orthogonal idempotents of R corresponding to the above decomposition of $\operatorname{Spec} R$, so that $X_i = \operatorname{Spec} R \setminus V(e_i)$ for all i, then

$$\mathcal{I}_{u^{-1}}(f) = \sum_{i=1}^n a_i e_i \in R.$$

Theorem 2.19 *The Hattori-Stallings rank r_{HS} is the composition*

$$K_0(R) \xrightarrow{\ r\ } [Spec\, R, \mathbf{Z}] \xrightarrow{\ \mathcal{I}_{u^{-1}}\ } R,$$

where r is the geometric rank associated with R.

Proof. Let E be an idempotent $N \times N$ matrix with entries in R, $\widetilde{E}$ the corresponding endomorphism of the free R-module R^N and $P = \operatorname{im} \widetilde{E}$ the associated finitely generated projective R-module. The geometric rank of P is a locally constant $\mathbf{N}$-valued function f on $\operatorname{Spec} R$. We know that there is a covering of $\operatorname{Spec} R$ by disjoint clopen subsets $X_1, \ldots, X_n$ and non-negative integers $a_1, \ldots, a_n$, such that $f = \sum_{i=1}^{n} a_i \chi_{X_i}$. The covering of $\operatorname{Spec} R$ by the X_i's corresponds to a family of orthogonal idempotents $e_1, \ldots, e_n \in R$ which sum up to 1. Hence, that covering determines a decomposition of the ring R into the direct product of ideals $R_i = Re_i$, $i = 1, \ldots, n$. With this notation, we have to show that

$$\operatorname{tr}(E) = \sum_{i=1}^{n} a_i e_i \in R . \tag{2.2}$$

Since $K_0(R)$ is generated by the classes of finitely generated projective R-modules, this will finish the proof. In order to prove (2.2), we need the principle described in the following lemma.

Lemma 2.20 *An element* $r \in R$ *is equal to zero if and only the element* $r/1 \in R_\wp$ *is equal to zero for any prime ideal* $\wp \in \operatorname{Spec} R$.

Proof of Lemma 2.20. We assume that r is zero locally and try to show that it is actually zero. Let $I = \{x \in R : xr = 0\}$ be the annihilator of r in R. If $I \neq R$, then I must be contained in a maximal ideal $\mathbf{m} \subseteq R$. Since $r/1 = 0/1 \in R_{\mathbf{m}}$, there exists $s \in R \setminus \mathbf{m}$ with $sr = 0$. But then $s \in I$ and this contradicts the assumption that $I \subseteq \mathbf{m}$. It follows that $I = R$ and hence $r = 0$, as needed. □

Proof of Theorem 2.19 (cont.). Let $\wp$ be a prime ideal of R, $R' = R_\wp$ and $E' \in \mathbf{M}_N(R')$ the matrix whose entries are the canonical images of the entries of E in R'. Then, we have to show that

$$\operatorname{tr}(E') = \sum_{i=1}^{n} a_i e_i/1 \in R' .$$

Without loss of generality, we may assume that $\wp \in X_1 = \operatorname{Spec} R \setminus V(e_1)$. Then, $e_1 \notin \wp$ and hence Lemma 2.17 implies that $e_1/1 = 1/1 \in R'$ and $e_i/1 = 0/1 \in R'$ for all $i > 1$. Therefore, the equality above reduces to

$$\operatorname{tr}(E') = a_1/1 \in R' .$$

We now consider the R'-module $P' = P \otimes_R R'$. Being a finitely generated projective module over the local ring R', P' is free; in fact, it is free with rank the value of the geometric rank function $f = r(P)$ at the point $\wp \in \operatorname{Spec} R$. But $\wp \in X_1$ and f is constant on X_1 with value a_1. It follows that $f(\wp) = a_1$ and hence $P' \simeq R'^{a_1}$. The endomorphism $\widetilde{E'}$ of R'^N associated with the matrix E' is that induced from the endomorphism $\widetilde{E}$ of R^N by localization. Therefore,

$$P' = P \otimes_R R' = \operatorname{im} \widetilde{E} \otimes_R R' = \operatorname{im}\left(\widetilde{E} \otimes 1\right) = \operatorname{im} \widetilde{E'} .$$

We also consider the complementary submodule $Q' = \text{im}\left(1 - \widetilde{E'}\right)$ of P' in R'^N. Being a finitely generated projective module over the local ring R', Q' is free. Since $P' \oplus Q' = R'^N$, we must have $Q' \simeq R'^{N-a_1}$. We can construct a basis of R'^N combining a basis of P' and a basis of Q'. The matrix of the endomorphism $\widetilde{E'}$ with respect to that basis is diagonal with a_1 ones and $N - a_1$ zeroes along the diagonal. Since E' is conjugate to that matrix, we have $\text{tr}(E') = a_1 \cdot (1/1) + (N - a_1) \cdot (0/1) = a_1/1 \in R'$, as needed.   □

In the sequel, we use Theorems 2.16 and 2.19 in the form of the following corollaries.

Corollary 2.21 *The Hattori-Stallings rank* $r_{HS} : K_0(R) \longrightarrow R$ *is a ring homomorphism with image the additive subgroup of R generated by the subset* $Idem(R) \subseteq R$.

Proof. Being a composition of ring homomorphisms (Theorem 2.19), r_{HS} is a ring homomorphism as well. Since the geometric rank r is surjective (Theorem 2.16), it follows that $\text{im}\, r_{HS} = \text{im}\, \mathcal{I}_{u^{-1}}$. Since the image of $\mathcal{I}_{u^{-1}}$ is easily seen to be the subgroup of R generated by its idempotents, the proof is finished.   □

Corollary 2.22 *Assume that the commutative ring R has characteristic 0. Then, the following assertions are equivalent:*

(i) R has no non-trivial idempotents.

(ii) The Hattori-Stallings rank of any finitely generated projective R-module is equal to $n \cdot 1 \in R$, for some $n \in \mathbf{Z}$.

Proof. (i)→(ii): If R has no non-trivial idempotents, then Corollary 2.21 implies that $\text{im}\, r_{HS} = \mathbf{Z} \cdot 1$.

(ii)→(i): Assume that $\text{im}\, r_{HS} = \mathbf{Z} \cdot 1$ and consider an idempotent $e \in R$. Then, $e = n \cdot 1$ for some $n \in \mathbf{Z}$. In view of the assumption about the characteristic of R, the equality $n^2 \cdot 1 = n \cdot 1 \in R$ occurs only when $n = 0, 1$; hence, the idempotent e is trivial.   □

2.1.3 The Connectedness of Spec kG

In this subsection, we specialize the previous discussion and consider the case where $R = kG$ is the group ring of an abelian group G with coefficients in a subring k of the field $\mathbf{C}$ of complex numbers, such that $k \cap \mathbf{Q} = \mathbf{Z}$. We prove that R has no non-trivial idempotents and hence conclude that the Hattori-Stallings rank of any finitely generated projective R-module is an integer multiple of $1 \in R$. It will follow that the abelian group G satisfies Bass' conjecture.

Proposition 2.23 *Let k be a subring of $\mathbf{C}$ with $k \cap \mathbf{Q} = \mathbf{Z}$, G an abelian group and $R = kG$ the corresponding group ring. Then, R has no non-trivial idempotents.*

Proof. Clearly, it suffices to consider the case where the group G is finitely generated. Then, $G = T \times F$, where T is a finite group and F is free abelian. Let $e = \sum_{i=1}^{n} e_i g_i \in kG$ be a non-trivial idempotent, where $e_i \in k \setminus \{0\}$ for all i and $g_i \neq g_j$ for all $i \neq j$. Since e is non-trivial, there is at least one i such that $g_i \neq 1$; without loss of generality, we assume that $g_1 \neq 1$. We now choose an integer $n \gg 0$, in such a way that the images $\overline{g_1}, \ldots, \overline{g_n}$ of the g_i's in the quotient group $\overline{G} = T \times F/nF$ satisfy:

(i) $\overline{g_i} \neq \overline{g_j}$ for all $i \neq j$ and
(ii) $\overline{g_1} \neq \overline{1}$.

Then, $\overline{e} = \sum_{i=1}^{n} e_i \overline{g_i}$ is a non-trivial idempotent in $k\overline{G}$. Therefore, it suffices to consider the case where the group G is finite. This case is taken care of by the following result.   □

Lemma 2.24 *Let k be a subring of $\mathbf{C}$ with $k \cap \mathbf{Q} = \mathbf{Z}$, G a finite group (not necessarily abelian) and $R = kG$ the corresponding group ring. Then, R has no non-trivial idempotents.*

Proof. Let n be the order of G and

$$L : kG \longrightarrow \text{End}_k(kG) \simeq \mathbf{M}_n(k)$$

the left regular representation. Then, $\text{tr}(L(a)) = na_1$ for all $a = \sum_g a_g g \in kG$. Indeed, by linearity it suffices to consider the case where $a = g$ is an element of G. In that case, $L(a)$ permutes the standard basis of kG and hence the trace $\text{tr}(L(a))$ counts the fixed points of the permutation. Since that permutation is fixed-point-free if $g \neq 1$ and the identity if $g = 1$, the formula follows. In particular, if $e = \sum_g e_g g \in kG$ is an idempotent then $\text{tr}(L(e)) = ne_1$. Viewing $L(e)$ as an $n \times n$ idempotent matrix with complex entries, we conclude that its trace is an integer i with $0 \leq i \leq n$. Therefore, $e_1 = \frac{1}{n}\text{tr}(L(e))$ is a rational number of the form $\frac{i}{n}$, for some integer i between 0 and n. Since $e_1 \in k$, our assumption shows that e_1 can be either 0 or 1. In the former case, $\text{tr}(L(e)) = 0$ and hence $L(e)$ is the zero matrix; therefore, $e = 0$. In the latter case, $\text{tr}(L(e)) = n$ and hence $L(e)$ is the identity matrix; therefore, $e = 1$. □

Remark 2.25 The proof of Proposition 2.23 works more generally in the case where the group G is locally residually finite. In fact, we shall prove in the following chapter that for *any* group G the group ring kG has no non-trivial idempotents, whenever k is a subring of $\mathbf{C}$ with $k \cap \mathbf{Q} = \mathbf{Z}$ (cf. Corollary 3.21).

We are now ready to state and prove the main result of this section.

Theorem 2.26 *Abelian groups satisfy Bass' conjecture.*

Proof. Let k be a subring of $\mathbf{C}$ with $k \cap \mathbf{Q} = \mathbf{Z}$, G an abelian group and $R = kG$ the corresponding group ring. Bass' conjecture for G asserts that the Hattori-Stallings rank of any finitely generated projective R-module is contained in $k \cdot 1 \subseteq R$. Since R has no idempotents other than 0 and 1 (cf. Proposition 2.23), the proof follows from Corollary 2.22.   □

Remark 2.27 Let k be a subring of $\mathbf{C}$ with $k \cap \mathbf{Q} = \mathbf{Z}$, G a group and $R = kG$ the corresponding group ring. We already know that the Hattori-Stallings rank of a finitely generated projective R-module is contained in $k \cdot [1] \subseteq T(R)$ if and only if it is an integer multiple of $[1] \in T(R)$ (cf. Corollary 1.48). Therefore, it follows from Corollaries 2.15 and 2.22 that Bass' conjecture for an abelian group G is equivalent to the assertion that R has no non-trivial idempotents and hence to the connectedness of the prime spectrum Spec R.

2.2 The Case of Finite Groups

Let G be a finite group, k a subring of the field $\mathbf{C}$ of complex numbers such that $k \cap \mathbf{Q} = \mathbf{Z}$ and E an idempotent matrix with entries in kG. In this section, we prove that $r_g(E) = 0$ whenever $g \neq 1$, i.e. that the finite group G satisfies Bass' conjecture. To that end, we consider the cyclic subgroup $H \leq G$ generated by g and show that E induces a certain idempotent matrix E' with entries in kH, in such a way that $r_g(E) = 0$ if and only if $r_g(E') = 0$. On the other hand, we know that H satisfies Bass' conjecture (being abelian) and hence $r_g(E')$ is indeed zero. We also establish the equivalence between the assertion that finite groups satisfy Bass' conjecture and a result of Swan on induced representations.

In §2.2.1, we consider a ring homomorphism $\varphi : A \longrightarrow B$, making B a finitely generated free right A-module, and construct a transfer homomorphism $\varphi^* : T(B) \longrightarrow T(A)$. Specializing the discussion, we consider in §2.2.2 the case of the pair of group rings associated with a group G and a subgroup H of it of finite index. The relation between the Hattori-Stallings rank of a finitely generated projective module over the group ring of G and that of the associated finitely generated projective module over the group ring of H will immediately yield Bass' conjecture for finite groups. In §2.2.3, we explain how the Hattori-Stallings rank of a finitely generated projective module over the group ring of a finite group determines and is, in fact, determined by the character of the associated finite dimensional representation of the group. In this way, we obtain Swan's theorem on induced representations of finite groups. It was precisely this result of Swan that Bass attempted to generalize, by formulating the conjecture we are studying for arbitrary groups.

2.2.1 The Transfer Homomorphism

Let us consider a ring homomorphism $\varphi : A \longrightarrow B$ and the induced right A-module structure on B. Since multiplication to the left by elements of B commutes with the right A-action, there is a ring homomorphism

$$L : B \longrightarrow \text{End}_A B \, ,$$

where $L(b) = L_b$ is the left multiplication with b for all $b \in B$. We assume that there is a positive integer n, such that $B \simeq A^n$ as right A-modules. In that case, we may fix an A-basis of B and obtain a ring homomorphism

$$B \xrightarrow{L} \text{End}_A B \simeq \text{End}_A(A^n) = \mathbf{M}_n(A) \, ,$$

which we denote by φ_1^*. For any $t \geq 1$ we consider the homomorphism of matrix rings

$$\varphi_t^* : \mathbf{M}_t(B) \longrightarrow \mathbf{M}_t(\mathbf{M}_n(A)) = \mathbf{M}_{tn}(A) \, ,$$

which is induced by φ_1^*.

Example 2.28 Let us consider the inclusion $\mathbf{R} \hookrightarrow \mathbf{C}$. Then, $\mathbf{C} \simeq \mathbf{R}^2$ as real vector spaces and the associated 2-dimensional real representation of $\mathbf{C}$ corresponding to the choice of the basis $\{1, i\}$ of $\mathbf{C}$ over $\mathbf{R}$, maps a complex number $a + bi$ $(a, b \in \mathbf{R})$ onto the matrix $\begin{pmatrix} a & -b \\ b & a \end{pmatrix}$.

Proposition 2.29 *Let $\varphi : A \longrightarrow B$ be a ring homomorphism, such that $B \simeq A^n$ as right A-modules for some $n \geq 1$. Then, there is a unique group homomorphism $\varphi^* : T(B) \longrightarrow T(A)$, which makes the following diagram commutative for all $t \geq 1$*

$$\begin{array}{ccc} \mathbf{M}_t(B) & \xrightarrow{\varphi_t^*} & \mathbf{M}_{tn}(A) \\ {\scriptstyle r^B} \downarrow & & \downarrow {\scriptstyle r^A} \\ T(B) & \xrightarrow{\varphi^*} & T(A) \end{array}$$

Here, $r^A = r^A_{HS}$ and $r^B = r^B_{HS}$ are the Hattori-Stallings trace maps associated with A and B respectively.

Proof. Let us denote by $\mathcal{D}_t$ the above diagram. Since $\varphi_1^* : B \longrightarrow \mathbf{M}_n(A)$ is a ring homomorphism, the composition

$$B \xrightarrow{\varphi_1^*} \mathbf{M}_n(A) \xrightarrow{r^A} T(A)$$

vanishes on the set $\{bb' - b'b : b, b' \in B\}$. Hence, this composition induces by passage to the quotient a group homomorphism $\varphi^* : T(B) \longrightarrow T(A)$. By the very definition of φ^* the following diagram is commutative

$$\begin{array}{ccc} B & \xrightarrow{\varphi_1^*} & \mathbf{M}_n(A) \\ \downarrow & & \downarrow {\scriptstyle r^A} \\ T(B) & \xrightarrow{\varphi^*} & T(A) \end{array}$$

Since the above diagram is precisely $\mathcal{D}_1$, we have established the uniqueness assertion of the Proposition. In order to show that φ^* makes $\mathcal{D}_t$ commutative for all $t > 1$ as well, we consider a matrix $X = (b_{ij}) \in \mathbf{M}_t(B)$ and the corresponding matrix $X' = \varphi_t^*(X) = (\varphi_1^*(b_{ij})) \in \mathbf{M}_t(\mathbf{M}_n(A)) = \mathbf{M}_{tn}(A)$. We have $\mathrm{tr}(X') = \sum_{i=1}^t \mathrm{tr}(\varphi_1^*(b_{ii}))$ and hence

$$r^A(X') = \overline{\mathrm{tr}(X')} = \sum_{i=1}^t \overline{\mathrm{tr}(\varphi_1^*(b_{ii}))} = \sum_{i=1}^t r^A(\varphi_1^*(b_{ii})) \ .$$

On the other hand, we have

$$\varphi^*\left(r^B(X)\right) = \varphi^*\left(\overline{\mathrm{tr}(X)}\right) = \varphi^*\left(\sum_{i=1}^t \overline{b_{ii}}\right) = \sum_{i=1}^t \varphi^*\left(\overline{b_{ii}}\right) \ .$$

In view of the commutativity of $\mathcal{D}_1$, we have $\varphi^*(\overline{b_{ii}}) = r^A(\varphi_1^*(b_{ii}))$ for all i; hence, we conclude that $r^A(X') = \varphi^*\left(r^B(X)\right)$, as needed.   □

Definition 2.30 *Let* $\varphi : A \longrightarrow B$ *be a ring homomorphism, such that* $B \simeq A^n$ *as right A-modules for some $n \geq 1$. The group homomorphism* $\varphi^* : T(B) \longrightarrow T(A)$, *defined in Proposition 2.29, is called the transfer homomorphism associated with* φ.

Remarks 2.31 (i) Let $\varphi : A \longrightarrow B$ be a ring homomorphism, which makes B a finitely generated free right A-module. Even though the ring homomorphisms φ_t, $t \geq 1$, depend upon the choice of a basis of B as a right A-module, the transfer φ^* is independent of that choice. This is clear, since a different choice of a basis changes φ_1^* by an inner automorphism of the matrix algebra $\mathbf{M}_n(A)$, whereas

$$r^A : \mathbf{M}_n(A) \longrightarrow T(A)$$

is invariant under conjugation.

(ii) If $\imath$ is the inclusion of $\mathbf{R}$ into $\mathbf{C}$ (cf. Example 2.28), then $\imath^* : \mathbf{C} \longrightarrow \mathbf{R}$ is the map $z \mapsto 2\mathrm{Re}\,z$, $z \in \mathbf{C}$.

2.2.2 Subgroups of Finite Index

Let G be a group, $H \leq G$ a subgroup, k a commutative ring and

$$\iota : kH \hookrightarrow kG$$

the inclusion of the corresponding group rings. If S is a set of representatives of the right H-cosets $\{gH : g \in G\}$, then there is a decomposition $kG = \bigoplus_{s \in S} s \cdot kH$ of right kH-modules. As every summand in this decomposition is isomorphic with the free right kH-module kH, the right kH-module kG is free. If, in addition, the index $[G : H] = \mathrm{card}\,S$ is finite then the ring homomorphism ι does satisfy the hypothesis of §2.2.1. In the present subsection, we study the specific properties of the transfer homomorphism ι^* associated with such a subgroup $H \leq G$ of finite index. In order to make explicit the dependence upon the containing group, we shall denote for any $h \in H$ (resp. $g \in G$) its conjugacy class in H (resp. in G) by $[h]_H$ (resp. by $[g]_G$).

Proposition 2.32 *Let G be a group and $H \leq G$ a subgroup of finite index. Then, there is a function*

$$\mu = \mu(G, H) : H \longrightarrow \mathbf{N} \setminus \{0\} ,$$

having the following two properties:

(i) μ is constant on H-conjugacy classes and

(ii) For any commutative ring k the transfer $\iota^* : T(kG) \longrightarrow T(kH)$, associated with the inclusion ι of the corresponding group rings, is such that $\iota^*([g]_G) = \sum \{\mu(h)[h]_H : [h]_H \subseteq [g]_G\}$ for all $g \in G$.

Proof. Let $[G : H] = n$ and fix a set $S = \{s_1, \ldots, s_n\}$ of representatives of the right H-cosets in G. An element $g \in G$ induces the left multiplication map L_g on the right kH-module $kG = \bigoplus_{i=1}^n s_i \cdot kH$. In order to represent L_g by a matrix with entries in kH, we begin by noting that $L_g(s_i) = gs_i$ for all i. The element $gs_i \in G$ can be written in the form $s_j h_i$ with $h_i \in H$ for a unique $j \in \{1, \ldots, n\}$; let σ be the self-map of the set $\{1, \ldots, n\}$, defined by $i \mapsto j$.[2] Therefore, $L_g(s_i) = s_{\sigma(i)} h_i$ and hence the i-th column of the matrix corresponding to L_g has h_i in the $\sigma(i)$-th row and 0's in all other rows. In particular, the trace of L_g is the sum of those h_i's for which $\sigma(i) = i$. But $\sigma(i) = i$ if and only if $s_i^{-1} g s_i \in H$, in which case $s_i^{-1} g s_i = h_i$. It follows that

$$\mathrm{tr}(L_g) = \sum \{s_i^{-1} g s_i : 1 \leq i \leq n \text{ and } s_i^{-1} g s_i \in H\} \in kH$$

and hence by the very definition of ι^* (diagram $\mathcal{D}_1$ in the proof of Proposition 2.29) we have

$$\iota^*([g]_G) = \sum \{[s_i^{-1} g s_i]_H : 1 \leq i \leq n \text{ and } s_i^{-1} g s_i \in H\} \in T(kH) . \quad (2.3)$$

Since any $s_i^{-1} g s_i$ is conjugate to g in G, we conclude that ι^* maps $[g]_G$ onto a sum (with multiplicities) of H-conjugacy classes $[h]_H$ that are contained in $[g]_G$. Moreover, any H-conjugacy class $[h]_H$ contained in $[g]_G$ is of the form $[s_i^{-1} g s_i]_H$ for some i and hence occurs with a positive multiplicity in $\iota^*([g]_G)$. In order to complete the proof, we define for any $h \in H$ the integer $\mu(h)$ as the multiplicity considered above. More formally, for any $h \in H$ we choose an element $g \in G$ in the conjugacy class $[h]_G$ of h in G and define $\mu(h)$ as the cardinality of the set of those integers i with $1 \leq i \leq n$, for which $s_i^{-1} g s_i \in H$ and $[h]_H = [s_i^{-1} g s_i]_H$.[3] It is clear that μ is constant on H-conjugacy classes and hence satisfies condition (i). By its very definition and (2.3), μ satisfies condition (ii) as well. ∎

---

[2] It is easily seen that σ is, in fact, a permutation of $\{1, \ldots, n\}$.

[3] The reader can easily provide a direct argument showing that $\mu(h)$ does not depend upon the choice of $g \in [h]_G$ and is, in fact, strictly positive; cf. Exercise 2.3.8.

Corollary 2.33 *Let G be a group and $H \leq G$ a subgroup of finite index. Then, there is a function*

$$\mu = \mu(G, H) : H \longrightarrow \mathbf{N} \setminus \{0\} ,$$

having the following two properties:

(i) μ is constant on H-conjugacy classes and

(ii) For any commutative ring k the transfer $\iota^* : T(kG) \longrightarrow T(kH)$, associated with the inclusion ι of the corresponding group rings, maps any element $\rho = \sum_{[g]_G} \rho_g [g]_G \in T(kG)$ (viewed as a function on G which is constant on the G-conjugacy classes and vanishes in all but finitely many of them) onto $\sum_{[h]_H} \mu(h) \rho_h [h]_H \in T(kH)$. □

Corollary 2.34 *Let G be a group, $H \leq G$ a subgroup of finite index, k a commutative ring and $\iota : kH \hookrightarrow kG$ the inclusion of the corresponding group rings. Then, for any $h \in H$ there is a positive integer $\mu(h)$ inducing a commutative diagram*

$$
\begin{array}{ccc}
\mathbf{M}_t(kG) & \xrightarrow{\iota_t^*} & \mathbf{M}_{tn}(kH) \\
r_h^{(G)} \downarrow & & \downarrow r_h^{(H)} \\
k & \xrightarrow{\mu(h)} & k
\end{array}
$$

Here, the bottom arrow is multiplication by $\mu(h)$, whereas $r_h^{(G)}$ (resp. $r_h^{(H)}$) denotes the map defined in §1.2.1 just before Proposition 1.46, corresponding to G and its element h (resp. to H and its element h).

Proof. Let $X \in \mathbf{M}_t(kG)$ and $X' = \iota_t^*(X) \in \mathbf{M}_{tn}(kH)$. Then,

$$
\begin{aligned}
\sum_{[h]_H} r_h^{(H)}(X')[h]_H &= r^{(H)}(X') \\
&= r^{(H)}(\iota_t^*(X)) \\
&= \iota^*\big(r^{(G)}(X)\big) \\
&= \iota^*\Big(\sum_{[g]_G} r_g^{(G)}(X)[g]_G\Big) \\
&= \sum_{[h]_H} \mu(h) r_h^{(G)}(X)[h]_H
\end{aligned}
$$

where the third (resp. fifth) equality follows invoking Proposition 2.29 (resp. Corollary 2.33). It follows that $r_h^{(H)}(X') = \mu(h)\, r_h^{(G)}(X)$ for all $h \in H$, as needed. □

2.2.3 Swan's Theorem

We are now ready to prove that finite groups satisfy Bass' conjecture. This will be a consequence of the following result.

Proposition 2.35 *Let G be a group whose center C is a subgroup of finite index. Then, G satisfies Bass' conjecture.*

Proof. Let k be a subring of the field $\mathbf{C}$ of complex numbers with $k \cap \mathbf{Q} = \mathbf{Z}$, $E \in \mathbf{M}_t(kG)$ an idempotent $t \times t$ matrix and $h \in G \setminus \{1\}$. The subgroup $H \leq G$ generated by C and h is abelian and has finite index in G, say $[G : H] = n$. We let ι denote the inclusion of kH into kG and consider the idempotent matrix $E' = \iota_t^*(E) \in \mathbf{M}_{tn}(kH)$. Then, Corollary 2.34 implies that $r_h^{(H)}(E') = \mu(h) \, r_h^{(G)}(E)$ for a suitable positive integer $\mu(h)$. Being abelian, the group H satisfies Bass' conjecture (cf. Theorem 2.26) and hence $r_h^{(H)}(E') = 0$. It follows that $r_h^{(G)}(E) = 0$ as well. $\qquad\square$

Corollary 2.36 *Finite groups satisfy Bass' conjecture.* $\qquad\square$

Remark 2.37 Let G be a group and E an idempotent matrix with entries in kG, where k is a subring of $\mathbf{C}$ with $k \cap \mathbf{Q} = \mathbf{Z}$. The argument in the proof of Proposition 2.35 shows that $r_h(E) = 0$, whenever $h \in G \setminus \{1\}$ is an element of a subgroup $H \leq G$ that satisfies the following two conditions:
  (i) The index $[G : H]$ is finite and
  (ii) H itself satisfies Bass' conjecture.

Using some basic representation theory, we may reformulate Corollary 2.36, in order to obtain the following result, that served as the primary motivation for Bass to formulate the conjecture we are studying.

Theorem 2.38 *(Swan) Let k be a subring of $\mathbf{C}$ with $k \cap \mathbf{Q} = \mathbf{Z}$, K its field of fractions and G a finite group. Then, for any finitely generated projective kG-module P, the KG-module $P \otimes_k K$ is free.*

In order to prove Swan's theorem, we examine the relation between the Hattori-Stallings rank of P and the character of the associated representation of the group G over K.

Let K be any field of characteristic 0, G a finite group and V a finite dimensional K-representation of G; equivalently, one may describe V as a finitely generated KG-module. We denote by α the corresponding homomorphism of G into $GL(V)$. In view of Maschke's theorem (Theorem 1.9), the KG-module V is projective and hence we may consider its Hattori-Stallings rank

$$ r_{HS}(V) = \sum\nolimits_{[g] \in \mathcal{C}(G)} r_g(V)[g] \in T(KG) = \bigoplus\nolimits_{[g] \in \mathcal{C}(G)} K \cdot [g] \, . $$

The character $\chi = \chi_V : G \longrightarrow K$ of V maps any element $g \in G$ onto the trace of the endomorphism $\alpha(g) \in GL(V)$.

Lemma 2.39 *Let K be a field of characteristic 0, G a finite group and V a finite dimensional K-representation of G. Then, the Hattori-Stallings rank $r_{HS}(V)$ of V is equal to $\dfrac{1}{|G|} \sum_{g \in G} \chi(g^{-1})[g]$ and hence $r_g(V) = \dfrac{\chi(g^{-1})}{|C_g|}$ for all $g \in G$, where C_g denotes the centralizer of g in G.*

Proof. We consider the induced KG-module $KG \otimes_K V_0$, where V_0 is the K-vector space obtained from the KG-module V by restriction of scalars, and define the K-linear maps

$$i : V \longrightarrow KG \otimes_K V_0 \quad \text{and} \quad \pi : KG \otimes_K V_0 \longrightarrow V \,,$$

by letting

$$i(v) = \frac{1}{|G|} \sum_{g \in G} g \otimes g^{-1}v \quad \text{and} \quad \pi(g \otimes v) = gv$$

for all $v \in V$ and $g \in G$. It is easily seen that both i and π are KG-linear, whereas $\pi \circ i$ is the identity map on V (cf. the proof of Maschke's theorem). Therefore, the endomorphism

$$i \circ \pi \in \operatorname{End}_{KG}(KG \otimes_K V_0)$$

is an idempotent which identifies the projective KG-module V with a direct summand of the free KG-module $KG \otimes_K V_0$. It follows that the Hattori-Stallings rank $r_{HS}(V)$ of V is equal to the class of $\operatorname{tr}(i \circ \pi) \in KG$ in the quotient group $T(KG)$. Let $\dim_K V_0 = n$ and choose a K-basis $v_1, \ldots, v_n$ of V_0; then, $1 \otimes v_1, \ldots, 1 \otimes v_n$ is a KG-basis of $KG \otimes_K V_0$. For any $l = 1, \ldots, n$ we compute

$$
\begin{aligned}
(i \circ \pi)(1 \otimes v_l) &= i(v_l) \\
&= \frac{1}{|G|} \sum_{g \in G} g \otimes g^{-1}v_l \\
&= \frac{1}{|G|} \sum_{g \in G} g \otimes \sum_{k=1}^{n} \alpha_{kl}(g^{-1})v_k \\
&= \sum_{k=1}^{n} \left[\frac{1}{|G|} \sum_{g \in G} \alpha_{kl}(g^{-1})g \right](1 \otimes v_k) \,.
\end{aligned}
$$

Here, for any $g \in G$ we denote by $\alpha_{kl}(g) \in K$ the (k,l)-entry of the matrix of the endomorphism $\alpha(g) \in \operatorname{End}_K V$, with respect to the basis $v_1, \ldots, v_n$. It follows that the (k,l)-entry of the matrix of $i \circ \pi$ with respect to the basis $1 \otimes v_1, \ldots, 1 \otimes v_n$ of $KG \otimes_K V_0$ is $\frac{1}{|G|} \sum_{g \in G} \alpha_{kl}(g^{-1})g$ and hence

$$
\begin{aligned}
\operatorname{tr}(i \circ \pi) &= \sum_{k=1}^{n} \frac{1}{|G|} \sum_{g \in G} \alpha_{kk}(g^{-1})g \\
&= \frac{1}{|G|} \sum_{g \in G} \left(\sum_{k=1}^{n} \alpha_{kk}(g^{-1}) \right)g \\
&= \frac{1}{|G|} \sum_{g \in G} \chi(g^{-1})g \,.
\end{aligned}
$$

It follows that $r_{HS}(V) = \frac{1}{|G|} \sum_{g \in G} \chi(g^{-1})[g] = \sum_{[g] \in C(G)} \frac{\chi(g^{-1})}{|C_g|}[g]$ and this finishes the proof. $\qquad\square$

Corollary 2.40 *Let K be a field of characteristic 0, G a finite group and V, V' two finite dimensional K-representations of G. Then, $V \simeq V'$ as KG-modules if and only if $r_{HS}(V) = r_{HS}(V') \in T(KG)$.*

Proof. The KG-modules V, V' are isomorphic if and only if their characters χ and χ' are equal (cf. [41, Theorem 7.19]). In view of Lemma 2.39, this latter condition is equivalent to the equality $r_{HS}(V) = r_{HS}(V')$.   □

Proof of Theorem 2.38. Let P be a finitely generated projective kG-module and $V = P \otimes_k K$. In view of Corollary 2.36, there is an element $r \in k$, such that the Hattori-Stallings rank of P is equal to $r[1] \in T(kG)$; in fact, Corollary 1.48 implies that r must be a non-negative integer. Invoking the naturality of the Hattori-Stallings rank with respect to coefficient ring homomorphisms (cf. Proposition 1.46), it follows that $r_{HS}(V) = r[1] \in T(KG)$. Therefore, the Hattori-Stallings rank of V is equal to that of the free KG-module $(KG)^r$. In view of Corollary 2.40, this implies that $V \simeq (KG)^r$ and hence the KG-module V is indeed free.   □

Remark 2.41 We proved Swan's theorem (Theorem 2.38) using the fact that finite groups satisfy Bass' conjecture (Corollary 2.36). Conversely, it is easy to prove that finite groups satisfy Bass' conjecture using Swan's theorem. Indeed, let G be a finite group, k a subring of $\mathbf{C}$ with $k \cap \mathbf{Q} = \mathbf{Z}$, K its field of fractions and P a finitely generated projective kG-module. Swan's theorem asserts that the KG-module $V = P \otimes_k K$ is free; say $V \simeq (KG)^r$ for some $r \geq 0$. Then, $r_{HS}(V) = r[1] \in T(KG)$ and hence $r_g(V) = 0$ for all $g \neq 1$. Taking into account Proposition 1.46, it follows that $r_g(P) = 0$ for all $g \neq 1$ as well; therefore, G satisfies Bass' conjecture.

2.3 Exercises

1. Let $\varphi : R \longrightarrow R'$ be a homomorphism of commutative rings.
   (i) Show that the map $J \mapsto \varphi^{-1}(J)$, where $J \subseteq R'$ is an ideal, restricts to a continuous map $\Phi : \operatorname{Spec} R' \longrightarrow \operatorname{Spec} R$.
   (ii) Consider a finitely generated projective R-module P and let $P' = P \otimes_R R'$ be the induced finitely generated projective R'-module. Show that the geometric rank $r(P')$ of P' is the composition $r(P) \circ \Phi$, where $r(P)$ is the geometric rank of P.
   (iii) (naturality of the geometric rank) Show that the diagram

$$
\begin{array}{ccc}
K_0(R) & \xrightarrow{r^R} & [\operatorname{Spec} R, \mathbf{Z}] \\
{\scriptstyle K_0(\varphi)} \downarrow & & \downarrow {\scriptstyle [\Phi, \mathbf{Z}]} \\
K_0(R') & \xrightarrow{r^{R'}} & [\operatorname{Spec} R', \mathbf{Z}]
\end{array}
$$

   is commutative, where r^R (resp. $r^{R'}$) is the geometric rank associated with R (resp. R') and $[\Phi, \mathbf{Z}]$ the map $f \mapsto f \circ \Phi$, $f \in [\operatorname{Spec} R, \mathbf{Z}]$.
2. Let $\varphi : R \longrightarrow R'$ be a homomorphism of commutative rings and Φ the continuous map defined in Exercise 1(i) above.

(i) (naturality of u) Show that the diagram

$$
\begin{array}{ccc}
\mathrm{Idem}(R) & \xrightarrow{u^R} & L(\mathrm{Spec}\,R) \\
\scriptstyle{Idem(\varphi)}\downarrow & & \downarrow \scriptstyle{L(\Phi)} \\
\mathrm{Idem}(R') & \xrightarrow{u^{R'}} & L(\mathrm{Spec}\,R')
\end{array}
$$

is commutative, where u^R (resp. $u^{R'}$) is the Boolean algebra isomorphism corresponding to R (resp. R'), as defined in Proposition 2.14, and $L(\Phi)$ the Boolean algebra morphism $X \mapsto \Phi^{-1}(X)$, $X \in L(\mathrm{Spec}\,R)$ (cf. Example 1.5(iii)).

(ii) (naturality of v) Conclude that the diagram

$$
\begin{array}{ccc}
L(\mathrm{Spec}\,R) & \xrightarrow{v^R} & \mathrm{Idem}(K_0(R)) \\
\scriptstyle{L(\Phi)}\downarrow & & \downarrow \scriptstyle{Idem(K_0(\varphi))} \\
L(\mathrm{Spec}\,R') & \xrightarrow{v^{R'}} & \mathrm{Idem}(K_0(R'))
\end{array}
$$

is commutative, where v^R (resp. $v^{R'}$) is the Boolean algebra morphism corresponding to R (resp. R'), as defined in the text following Corollary 2.15.

(iii) (naturality of $\mathcal{I}_v$) Conclude that the diagram

$$
\begin{array}{ccc}
[\mathrm{Spec}\,R, \mathbf{Z}] & \xrightarrow{\mathcal{I}_v^R} & K_0(R) \\
\scriptstyle{[\Phi,\mathbf{Z}]}\downarrow & & \downarrow \scriptstyle{K_0(\varphi)} \\
[\mathrm{Spec}\,R', \mathbf{Z}] & \xrightarrow{\mathcal{I}_v^{R'}} & K_0(R')
\end{array}
$$

is commutative, where $\mathcal{I}_v^R$ (resp. $\mathcal{I}_v^{R'}$) is the right inverse of the geometric rank, that was defined in the text preceding Theorem 2.16, and $[\Phi, \mathbf{Z}]$ the map $f \mapsto f \circ \Phi$, $f \in [\mathrm{Spec}\,R, \mathbf{Z}]$.

(*Hint:* This is a formal consequence of (ii) and Exercises B.3.1(iii) and B.3.2(iii).)

3. Let R be a commutative ring, $r : K_0(R) \longrightarrow [\mathrm{Spec}\,R, \mathbf{Z}]$ its geometric rank and $\overline{K}_0(R) = \ker r$.

(i) Show that if R is a field then r is bijective.

(ii) Let $\mathcal{F}(R)$ be the set of all homomorphisms $\varphi : R \longrightarrow F$, where F is a field. Show that $\overline{K}_0(R) = \bigcap_{\varphi \in \mathcal{F}(R)} \ker \left[K_0(R) \xrightarrow{K_0(\varphi)} K_0(F) \right]$.

4. Let A be a commutative ring and assume that $a_0, a_1, \ldots, a_n \in A$ are such that the polynomial $f(T) = \sum_{i=0}^{n} a_i T^i$ is invertible in the polynomial ring $A[T]$. Then, show that a_i is nilpotent for all $i = 1, \ldots, n$.

(*Hint:* If A is an integral domain, then a_i must vanish for all $i = 1, \ldots, n$. Now use Lemma 2.3(v).)

5. Let R be a commutative ring and $\overline{K}_0(R)$ the kernel of the geometric rank $r : K_0(R) \longrightarrow [\mathrm{Spec}\,R, \mathbf{Z}]$. The goal of this Exercise is to prove that $\overline{K}_0(R)$

is the nil radical of the ring $K_0(R)$. To that end, we define for any finitely generated projective R-module P and any $n \in \mathbf{N}$ the power series

$$\gamma(P, n) = \sum_{i=0}^{\infty} [\Lambda_R^i P] T^i (1 - T)^{n-i} \in K_0(R)[[T]] \ .$$

Here, we denote by $\Lambda_R^i P$ the i-th exterior power of P for all $i \geq 0$. Show that:

(i) $[\Lambda_R^i(P \oplus R)] = [\Lambda_R^i P] + [\Lambda_R^{i-1} P] \in K_0(R)$ for all $i \geq 1$.

(ii) $\gamma(P, n) = \gamma(P \oplus R, n + 1)$.

(iii) The power series $\gamma(P, n)$ depends only upon the class $x = [P] - [R^n] \in K_0(R)$; let us denote it by $\gamma(x)$.

(iv) If $x = [P] - [R^n] \in K_0(R)$ and $\gamma(x) = \sum_{i=0}^{\infty} \gamma_i(x) T^i$, then $\gamma_0(x) = 1$, $\gamma_1(x) = x$ and $\gamma_2(x) = [\Lambda_R^2 P] - (n - 1)[P] + \frac{1}{2} n(n - 1)$.

(v) $\gamma : K_0(R) \longrightarrow 1 + T \cdot K_0(R)[[T]] \subseteq U\{K_0(R)[[T]]\}$ is a homomorphism of groups.

(vi) If $x \in K_0(R)$ and $\gamma(x), \gamma(-x) \in K_0(R)[T]$, then $x \in \operatorname{nil} K_0(R)$. (*Hint:* Use parts (iv), (v) and the result of Exercise 4 above.)

(vii) $\gamma\big(\overline{K}_0(R)\big) \subseteq K_0(R)[T]$. (*Hint:* If P is a finitely generated projective R-module, such that $r(P)$ is the constant function with value n, then $\Lambda_R^i P = 0$ for all $i > n$.)

(viii) Conclude that $\overline{K}_0(R) = \operatorname{nil} K_0(R)$.

6. Let R, T be commutative rings.

(i) For any ring homomorphism $\varphi : K_0(R) \longrightarrow T$, we consider the Boolean algebra morphism $\widetilde{\varphi} : \operatorname{Idem}(R) \longrightarrow \operatorname{Idem}(T)$, which is defined as the composition

$$\operatorname{Idem}(R) \xrightarrow{\sigma} \operatorname{Idem}(K_0(R)) \xrightarrow{Idem(\varphi)} \operatorname{Idem}(T) \ .$$

Here, σ is the map defined in Proposition 1.28. We now define a map

$$\lambda : \operatorname{Hom}_{Ring}(K_0(R), T) \longrightarrow \operatorname{Hom}_{Boole}(\operatorname{Idem}(R), \operatorname{Idem}(T)) \ ,$$

by letting $\lambda(\varphi) = \widetilde{\varphi}$ for all $\varphi \in \operatorname{Hom}_{Ring}(K_0(R), T)$. If the ring T is reduced, then show that λ is bijective. (*Hint:* Use Exercise 5(viii) above and Remark B.8(i) of Appendix B, combined with Proposition 2.14.)

(ii) Assume that R has no non-trivial idempotents and T is reduced. Then, show that there is a unique ring homomorphism from $K_0(R)$ into T.

(iii) For any prime ideal $\wp \in \operatorname{Spec} R$ consider the ring homomorphism $\varphi_\wp : K_0(R) \longrightarrow T$, which is defined as the composition

$$K_0(R) \xrightarrow{r} [\operatorname{Spec} R, \mathbf{Z}] \xrightarrow{ev_\wp} \mathbf{Z} \xrightarrow{\imath} T \ ,$$

where r is the geometric rank, $ev_\wp$ the evaluation map at $\wp$ and $\imath$ the map $n \mapsto n \cdot 1$, $n \in \mathbf{Z}$. If R has no non-trivial idempotents, then show that $\varphi_\wp = \varphi_{\wp'}$ for all $\wp, \wp' \in \operatorname{Spec} R$.

7. This Exercise refers to the transfer homomorphism φ^* associated with a ring homomorphism φ, as described in §2.2.1.

(i) Let $\varphi : A \longrightarrow B$ be a ring homomorphism, such that $B \simeq A^n$ as (A, A)-bimodules (cf. Exercise 1.3.1). Show that the composition

$$T(A) \xrightarrow{T(\varphi)} T(B) \xrightarrow{\varphi^*} T(A)$$

is multiplication by n.

(ii) Let R be a ring, $n \in \mathbf{N}$ and $\varphi : R \longrightarrow \mathbf{M}_n(R)$ the ring homomorphism with $\varphi(r) = rI_n$ for all $r \in R$. Show that the composition

$$T(R) \xrightarrow{\bar{\imath}} T(\mathbf{M}_n(R)) \xrightarrow{\varphi^*} T(R) ,$$

where $\bar{\imath}$ is the isomorphism of Lemma 1.37, is multiplication by n.

(iii) Let k be a commutative ring, G a finite group of order n and φ the natural inclusion of k into kG. Show that $\varphi^* : T(kG) \longrightarrow k$ maps any element $\sum_{[g]\in\mathcal{C}(G)} a_g[g] \in T(kG)$ onto $na_1 \in k$.

8. Let G be a group, $H \leq G$ a subgroup and S a set of representatives of the right H-cosets in G. For any pair $(h, g) \in H \times G$ we consider the subset $S(h, g) \subseteq S$, which consists of those elements $s \in S$ for which $s^{-1}gs \in H$ and $[h]_H = [s^{-1}gs]_H \in \mathcal{C}(H)$.

(i) Show that for any $x \in H$ the sets $S(h, g)$ and $S(x^{-1}hx, g)$ coincide.

(ii) Show that for any $x \in G$ the sets $S(h, g)$ and $S(h, x^{-1}gx)$ have the same cardinality.

(*Hint:* Consider the permutation β of S, which is defined by letting $\beta(s)H = xsH$ for all $s \in S$.)

(iii) If $[h]_G = [g]_G \in \mathcal{C}(G)$, then show that $S(h, g) \neq \emptyset$.

9. Let G be a group, $\Lambda_G \subseteq \mathbf{Q}$ the subring defined in Exercise 1.3.10 and k a subring of the field $\mathbf{C}$ of complex numbers, such that $k \cap \Lambda_G = \mathbf{Z}$.

(i) Assume that the group G is finite. Then, show that the group algebra kG has no non-trivial idempotents.

(ii) Assume that the group G is abelian. Then, show that the group algebra kG has no non-trivial idempotents and hence conclude that the pair (k, G) satisfies Bass' conjecture.

(iii) Assume that the center of G is a subgroup of finite index. Then, show that the pair (k, G) satisfies Bass' conjecture.

(iv) Assume that the group G is finite. Then, show that the pair (k, G) satisfies Bass' conjecture.

Notes and Comments on Chap. 2. The analogy between finitely generated projective modules P over a commutative ring R and vector bundles on a manifold is based on a result of R. Swan [71], whereas the basic properties of the geometric rank $r(P)$ can be found in books on Commutative Algebra (e.g. in [8]). For further results on the K_0-group of a commutative ring, the reader may consult C. Weibel's book-in-progress [73]. The validity of Bass' conjecture for abelian groups was already noted in [4]. The construction of the transfer homomorphism associated with an extension of rings, as presented in §2.2.1, is also due to H. Bass who reformulated in [loc.cit.] R. Swan's result [70] on induced representations, in terms of the Hattori-Stallings rank.

3

Reduction to Positive Characteristic

3.1 The Rationality of the Canonical Trace

Let E be an idempotent matrix with entries in the group algebra kG of a group G with coefficients in a field k. If $r_g(E) = 0$ for all $g \in G \setminus \{1\}$, then Corollary 1.48 implies that $r_1(E) = n \cdot 1 \in k$ for a suitable non-negative integer n. In this section, we study the value of the trace functional r_1 on E and show that $r_1(E)$ is always, i.e. without any assumption on the $r_g(E)$'s, an element of the prime field of k (Zaleskii's theorem). Even though we are mainly interested in the case where $k = \mathbf{C}$, we follow closely Zaleskii's argument and consider initially the case where k is a field of positive characteristic. In that case, the result is a consequence of certain basic properties of the Frobenius operator acting on the corresponding matrix algebra. In order to lift this result to the characteristic 0 case, we proceed in two steps and study first the case where k is the field $\overline{\mathbf{Q}}$ of algebraic numbers. Then, we consider the general case where $k = \mathbf{C}$, by associating with any idempotent matrix in $\mathbf{M}_n(\mathbf{C}G)$ a certain idempotent matrix in $\mathbf{M}_n(\overline{\mathbf{Q}}G)$. In the meantime, we prove a positivity result of Kaplansky, which is to play an important role in the homological approach to the idempotent conjecture that will be presented in Chap. 4.

In §3.1.1, we consider idempotent matrices E with entries in the group algebra kG of a group G with coefficients in a field k of characteristic p. We study the behavior of the p-th power map on the matrix algebra $\mathbf{M}_n(kG)$ with respect to traces and show that $r_1(E) \in \mathbf{F}_p$. In the following subsection, we lift this result to the case where $k = \overline{\mathbf{Q}}$ and show that the trace functional r_1 takes rational values on idempotent matrices in $\mathbf{M}_n(\overline{\mathbf{Q}}G)$. In order to reduce to the positive characteristic case, we use certain basic properties of the ring of algebraic integers. In §3.1.3, we prove Kaplansky's theorem on the positivity of the trace r_1 on non-zero idempotent matrices with entries in $\mathbf{C}G$, by embedding the group algebra $\mathbf{C}G$ in the von Neumann algebra $\mathcal{N}G$. As an immediate consequence of Kaplansky's positivity theorem, we conclude that $r_1(E)$ is an algebraic number for all idempotent matrices $E \in \mathbf{M}_n(\mathbf{C}G)$.

We use this information in §3.1.4, combined with the existence of $\overline{\mathbf{Q}}$-algebra homomorphisms from finitely generated commutative $\overline{\mathbf{Q}}$-algebras back to $\overline{\mathbf{Q}}$, in order to complete the proof of the rationality of $r_1(E)$ for any idempotent matrix $E \in \mathbf{M}_n(\mathbf{C}G)$.

3.1.1 Coefficient Fields of Positive Characteristic

As explained above, even though we are primarily interested in the case of idempotent matrices with entries in a group algebra with coefficients in a subring k of the field $\mathbf{C}$ of complex numbers, we first consider the case where the ring k is a field of positive characteristic. The goal of this subsection is to prove the following result.

Theorem 3.1 *(Zaleskii; the positive characteristic case) Let k be a field of characteristic $p > 0$, G a group and E an idempotent matrix with entries in the group algebra kG. Then, $r_1(E)$ is an element of the prime field $\mathbf{F}_p$ of k. In particular, if $e = \sum_{g \in G} e_g g \in kG$ is an idempotent, where $e_g \in k$ for all $g \in G$, then $e_1 \in \mathbf{F}_p \subseteq k$.*

In order to prove Zaleskii's theorem, we study the p-th power map (Frobenius operator) on the ring of matrices with entries in kG.

Let R be a ring and p a prime number. We consider the operator

$$F : R \longrightarrow R \,,$$

which is given by $x \mapsto x^p$, $x \in R$. If R is commutative and has characteristic p, then F is well-known to be an additive group homomorphism (cf. Lemma 3.2(ii) below). We examine the behavior of F with respect to addition, in the more general case where R is a not necessarily commutative ring of characteristic p.

Lemma 3.2 *Let p be a prime number and R a ring of characteristic p.*
(i) If $(x_i)_{i \in I}$ is a finite family of elements of R and $x = \sum_i x_i$, then $F(x) \equiv \sum_i F(x_i) \pmod{[R, R]}$.
(ii) If R is commutative then F is an additive group homomorphism.

Proof. (i) It suffices to consider the universal case, where R is the free noncommutative $\mathbf{F}_p$-algebra on the letters $(x_i)_{i \in I}$. Then, x^p is the sum of monomials of the form $x_{i_1} x_{i_2} \cdots x_{i_p}$, where $i_t \in I$ for all $t = 1, 2, \ldots, p$; let X be the set of these monomials. We consider the action of the cyclic group $\Lambda = <\lambda>$ of order p on X, which is defined by

$$x_{i_1} x_{i_2} \cdots x_{i_p} \overset{\lambda}{\mapsto} x_{i_2} \cdots x_{i_p} x_{i_1} \,.$$

It is clear that $\mathbf{x} - \lambda \cdot \mathbf{x} \in [R, R]$ for any element $\mathbf{x} = x_{i_1} x_{i_2} \cdots x_{i_p} \in X$. Any Λ-orbit for this action, being Λ-equivalent to a coset space of Λ, has either

p elements or else is a singleton. The contribution to the sum of an orbit consisting of p elements with a representative $\mathbf{x}$ is

$$\sum_{j=0}^{p-1} \lambda^j \cdot \mathbf{x} \equiv \sum_{j=0}^{p-1} \mathbf{x} \equiv p\mathbf{x} \equiv 0 \pmod{[R,R]} .$$

An orbit with one element is of the form x_i^p for some i and hence the contribution of the singleton orbits is $\sum_i x_i^p$.

(ii) This is an immediate consequence of (i), since the commutator subgroup $[R,R]$ is trivial if R is commutative. $\qquad\square$

Corollary 3.3 *Let p be a prime number, R a ring of characteristic p and τ a trace on R with values in an abelian group. If $(x_i)_{i\in I}$ is a finite family of elements of R and $x = \sum_i x_i$, then $\tau(x^p) = \sum_i \tau(x_i^p)$.* $\qquad\square$

We shall apply Corollary 3.3 in the case where R is a ring of matrices with entries in the group algebra of a group G with coefficients in a field of positive characteristic and proceed in the description of the relevant traces. We let $\mathcal{C}(G)$ be the set of G-conjugacy classes and consider for any positive integer m the subset $\mathcal{C}_m \subseteq \mathcal{C}(G)$, which consists of the conjugacy classes of elements of order m; in other words,

$$\mathcal{C}_m = \{[g] \in \mathcal{C}(G) : o(g) = m\} .$$

We fix a commutative ring k and recall that the k-module $T(kG)$ is free with basis the set $\mathcal{C}(G)$; let $(\pi_{[g]})_{[g]}$ be the corresponding dual basis. For any $m \geq 1$ we consider the k-linear functional

$$f_m : T(kG) \longrightarrow k ,$$

which is defined by $\rho \mapsto \sum_{[g]\in\mathcal{C}_m} \pi_{[g]}(\rho)$, $\rho \in T(kG)$. Since $\mathcal{C}_1 = \{[1]\}$, it follows that $f_1 = \pi_{[1]}$.

Lemma 3.4 *For any $\rho \in T(kG)$ there exists an integer $m_0 = m_0(\rho) \geq 1$, such that $f_m(\rho) = 0$ for all $m \geq m_0$.*

Proof. Taking into account the linearity of the f_m's, it suffices to consider the case where $\rho = [g]$, for some element $g \in G$. In that case, the result is clear since $f_m([g]) = 0$ for all m if g has infinite order, whereas $f_m([g]) = 0$ for all $m > o(g)$ if g has finite order. $\qquad\square$

We now fix a positive integer n and consider the k-algebra $R = \mathbf{M}_n(kG)$ of $n \times n$ matrices with entries in kG. For any $m \geq 1$, we consider the k-linear functional τ_m on R, which is defined as the composition

$$\mathbf{M}_n(kG) \xrightarrow{r} T(kG) \xrightarrow{f_m} k ,$$

where $r = r_{HS}$ is the Hattori-Stallings trace.

Lemma 3.5 *Let $(\tau_m)_{m \geq 1}$ be the functionals on the k-algebra $\mathbf{M}_n(kG)$ defined above. Then:*

(i) τ_m is a k-linear trace for all $m \geq 1$.

(ii) $\tau_1 = r_1$.

(iii) For any matrix $X \in \mathbf{M}_n(kG)$ there exists a non-negative integer $m_0 = m_0(X)$, such that $\tau_m(X) = 0$ for all $m \geq m_0$.

Proof. Assertion (i) is clear, since τ_m is the composition of the k-linear trace r followed by the k-linear functional f_m. Assertion (ii) follows since $f_1 = \pi_{[1]}$, whereas (iii) is an immediate consequence of Lemma 3.4. $\qquad\square$

We now examine the behavior of the trace functionals defined above with respect to idempotent matrices, in the positive characteristic case.

Proposition 3.6 *Let p be a prime number, k a commutative ring of characteristic p, G a group and $\mathbf{M}_n(kG)$ the k-algebra of $n \times n$ matrices with entries in kG. If $E \in \mathbf{M}_n(kG)$ is an idempotent matrix, then:*

(i) $\tau_{p^t}(E) = 0$ for all $t \geq 1$ and

(ii) $\tau_1(E) = (\tau_1(E))^p$.

Proof. Let $E = (e_{ij})_{i,j} = \sum_{i,j} e_{ij} E_{ij}$, where E_{ij} is the (i,j)-th matrix unit and $e_{ij} \in kG$ for all $i, j = 1, \ldots, n$. We write $e_{ij} = \sum_g e_{ij,g} g$, where $e_{ij,g} \in k$ for all $i, j = 1, \ldots, n$ and $g \in G$. Then, $E = \sum_{i,j,g} e_{ij,g} g E_{ij}$ and hence

$$
\begin{aligned}
\tau_m(E) &= \sum_{i,j,g} \tau_m(e_{ij,g} g E_{ij}) \\
&= \sum_{i,g} f_m(e_{ii,g}[g]) \\
&= \sum_{i,g} e_{ii,g} f_m([g]) \\
&= \sum \{e_{ii,g} : 1 \leq i \leq n, o(g) = m\}
\end{aligned}
$$

for all $m \geq 1$. In particular,

$$
\tau_{p^t}(E) = \sum \{e_{ii,g} : 1 \leq i \leq n, o(g) = p^t\} \tag{3.1}
$$

for all $t \geq 0$. On the other hand, since $E = E^p$, we may apply Corollary 3.3 for the trace τ_m on the ring $R = \mathbf{M}_n(kG)$, in order to conclude that

$$
\begin{aligned}
\tau_m(E) = \tau_m(E^p) &= \sum_{i,j,g} \tau_m((e_{ij,g} g E_{ij})^p) \\
&= \sum_{i,g} \tau_m(e_{ii,g}^p g^p E_{ii}) \\
&= \sum_{i,g} f_m(e_{ii,g}^p[g^p]) \\
&= \sum_{i,g} e_{ii,g}^p f_m([g^p]) \\
&= \sum \{e_{ii,g}^p : 1 \leq i \leq n, o(g^p) = m\}
\end{aligned}
$$

for all $m \geq 1$. In particular,

$$\tau_{p^t}(E) = \sum \{e_{ii,g}^p : 1 \le i \le n, o(g^p) = p^t\} \tag{3.2}$$

for all $t \ge 0$.

(i) If $t \ge 1$ then $o(g^p) = p^t$ if and only if $o(g) = p^{t+1}$. Hence, it follows from (3.2) that

$$
\begin{aligned}
\tau_{p^t}(E) &= \sum \{e_{ii,g}^p : 1 \le i \le n, o(g) = p^{t+1}\} \\
&= \Big(\sum \{e_{ii,g} : 1 \le i \le n, o(g) = p^{t+1}\} \Big)^p \\
&= \big(\tau_{p^{t+1}}(E) \big)^p .
\end{aligned}
$$

Here, the second equality follows from Lemma 3.2(ii) and the third one from (3.1). Since $\tau_{p^t}(E) = 0$ for $t \gg 0$ (cf. Lemma 3.5(iii)), the equalities $\tau_{p^t}(E) = \big(\tau_{p^{t+1}}(E) \big)^p$, $t \ge 1$, that we have just established, imply that $\tau_{p^t}(E) = 0$ for all $t \ge 1$.

(ii) Since $o(g^p) = 1$ if and only if $g = 1$ or $o(g) = p$, (3.2) for $t = 0$ reduces to

$$
\begin{aligned}
\tau_1(E) &= \sum_{i=1}^n e_{ii,1}^p + \sum \{e_{ii,g}^p : 1 \le i \le n, o(g) = p\} \\
&= \Big(\sum_{i=1}^n e_{ii,1} \Big)^p + \Big(\sum \{e_{ii,g} : 1 \le i \le n, o(g) = p\} \Big)^p \\
&= (\tau_1(E))^p + (\tau_p(E))^p \\
&= (\tau_1(E))^p .
\end{aligned}
$$

In the above chain of equalities, the second one is an application of Lemma 3.2(ii), the third one follows from (3.1) and the last one results from the vanishing of $\tau_p(E)$, that was established in part (i). □

We are now ready to prove Zaleskii's theorem.

Proof of Theorem 3.1. It follows from Proposition 3.6(ii) that the element $x = r_1(E) = \tau_1(E) \in k$ (cf. Lemma 3.5(ii)) satisfies the polynomial equation $x^p - x = 0$. This finishes the proof, since the roots of that equation are precisely the elements of the prime field $\mathbf{F}_p \subseteq k$. □

3.1.2 Lifting to the Field of Algebraic Numbers

Our next goal is to prove the analogue of Theorem 3.1 for idempotent matrices with entries in the complex group algebra $\mathbf{C}G$ of a group G. As a first step in that direction, we consider in the present subsection idempotent matrices E with entries in the group algebra $\overline{\mathbf{Q}}G$, where $\overline{\mathbf{Q}}$ is the field of algebraic numbers, and prove the following result:

Theorem 3.7 *(Zaleskii; the case of the field of algebraic numbers) Let G be a group and E an idempotent matrix with entries in the group algebra $\overline{\mathbf{Q}}G$. Then, $r_1(E)$ is a rational number. In particular, if $e = \sum_{g \in G} e_g g \in \overline{\mathbf{Q}}G$ is an idempotent, where $e_g \in \overline{\mathbf{Q}}$ for all $g \in G$, then $e_1 \in \mathbf{Q}$.*

In order to reduce the proof of Theorem 3.7 to the case of fields of positive characteristic, we need a few basic properties of the ring $\mathcal{R}$ of algebraic integers. The following properties of $\mathcal{R}$ are derived at the end of §A.2 of Appendix A from general facts about integral dependence:

(AI1) For any algebraic number $x \in \overline{\mathbf{Q}}$ there is a non-zero integer n, such that $nx \in \mathcal{R}$.

(AI2) Let x be an algebraic integer and $x_1(= x), x_2, \ldots, x_k$ its Galois conjugates. Then, all x_i's are algebraic integers, whereas the polynomial $\prod_{i=1}^{k}(X - x_i)$ has coefficients in $\mathbf{Z}$.

(AI3) For any prime number $p \in \mathbf{Z}$ there exists a maximal ideal $\mathcal{M} \subseteq \mathcal{R}$, such that $\mathbf{Z} \cap \mathcal{M} = p\mathbf{Z}$; then, the field $\mathcal{R}/\mathcal{M}$ has characteristic p.

Proof of Theorem 3.7. Assume that E is a matrix of size n. Since $x = r_1(E) \in \overline{\mathbf{Q}}$ is an algebraic number, property (AI1) implies that there exists a positive integer m, such that mx is an algebraic integer. Replacing the matrix E by the block diagonal $nm \times nm$ matrix that consists of m copies of E along the diagonal, we may assume that $x \in \mathcal{R}$. Let us consider the Galois conjugates $x_1(= x), x_2, \ldots, x_k$ of x. Then, there are automorphisms $\sigma_1, \ldots, \sigma_k$ of the field $\overline{\mathbf{Q}}$ of algebraic numbers, such that $x_i = \sigma_i(x)$ for all $i = 1, \ldots, k$. We know that $x_i \in \mathcal{R}$ for all $i = 1, \ldots, k$, whereas the polynomial $f(X) = \prod_{i=1}^{k}(X - x_i)$ has integer coefficients (property (AI2)). We also know from Galois theory that $f(X)$ is the minimum polynomial of x over $\mathbf{Q}$; therefore, the monic polynomial $f(X)$ is irreducible in $\mathbf{Q}[X]$ and hence in $\mathbf{Z}[X]$ as well.

Claim 3.8 *There is a finite set Π of prime numbers, such that for any prime number $p \notin \Pi$ the reduction $f_p(X) \in \mathbf{F}_p[X]$ of $f(X)$ modulo p decomposes into a product of linear factors.*

Having established the above claim, we may invoke Corollary C.4 of Appendix C and conclude that the degree k of the irreducible polynomial $f(X) \in \mathbf{Z}[X]$ is 1. But then $f(X) = X - x_1$ and hence $x = x_1 \in \mathbf{Z}$. □

Proof of Claim 3.8. Any element $a \in \overline{\mathbf{Q}}G$ can be written uniquely as a sum of elements of the form $a_i g_i$, $i = 1, \ldots, t$, where t is a non-negative integer, the g_i's are distinct elements of G and the a_i's non-zero algebraic numbers. We denote by S_a the (finite) subset of $\overline{\mathbf{Q}}$ consisting of the a_i's. If $A = (a_{ij})_{i,j}$ is an $n \times n$ matrix with entries in $\overline{\mathbf{Q}}G$, we denote by S_A the union $\bigcup_{i,j} S_{a_{ij}}$; this is still a finite set of algebraic numbers.

Any field automorphism σ of $\overline{\mathbf{Q}}$ can be extended to an automorphism $\widetilde{\sigma}$ of the group algebra $\overline{\mathbf{Q}}G$, by letting $\widetilde{\sigma}(g) = g$ for all $g \in G$. By an obvious abuse of notation, we also denote by $\widetilde{\sigma}$ the associated automorphism of the matrix algebra $\mathbf{M}_n(\overline{\mathbf{Q}}G)$. In particular, let us consider the idempotent matrices $\widetilde{\sigma}_1(E), \ldots, \widetilde{\sigma}_k(E)$ and the finite set $S = \bigcup_{i=1}^{k} S_{\widetilde{\sigma}_i(E)}$. In view of property AI1, we may choose a (suitably large) integer l, such that $ly \in \mathcal{R}$ for all $y \in S$. Then, any $y \in S$ is an element of the subring $R = \mathcal{R}[l^{-1}]$ of $\overline{\mathbf{Q}}$. Therefore, the entries of $\widetilde{\sigma}_i(E)$ are elements of the group ring RG for all $i = 1, \ldots, k$ and hence each one of these matrices may be viewed as an idempotent in $\mathbf{M}_n(RG)$.

Let Π be the set consisting of those prime numbers that divide l and consider a prime number $p \notin \Pi$. In view of property (AI3), we can find a maximal ideal $\mathcal{M} \subseteq \mathcal{R}$, such that the residue field $\mathbf{F} = \mathcal{R}/\mathcal{M}$ has characteristic p. Since l is not a multiple of p, the residue class $\bar{l}$ of l in $\mathbf{F}$ is invertible and hence the ideal $\mathbf{m} = \mathcal{M}[l^{-1}] \subseteq \mathcal{R}[l^{-1}] = R$ is such that

$$R/\mathbf{m} = \mathcal{R}[l^{-1}]/\mathcal{M}[l^{-1}] = (\mathcal{R}/\mathcal{M})[\bar{l}^{-1}] = \mathbf{F}[\bar{l}^{-1}] = \mathbf{F} .$$

In order to justify formally the second equality above, we note that $\mathcal{R}[l^{-1}]$ (resp. $(\mathcal{R}/\mathcal{M})[\bar{l}^{-1}]$) is isomorphic to the quotient of the polynomial ring $\mathcal{R}[X]$ (resp. $(\mathcal{R}/\mathcal{M})[X]$) by the principal ideal generated by the polynomial $lX - 1$ (resp. $\bar{l}X - \bar{1}$). The natural ring homomorphism $R \longrightarrow R/\mathbf{m} = \mathbf{F}$ induces a homomorphism of group rings $RG \longrightarrow \mathbf{F}G$ and an associated homomorphism of matrix rings $\mathbf{M}_n(RG) \longrightarrow \mathbf{M}_n(\mathbf{F}G)$. We fix an integer $i \in \{1, \ldots, k\}$ and consider the automorphism σ_i of $\bar{\mathbf{Q}}$, the matrix $\widetilde{\sigma}_i(E) \in \mathbf{M}_n(RG)$ and the associated matrix $\overline{\widetilde{\sigma}_i(E)} \in \mathbf{M}_n(\mathbf{F}G)$. In view of the naturality of the trace functional r_1 with respect to coefficient ring homomorphisms (cf. Proposition 1.46), we have

$$r_1(\widetilde{\sigma}_i(E)) = \sigma_i(r_1(E)) = \sigma_i(x) = x_i \in R ,$$

whereas $r_1\left(\overline{\widetilde{\sigma}_i(E)}\right)$ is the residue class $\overline{x_i}$ of $x_i = r_1(\widetilde{\sigma}_i(E))$ in $\mathbf{F}$. Since $\mathbf{F}$ is a field of characteristic p, we may apply Theorem 3.1 for the idempotent matrix $\overline{\widetilde{\sigma}_i(E)} \in \mathbf{M}_n(\mathbf{F}G)$, in order to conclude that

$$\overline{x_i} = r_1\left(\overline{\widetilde{\sigma}_i(E)}\right) \in \mathbf{F}_p \subseteq \mathbf{F} .$$

Of course, this is true for all $i = 1, \ldots, k$. Let us now consider the commutative diagram

$$\begin{array}{ccc} \mathbf{Z} & \longrightarrow & R \\ \downarrow & & \downarrow \\ \mathbf{F}_p & \longrightarrow & \mathbf{F} \end{array}$$

where the horizontal (resp. vertical) arrows are the natural inclusions (resp. projections), and the induced diagram of polynomial rings

$$\begin{array}{ccc} \mathbf{Z}[X] & \longrightarrow & R[X] \\ \downarrow & & \downarrow \\ \mathbf{F}_p[X] & \longrightarrow & \mathbf{F}[X] \end{array}$$

The polynomial $f(X) \in \mathbf{Z}[X]$ decomposes in the ring $R[X]$ into the product $\prod_{i=1}^{k}(X - x_i)$ and hence its reduction $f_p(X) \in \mathbf{F}_p[X]$ decomposes in the ring $\mathbf{F}[X]$ into the product $\prod_{i=1}^{k}(X - \overline{x_i})$. Since $\overline{x_i} \in \mathbf{F}_p$ for all $i = 1, \ldots, k$, this decomposition takes place in $\mathbf{F}_p[X]$, i.e. $f_p(X)$ decomposes into a product of linear factors in $\mathbf{F}_p[X]$. $\qquad\square$

3.1.3 The Kaplansky Positivity Theorem

Having proved that the trace functional r_1 takes rational values on idempotent matrices E with entries in the group algebra $\overline{\mathbf{Q}}G$ of a group G, we now consider the general case where the entries of E are elements of the complex group algebra $\mathbf{C}G$. Our main goal in the present subsection is to prove that, for such an idempotent matrix E, $r_1(E)$ is a non-negative real number. This result will be strengthened in §3.1.4, where we prove that $r_1(E)$ is, in fact, a non-negative rational number.

In order to prove that $r_1(E) \geq 0$, we work with a suitable completion of the group algebra $\mathbf{C}G$; surprisingly enough, the only known proofs of the positivity of $r_1(E)$ involve analytic techniques. More precisely, we let the group algebra $\mathbf{C}G$ act on the Hilbert space $\ell^2 G$ by the left regular representation

$$L : \mathbf{C}G \longrightarrow \mathcal{B}(\ell^2 G)$$

and consider the von Neumann algebra $\mathcal{N}G$, which is defined as the closure of the image $L(\mathbf{C}G)$ in the weak operator topology (cf. §1.1.2.III). Recall that the weak operator topology (WOT) on the algebra of bounded linear operators on $\ell^2 G$ is the locally convex topology induced by the family of semi-norms $(P_{\xi,\eta})_{\xi,\eta \in \ell^2 G}$, where

$$P_{\xi,\eta}(a) = |< a(\xi), \eta >|$$

for all $\xi, \eta \in \ell^2 G$ and $a \in \mathcal{B}(\ell^2 G)$. Then, $\mathcal{N}G$ is a C^*-algebra containing the group algebra $\mathbf{C}G \simeq L(\mathbf{C}G)$ as a WOT-dense subalgebra. The elements of $\mathbf{C}G$ are finite linear combinations of elements of G and hence look much simpler than those of $\mathcal{N}G$, which are operators on $\ell^2 G$ of a certain special form. Nevertheless, it will turn out that the passage to the von Neumann algebra level offers certain advantages for the study of our problem. In fact, we shall prove a more general positivity result for idempotent matrices with entries in $\mathcal{N}G$.

We begin with a lemma describing certain properties that are satisfied by the operators in the von Neumann algebra $\mathcal{N}G$.

Lemma 3.9 *Let G be a group and consider an operator $a \in \mathcal{N}G$.*

(i) If $(\delta_g)_{g \in G}$ denotes the canonical orthonormal basis of $\ell^2 G$, then we have $< a(\delta_g), \delta_{hg} > = < a(\delta_1), \delta_h >$ for all $g, h \in G$.[1]

(ii) For any vector $\xi \in \ell^2 G$ and any group element $g \in G$ the family of complex numbers $(< a(\delta_1), \delta_x > \cdot < \xi, \delta_{x^{-1}g} >)_x$ is summable and

$$\sum_{x \in G} < a(\delta_1), \delta_x > \cdot < \xi, \delta_{x^{-1}g} > = < a(\xi), \delta_g > .$$

(iii) If $a(\delta_1) = 0 \in \ell^2 G$ then a is the zero operator.

---

[1] In fact, this property characterizes the operators in $\mathcal{N}G$; cf. Exercise 3.3.1.

Proof. (i) First of all, let us consider the case where $a = L_x$ for some $x \in G$. In that case, we have to prove that $<\delta_{xg}, \delta_{hg}> = <\delta_x, \delta_h>$. But this equality is obvious, since $xg = hg$ if and only if $x = h$. Both sides of the formula to be proved are linear and WOT-continuous in a and hence the result follows from the special case considered above, since $\mathcal{N}G$ is the WOT-closure of the linear span of the set $\{L_x : x \in G\}$.

(ii) Since $\xi = \sum_x <\xi, \delta_x> \delta_x$, it follows that $a(\xi) = \sum_x <\xi, \delta_x> a(\delta_x)$. In view of the linearity and continuity of the inner product, we conclude that

$$
\begin{aligned}
<a(\xi), \delta_g> &= \sum_x <\xi, \delta_x> \cdot <a(\delta_x), \delta_g> \\
&= \sum_x <\xi, \delta_x> \cdot <a(\delta_1), \delta_{gx^{-1}}> \\
&= \sum_y <\xi, \delta_{y^{-1}g}> \cdot <a(\delta_1), \delta_y> ,
\end{aligned}
$$

where the second equality follows from (i) above.

(iii) If $a(\delta_1) = 0$, then the equality of (ii) above implies that the inner product $<a(\xi), \delta_g>$ vanishes for all vectors $\xi \in \ell^2 G$ and all group elements $g \in G$. It follows readily from this that $a = 0$. □

We note that the linear functional $r_1 : \mathbf{C}G \longrightarrow \mathbf{C}$, which maps an element $a \in \mathbf{C}G$ onto the coefficient of $1 \in G$ in a, extends to a linear functional

$$\tau : \mathcal{N}G \longrightarrow \mathbf{C}, \tag{3.3}$$

by letting $\tau(a) = <a(\delta_1), \delta_1>$ for all $a \in \mathcal{N}G$.

Remark 3.10 Let G be a group and τ the linear functional defined above. Then, the assertion of Lemma 3.9(i) implies that $\tau(a) = <a(\delta_g), \delta_g>$ for all $a \in \mathcal{N}G$ and $g \in G$.

Proposition 3.11 *Let G be a group and τ the linear functional defined above. Then:*

(i) τ is a WOT-continuous trace.

(ii) τ is positive and faithful, i.e. $\tau(a^*a) \geq 0$ for all $a \in \mathcal{N}G$, whereas $\tau(a^*a) = 0$ if and only if $a = 0$.

(iii) τ is normalized, i.e. $\tau(1) = 1$, where $1 \in \mathcal{N}G$ is the identity operator. The trace τ will be referred to as the canonical trace on the von Neumann algebra $\mathcal{N}G$.

Proof. (i) It is clear that τ is WOT-continuous. In order to show that τ is a trace, we fix an operator $a \in \mathcal{N}G$ and note that for any $g \in G$ we have

$$<aL_g(\delta_{g^{-1}}), \delta_{g^{-1}}> = <a(\delta_1), \delta_{g^{-1}}> = <a(\delta_1), L_g^*(\delta_1)> = <L_g a(\delta_1), \delta_1> ,$$

where the second equality follows since $L_g^* = L_{g^{-1}}$. Invoking Remark 3.10, we conclude that $\tau(aL_g) = \tau(L_g a)$. This being the case for all $g \in G$, it follows that $\tau(aa') = \tau(a'a)$ for all $a' \in L(\mathbf{C}G)$. Since multiplication in $\mathcal{B}(\ell^2 G)$ is separately WOT-continuous (cf. Remark 1.13(ii)), the WOT-continuity of τ implies that $\tau(aa') = \tau(a'a)$ for all $a' \in \mathcal{N}G$.

(ii) For any $a \in \mathcal{N}G$ we have

$$\tau(a^*a) = <a^*a(\delta_1), \delta_1> = <a(\delta_1), a(\delta_1)> = \|a(\delta_1)\|^2 \geq 0 .$$

In particular, $\tau(a^*a) = 0$ if and only if $a(\delta_1) = 0$; this proves the final assertion, in view of Lemma 3.9(iii).

(iii) We compute $\tau(1) = <\delta_1, \delta_1> = \|\delta_1\|^2 = 1$.   □

For any positive integer n we consider the matrix algebra $\mathbf{M}_n(\mathcal{N}G)$; it is a von Neumann algebra of operators acting on the direct sum of n copies of $\ell^2 G$. The canonical trace τ of (3.3) induces the trace

$$\tau_n : \mathbf{M}_n(\mathcal{N}G) \longrightarrow \mathbf{C} ,$$

which maps any matrix $A = (a_{ij})_{i,j} \in \mathbf{M}_n(\mathcal{N}G)$ onto the complex number $\sum_{i=1}^{n} \tau(a_{ii})$ (cf. Proposition 1.39(i)). The trace τ_n is positive, faithful and $\tau_n(I_n) = n$, where I_n is the identity $n \times n$ matrix (cf. Remark 1.42(ii),(iii)).

We are now ready to state Kaplansky's result:

Theorem 3.12 *(Kaplansky positivity theorem) Let G be a group and E an idempotent $n \times n$ matrix with entries in the von Neumann algebra $\mathcal{N}G$. Then:*

(i) The complex number $\tau_n(E)$ is, in fact, real and satisfies the inequalities $0 \leq \tau_n(E) \leq n$.

(ii) $\tau_n(E) = 0$ (resp. n) if and only if E is the zero (resp. the identity) $n \times n$ matrix.

In particular, if $e \in \mathcal{N}G$ is an idempotent then $\tau(e)$ is a real number contained in the interval $[0,1]$; moreover, $\tau(e) = 0$ (resp. 1) if and only if $e = 0$ (resp. 1).

The following lemma will be the main technical tool in the proof of the theorem. It uses some of the analytic properties of a von Neumann algebra, thereby justifying the use of $\mathcal{N}G$ in the study of our problem.

Lemma 3.13 *Let $\mathcal{N}$ be a von Neumann algebra of operators acting on the Hilbert space $\mathcal{H}$. For any idempotent $e \in \mathcal{N}$ there is a projection $f \in \mathcal{N}$, such that $ef = f$ and $fe = e$.*

Proof. Since $e \in \mathrm{Idem}(\mathcal{N})$, the subspace $V = \mathrm{im}\, e$ is easily seen to be closed and $\mathcal{N}'$-invariant. Therefore, Lemma 1.17(ii) implies that the orthogonal projection f onto V is contained in $\mathcal{N}''$. Invoking Theorem 1.18, we conclude that $f \in \mathcal{N}$. The equalities $ef = f$ and $fe = e$ follow since e and f are idempotent operators on $\mathcal{H}$ with the same image.   □

Proof of Theorem 3.12. In view of Lemma 3.13, we may choose a projection $F \in \mathbf{M}_n(\mathcal{N}G)$, such that $FE = E$ and $EF = F$. Then, we have

$$\tau_n(E) = \tau_n(FE) = \tau_n(EF) = \tau_n(F) = \tau_n(F^*F) \geq 0 \,,$$

where the second equality follows from the trace property of τ_n and the inequality from its positivity. Considering the idempotent matrix $I_n - E$, we may prove in a similar way that $\tau_n(I_n - E)$ is a non-negative real number as well. But $\tau_n(E) + \tau_n(I_n - E) = \tau_n(I_n) = n$ and hence both numbers $\tau_n(E)$ and $\tau_n(I_n - E)$ are contained in the interval $[0, n]$. If $\tau_n(E) = 0$ then $\tau_n(F^*F) = 0$ and hence $F = 0$, in view of the faithfulness of τ_n. But then $E = FE = 0$ as well. Working similarly with the idempotent matrix $I_n - E$, one can show that $\tau_n(E) = n$ if and only if $E = I_n$.   □

Corollary 3.14 *Let G be a group and E an idempotent $n \times n$ matrix with entries in the group algebra $\mathbf{C}G$. Then:*

   (i) The complex number $r_1(E)$ is, in fact, real and satisfies the inequalities $0 \leq r_1(E) \leq n$.

   (ii) $r_1(E) = 0$ (resp. n) if and only if E is the zero (resp. the identity) $n \times n$ matrix.
In particular, if $e = \sum_g e_g g \in \mathbf{C}G$ is an idempotent, where $e_g \in \mathbf{C}$ for all $g \in G$, then e_1 is a real number contained in the interval $[0, 1]$; moreover, $e_1 = 0$ (resp. 1) if and only if $e = 0$ (resp. 1).

Proof. This follows from Theorem 3.12, since the restriction of the trace functional τ_n to the subalgebra $\mathbf{M}_n(\mathbf{C}G) \simeq \mathbf{M}_n(L(\mathbf{C}G))$ of $\mathbf{M}_n(\mathcal{N}G)$ coincides with the trace functional r_1.   □

The following result will play an important role in the homological approach to the idempotent conjecture, that will be presented in Chap. 4.

Proposition 3.15 *Let G be a group. Then, the following conditions are equivalent for an idempotent $e \in \mathbf{C}G$:*

   (i) The idempotent e is trivial.
   (ii) For any $g \in G \setminus \{1\}$ we have $r_g(e) = 0$.

Proof. It is clear that (i)→(ii). In order to show that (ii)→(i), let us assume that (ii) holds and write $e = \sum_{g \in G} e_g g \in \mathbf{C}G$, where $e_g \in \mathbf{C}$ for all $g \in G$. We consider the augmentation homomorphism $\varepsilon : \mathbf{C}G \longrightarrow \mathbf{C}$. Since $\varepsilon(e) \in \mathbf{C}$ is an idempotent, we have $\varepsilon(e) = 0$ or 1. We now compute

$$\begin{aligned}
\varepsilon(e) &= \sum \{e_g : g \in G\} \\
&= e_1 + \sum \{e_g : g \in G, g \neq 1\} \\
&= e_1 + \sum \{r_g(e) : [g] \in \mathcal{C}(G), [g] \neq [1]\} \\
&= e_1 \,.
\end{aligned}$$

Therefore, $e_1 = 0$ or 1; this finishes the proof, in view of the last assertion of Corollary 3.14.   □

Our next goal is to prove that for any idempotent matrix E with entries in the complex group algebra $\mathbf{C}G$ of a group G, the real number $r_1(E)$ is algebraic. To that end, we need the following lemma.

Lemma 3.16 *Let K, L be fields and assume that L is algebraically closed.*

(i) Let K' be a monogenic algebraic field extension of K. Then, any embedding of K into L can be extended to an embedding of K' into L.

(ii) Assume that L is the algebraic closure of K. Let ς be an automorphism of K and σ an embedding of L into itself that extends ς. Then, σ is an automorphism of L.

(iii) Assume that L is the algebraic closure of K. Then, any automorphism of K can be extended to an automorphism of L.

(iv) Assume that K is a subfield of L. Then, any automorphism of K can be extended to an automorphism of L.

Proof. (i) We may regard K as a subfield of L, by means of the given embedding. Let $x \in K'$ be such that $K' = K(x)$ and consider the minimum polynomial $f(X) \in K[X]$ of x over K. Let y be a root of $f(X)$ in L; then, there is a unique K-algebra homomorphism from $K' = K(x) \simeq K[X]/(f(X))$ into L, which maps the class of X onto y. Since K' is a field, this homomorphism is an embedding.

(ii) We have to prove that σ is surjective. To that end, let us fix an element $x \in L$ and consider its minimum polynomial $f(X) \in K[X]$. We also consider the polynomial $g(X) \in K[X]$, whose coefficients are obtained from those of $f(X)$ by applying ς^{-1}. Since L is algebraically closed, there are elements $x_1, \ldots, x_n \in L$ such that $g(X) = \prod_{i=1}^{n}(X - x_i) \in L[X]$. Applying σ to the coefficients of the polynomials on both sides of that equation, we conclude that $f(X) = \prod_{i=1}^{n}(X - \sigma(x_i)) \in L[X]$. But x is a root of $f(X)$ and hence $x = \sigma(x_i)$ for some i.

(iii) Let ς be an automorphism of K and consider the embedding $\imath$ of K into L, which is defined as the composition

$$ K \xrightarrow{\varsigma} K \hookrightarrow L \,. $$

Let Λ be the set of pairs $(E, \jmath)$, where E is a subfield of L containing K and $\jmath$ an embedding of E into L extending $\imath$. We order Λ by letting $(E, \jmath) \leq (E', \jmath')$ if $E \subseteq E'$ and $\jmath'$ extends $\jmath$. By an application of Zorn's lemma, we may choose a maximal element $(E_0, \jmath_0)$ in Λ. The maximality of $(E_0, \jmath_0)$, combined with part (i) above, shows that $E_0 = L$. We now invoke part (ii) and conclude that the embedding $\jmath_0$ of L into itself is actually an automorphism of L.

(iv) Let $(X_i)_{i \in I}$ be a transcendency basis of L over K and consider the subfield $K' = K(X_i; i \in I)$ of L, which is a purely transcendental extension of K. It is clear that any automorphism ς of K can be extended to an automorphism ς' of K'. Since L is the algebraic closure of K', ς' can be further extended to an automorphism σ of L, in view of part (iii) above. $\square$

Corollary 3.17 *Let x be a complex number. If $\sigma(x)$ is a non-negative real number for any field automorphism σ of $\mathbf{C}$, then $x \in \overline{\mathbf{Q}}$.*

Proof. Letting σ be the identity of $\mathbf{C}$, we deduce that x is a non-negative real number. Assume that x is transcendental over $\mathbf{Q}$; then, x is non-zero and hence $x > 0$. The subfield $K = \mathbf{Q}(x)$ of $\mathbf{C}$ is isomorphic to the field of rational functions in one variable over $\mathbf{Q}$ and hence there is a unique field automorphism ς of K that maps x onto $-x$. Since $\mathbf{C}$ is algebraically closed, we may invoke Lemma 3.16(iv) in order to extend ς to an automorphism σ of $\mathbf{C}$. But then $\sigma(x) = \varsigma(x) = -x < 0$, a contradiction.   □

Let us now consider an automorphism σ of the field $\mathbf{C}$ of complex numbers and its extension $\widetilde{\sigma}$ to the complex group algebra $\mathbf{C}G$ of a group G, which is such that $\widetilde{\sigma}(g) = g$ for all $g \in G$. In view of the naturality of the trace r_1 with respect to coefficient ring homomorphisms (cf. Proposition 1.46), we have

$$r_1(\widetilde{\sigma}(E)) = \sigma(r_1(E)) \tag{3.4}$$

for any idempotent matrix E with entries in $\mathbf{C}G$, where $\widetilde{\sigma}(E)$ is the idempotent matrix whose entries are obtained from those of E by applying $\widetilde{\sigma}$.

Proposition 3.18 *Let G be a group and E an idempotent matrix with entries in the complex group algebra $\mathbf{C}G$; then, $r_1(E)$ is an algebraic number. In particular, if $e = \sum_g e_g g \in \mathbf{C}G$ is an idempotent, where $e_g \in \mathbf{C}$ for all $g \in G$, then e_1 is an algebraic number.*

Proof. Let $x = r_1(E) \in \mathbf{C}$. We consider a field automorphism σ of $\mathbf{C}$ and the induced automorphism $\widetilde{\sigma}$ of the complex group algebra $\mathbf{C}G$. If $E' = \widetilde{\sigma}(E)$ then $r_1(E') = \sigma(x)$, in view of (3.4). We now invoke Corollary 3.14, applied to the idempotent matrix E', in order to conclude that $\sigma(x)$ is real and non-negative. Since this is the case for any automorphism σ of $\mathbf{C}$, the proof is finished using Corollary 3.17.   □

3.1.4 Idempotent Matrices with Entries in the Complex Group Algebra

In this subsection, we complete Theorem 3.7 and consider idempotent matrices with entries in the complex group algebra of a group G. In order to prove that the trace functional r_1 takes rational values on the set of these matrices, we follow a very simple strategy and associate with any such matrix E an idempotent matrix with entries in $\overline{\mathbf{Q}}G$. To that end, we consider the finitely generated $\overline{\mathbf{Q}}$-subalgebra R of $\mathbf{C}$ generated by all complex numbers that are involved in E and use a $\overline{\mathbf{Q}}$-algebra homomorphism from R back to $\overline{\mathbf{Q}}$.

Theorem 3.19 *(Zaleskii) Let G be a group and E an idempotent matrix with entries in the complex group algebra $\mathbf{C}G$. Then, $r_1(E)$ is a rational number. In particular, if $e = \sum_{g \in G} e_g g \in \mathbf{C}G$ is an idempotent, where $e_g \in \mathbf{C}$ for all $g \in G$, then $e_1 \in \mathbf{Q}$.*

Proof. As in the proof of Claim 3.8, for any element $a = \sum_{i=1}^{t} a_i g_i \in \mathbf{C}G$, where the g_i's are distinct elements of G and the a_i's non-zero complex numbers, we denote by S_a the subset of $\mathbf{C}$ consisting of the a_i's. If $A = (a_{ij})_{i,j}$ is an $n \times n$ matrix with entries in $\mathbf{C}G$, we denote by S_A the finite subset $\bigcup_{i,j} S_{a_{ij}}$ of $\mathbf{C}$.

In particular, let us consider the subset $S_E \subseteq \mathbf{C}$ and the finitely generated $\overline{\mathbf{Q}}$-subalgebra $R = \overline{\mathbf{Q}}[S_E]$ of $\mathbf{C}$. It is clear that E may be viewed as an idempotent matrix with entries in RG. We fix a $\overline{\mathbf{Q}}$-algebra homomorphism $\pi : R \longrightarrow \overline{\mathbf{Q}}$; such a homomorphism exists in view of Corollary A.23 of Appendix A. Since the complex number $x = r_1(E) \in R$ is algebraic (cf. Proposition 3.18), we have $\pi(x) = x$. The homomorphism π can be extended to a homomorphism $\widetilde{\pi} : RG \longrightarrow \overline{\mathbf{Q}}G$, by letting $\widetilde{\pi}(g) = g$ for all $g \in G$. We also denote by $\widetilde{\pi}$ the associated homomorphism of matrix rings $\mathbf{M}_n(RG) \longrightarrow \mathbf{M}_n(\overline{\mathbf{Q}}G)$ and consider the idempotent matrix $\widetilde{\pi}(E) \in \mathbf{M}_n(\overline{\mathbf{Q}}G)$. In view of the naturality of the trace functional r_1 with respect to coefficient ring homomorphisms (cf. Proposition 1.46), we have

$$r_1(\widetilde{\pi}(E)) = \pi(r_1(E)) = \pi(x) = x \ .$$

We now invoke Theorem 3.7, applied to the idempotent matrix $\widetilde{\pi}(E)$, in order to conclude that x is a rational number, as needed. $\square$

The passage from $\mathbf{C}$ to any field of characteristic 0 is now routine.

Corollary 3.20 *Let k be a field of characteristic 0, G a group and E an idempotent matrix with entries in the group algebra kG. Then, $r_1(E)$ is an element of the prime field $\mathbf{Q} \subseteq k$. In particular, if $e = \sum_{g \in G} e_g g \in kG$ is an idempotent, where $e_g \in k$ for all $g \in G$, then $e_1 \in \mathbf{Q} \subseteq k$.*

Proof. Let $S_E \subseteq k$ be the finite set consisting of all elements of k that are involved in E (cf. the proof of Claim 3.8 and that of Theorem 3.19). Replacing it, if necessary, by its subfield $\mathbf{Q}(S_E)$, we may assume that the field k is finitely generated. In that case, there exists a subfield $k_0 \subseteq k$, which is purely transcendental over $\mathbf{Q}$ with tr.deg $(k_0/\mathbf{Q}) < \infty$, such that k is finite algebraic over it. It is easily seen that any field K of characteristic 0 with tr.deg $(K/\mathbf{Q}) < \infty$ is countable. Since the field $\mathbf{C}$ is uncountable, its transcendence degree over $\mathbf{Q}$ is infinite. Therefore, k_0 may be embedded in $\mathbf{C}$. We may extend this embedding to k, by a repeated application of Lemma 3.16(i), and hence view k as a subfield of $\mathbf{C}$. Then, the result follows from Theorem 3.19. $\square$

The following result is an immediate consequence of the theorems of Kaplansky and Zaleskii and supplements Proposition 2.23 and Lemma 2.24 (cf. Remark 2.25).

Corollary 3.21 *The following conditions are equivalent for a subring k of the field $\mathbf{C}$ of complex numbers.*

(i) $k \cap \mathbf{Q} = \mathbf{Z}$.

(ii) For any group G the group algebra kG has no idempotents $\neq 0, 1$.

Proof. We already know that (ii)→(i); cf. Remark 1.53(i). In order to prove the reverse implication, let us consider a subring k of $\mathbf{C}$ satisfying (i), a group G and an idempotent $e = \sum_g e_g g \in kG$, where $e_g \in k$ for all $g \in G$. In view of Theorem 3.19, the complex number e_1 is rational and hence $e_1 \in k \cap \mathbf{Q} = \mathbf{Z}$. Since $0 \leq e_1 \leq 1$ (cf. Corollary 3.14), it follows that $e_1 = 0$ or 1. But then e is trivial, in view of the last assertion of loc.cit.   □

3.2 The Support of the Hattori-Stallings Rank

In Chap. 2, we proved Bass' conjecture for groups that are finite extensions of their center (cf. Proposition 2.35). In this section, we prove a result of Linnell, describing certain conditions on a group that are necessary for it to be a counterexample to Bass' conjecture. As it turns out, these conditions are not satisfied by groups that are finite extensions of their center; therefore, Linnell's theorem generalizes the above mentioned result. In the spirit of the present chapter, the approach we follow consists in replacing the subring k of $\mathbf{C}$ (that satisfies the usual condition $k \cap \mathbf{Q} = \mathbf{Z}$) by a quotient ring R of prime power characteristic. Then, any idempotent matrix E with entries in the group algebra kG of a group G induces an idempotent matrix $\overline{E}$ with entries in RG. The idea is to study the Hattori-Stallings rank $r_{HS}(E)$ of E by examining the corresponding rank of $\overline{E}$. This latter rank can be analyzed by using the action of the Frobenius operator (more precisely, of iterates of the Frobenius operator) on the ring of matrices with entries in RG. By the same technique, we obtain a result of Bass that concerns idempotent matrices with entries in a complex group algebra and prove that torsion-free polycyclic-by-finite groups satisfy the idempotent conjecture.

In §3.2.1, we consider the action of the Frobenius operator on a ring R of prime power characteristic and then apply our results, studying the form of the Hattori-Stallings rank of idempotent matrices in $\mathbf{M}_n(R)$. In the following subsection, we prove the theorems of Bass and Linnell and state a few immediate consequences concerning the idempotent and Bass' conjectures. Finally, in §3.2.3, we apply Linnell's theorem to the study of Bass' conjecture for solvable groups and prove that the conjecture is satisfied by the solvable groups of finite Hirsch number.

3.2.1 Iterates of the Frobenius Operator

Let R be a ring, p a prime number and

$$F : R \longrightarrow R$$

the Frobenius operator, which is given by $x \mapsto x^p$, $x \in R$. If the ring R has characteristic p, then we know that F is additive modulo $[R, R]$ (cf. Lemma 3.2(i)). We examine the behavior of certain iterates of F with respect to addition in the more general case where R has characteristic a power of p and then apply our results to the study of the Hattori-Stallings rank of idempotent matrices with entries in R.

Lemma 3.22 *Let p be a prime number and R a ring of characteristic p^n, for some $n \geq 1$. Then, for any finite family $(x_i)_{i \in I}$ of elements of R and any $k \geq n - 1$ we have*

$$\left(\sum_i x_i\right)^{p^k} \equiv \sum (x_{i_1} x_{i_2} \cdots x_{i_s})^{p^{k-n+1}} \quad (\mathrm{mod} \ [R, R]) \ ,$$

where $s = p^{n-1}$ and the summation on the right is extended over all s-tuples of indices $(i_1, i_2, \ldots, i_s) \in I^s$.

Proof. We consider the universal case, where R is the free non-commutative $\mathbf{Z}/p^n\mathbf{Z}$-algebra on the letters $(x_i)_{i \in I}$. Then, $\left(\sum_i x_i\right)^{p^k}$ is the sum of monomials of the form $x_{i_1} x_{i_2} \cdots x_{i_{p^k}}$, where $i_t \in I$ for all $t = 1, 2, \ldots, p^k$; let X be the set of these monomials. We consider the action of the cyclic group $\Lambda = <\lambda>$ of order p^k on X, which is defined by

$$x_{i_1} x_{i_2} \cdots x_{i_{p^k}} \overset{\lambda}{\mapsto} x_{i_2} \cdots x_{i_{p^k}} x_{i_1} \ .$$

It is clear that $\mathbf{x} - \lambda \cdot \mathbf{x} \in [R, R]$ for any $\mathbf{x} = x_{i_1} x_{i_2} \cdots x_{i_{p^k}} \in X$. Let $X' \subseteq X$ be a Λ-orbit; being Λ-equivalent to a coset space of Λ, X' has p^m elements for some $m = 0, 1, \ldots, k$. If $\mathrm{card}(X') = p^m$ then $X' \simeq \Lambda/\Lambda'$ as Λ-sets, where $\Lambda' = <\lambda^{p^m}>$. Let us consider an orbit X' with p^m elements, where $m \geq n$. If $\mathbf{x} = x_{i_1} x_{i_2} \cdots x_{i_{p^k}} \in X'$ is a representative of the orbit, then the contribution of X' to the sum is

$$\sum_{j=0}^{p^m - 1} \lambda^j \cdot \mathbf{x} \equiv \sum_{j=0}^{p^m - 1} \mathbf{x} \equiv p^m \mathbf{x} \equiv 0 \quad (\mathrm{mod} \ [R, R]) \ .$$

We now consider an orbit X' with p^m elements, where $m \leq n-1$. In that case, any element $\mathbf{x} = x_{i_1} x_{i_2} \cdots x_{i_{p^k}} \in X'$ is invariant under λ^{p^m} and hence under $\lambda^{p^{n-1}} = \lambda^s$ as well. But then $\mathbf{x}$ is of the form $(x_{i_1} x_{i_2} \cdots x_{i_s})^{p^{k-n+1}}$. Conversely, any element $\mathbf{x} \in X$ of the above form is invariant under $\lambda^s = \lambda^{p^{n-1}}$ and hence its orbit has p^m elements for some $m \leq n - 1$. It follows that the contribution to the sum of those orbits X' that contain p^m elements, for some $m \leq n - 1$, is equal to

$$\sum (x_{i_1} x_{i_2} \cdots x_{i_s})^{p^{k-n+1}} \ ,$$

where the summation is extended over all s-tuples of indices $(i_1, i_2, \ldots, i_s) \in I^s$. □

Applying the result above to the case of a ring of matrices with entries in a group algebra, we obtain an interesting property of the Hattori-Stallings rank of idempotent matrices.

Proposition 3.23 *Let p be a prime number and R a commutative ring of characteristic p^n, for some $n \geq 1$. If G is a group and E an idempotent matrix with entries in the group ring RG, then there is an integer $N \in \mathbf{N}$ and elements $r_1, \ldots, r_N \in R$ and $g_1, \ldots, g_N \in G$, such that*

$$r_{HS}(E) = \sum_{i=1}^{N} r_i^{p^k} \left[g_i^{p^k} \right] \in T(RG)$$

for all $k \geq 0$.

Proof. Let $E = \sum_{i,j=1}^{d} e_{ij} E_{ij} \in \mathbf{M}_d(RG)$, where the E_{ij}'s are the matrix units and $e_{ij} \in RG$ for all $i, j = 1, \ldots, d$. We fix an integer $k \geq 2n - 2$ and note that Lemma 3.22, applied to the ring $\mathbf{M}_d(RG)$, shows that

$$E \equiv E^{p^k}$$
$$\equiv \left(\sum_{i,j=1}^{d} e_{ij} E_{ij} \right)^{p^k}$$
$$\equiv \sum (e_{i_1 j_1} e_{i_2 j_2} \cdots e_{i_s j_s} E_{i_1 j_1} E_{i_2 j_2} \cdots E_{i_s j_s})^{p^{k-n+1}}$$

modulo $[\mathbf{M}_d(RG), \mathbf{M}_d(RG)]$, where $s = p^{n-1}$ and the summation is extended over all $2s$-tuples of indices $(i_1, j_1, i_2, j_2, \ldots, i_s, j_s) \in \{1, \ldots, d\}^{2s}$. The trace of a commutator of matrices is zero modulo $[RG, RG]$, whereas the trace of a matrix of the form $(E_{i_1 j_1} E_{i_2 j_2} \cdots E_{i_s j_s})^{p^{k-n+1}}$ is 1 if $j_1 = i_2, j_2 = i_3, \ldots, j_{s-1} = i_s, j_s = i_1$ and 0 otherwise. Therefore, we conclude that

$$\mathrm{tr}(E) \equiv \sum (e_{i_1 i_2} e_{i_2 i_3} \cdots e_{i_s i_1})^{p^{k-n+1}} \quad (\mathrm{mod} \ [RG, RG]),$$

where the summation is extended over all s-tuples of indices $(i_1, i_2, \ldots, i_s) \in \{1, \ldots, d\}^s$. Being an element of RG, any product $e_{i_1 i_2} e_{i_2 i_3} \cdots e_{i_s i_1}$ can be written as a sum $\sum_l \varrho_l g_l$ for suitable elements $\varrho_l \in R$ and $g_l \in G$. Using Lemma 3.22 again, this time applied to the ring RG, we conclude that a typical summand in the summation above is equal to

$$(e_{i_1 i_2} e_{i_2 i_3} \cdots e_{i_s i_1})^{p^{k-n+1}} \equiv \left(\sum_l \varrho_l g_l \right)^{p^{k-n+1}}$$
$$\equiv \sum (\varrho_{l_1} \varrho_{l_2} \cdots \varrho_{l_s} g_{l_1} g_{l_2} \cdots g_{l_s})^{p^{k-2n+2}}$$
$$\equiv \sum (\varrho_{l_1} \varrho_{l_2} \cdots \varrho_{l_s})^{p^{k-2n+2}} (g_{l_1} g_{l_2} \cdots g_{l_s})^{p^{k-2n+2}}$$

modulo $[RG, RG]$, where the latter summations are extended over all s-tuples $(l_1, l_2, \ldots, l_s)$. Since $k - 2n + 2$ can be any non-negative integer, the proof is finished by letting $\varrho_{l_1} \varrho_{l_2} \cdots \varrho_{l_s}$ (resp. $g_{l_1} g_{l_2} \cdots g_{l_s}$) be one of the r_i's (resp. one of the g_i's) in the statement. $\qquad \square$

In order to obtain a simpler form for the expression of the Hattori-Stallings rank given in Proposition 3.23, in the special case where the ring R is finite, we need the following lemma.

Lemma 3.24 *Let k be a commutative ring and $\mathbf{m} \subseteq k$ a maximal ideal, such that the residue field $k/\mathbf{m}$ is finite of order p^f, for some prime number p and some $f \geq 1$.*

(i) *If $x \in 1 + \mathbf{m}$ then $x^{p^{f(n-1)}} \in 1 + \mathbf{m}^n$ for all $n \geq 1$.*

(ii) *If $x \in k$ then $x^{p^{fn}} - x^{p^{f(n-1)}} \in \mathbf{m}^n$ for all $n \geq 1$.*

Proof. (i) First of all, we note that $p \cdot 1 \in \mathbf{m}$, since the residue field $k/\mathbf{m}$ has characteristic p. Therefore, $p^f \cdot 1 \in \mathbf{m}$ as well and hence the result follows by induction on n, using the binomial formula.

(ii) We begin by observing that

$$x^{p^{fn}} - x^{p^{f(n-1)}} = x^{p^{f(n-1)}} \left(x^{(p^f-1)p^{f(n-1)}} - 1 \right). \tag{3.5}$$

If $x \in \mathbf{m}$ then $x^{p^{f(n-1)}} \in \mathbf{m}^{p^{f(n-1)}}$ and the result follows from (3.5), since $p^{f(n-1)} \geq p^{n-1} \geq n$. We now assume that $x \notin \mathbf{m}$. Then, the residue class $\overline{x} \in k/\mathbf{m} = \mathbf{F}_{p^f}$ is non-zero and hence $\overline{x}^{p^f-1} = \overline{1} \in k/\mathbf{m}$, i.e. $x^{p^f-1} \in 1 + \mathbf{m}$. Invoking assertion (i) above, we conclude that $x^{(p^f-1)p^{f(n-1)}} \in 1 + \mathbf{m}^n$. In view of (3.5), the result follows in this case as well. $\square$

Corollary 3.25 *Let k be a commutative ring and $\mathbf{m} \subseteq k$ a maximal ideal, such that the residue field $k/\mathbf{m}$ is finite of order p^f, for some prime number p and some $f \geq 1$. Let $R = k/\mathbf{m}^n$ for some $n \geq 1$. If G is a group and E an idempotent matrix with entries in RG, then there is an integer $N \in \mathbf{N}$ and elements $r_1, \ldots, r_N \in R$ and $g_1, \ldots, g_N \in G$, such that*

$$r_{HS}(E) = \sum_{i=1}^{N} r_i[g_i] = \sum_{i=1}^{N} r_i\left[g_i^{p^f}\right] \in T(RG).$$

Proof. Since the residue field $k/\mathbf{m}$ has characteristic p, it follows that $p \cdot 1 \in \mathbf{m}$. Therefore, $p^n \cdot 1 \in \mathbf{m}^n$ and hence $R = k/\mathbf{m}^n$ is a commutative ring whose characteristic is a p-th power. Invoking Proposition 3.23, we may choose an integer $N \in \mathbf{N}$ and elements $\widetilde{r}_1, \ldots, \widetilde{r}_N \in R$ and $\widetilde{g}_1, \ldots, \widetilde{g}_N \in G$, such that

$$r_{HS}(E) = \sum_{i=1}^{N} \widetilde{r}_i^{\,p^k}\left[\widetilde{g}_i^{\,p^k}\right] \in T(RG)$$

for all $k \geq 0$. For $k = f(n-1)$ and $k = fn$ this equation becomes

$$r_{HS}(E) = \sum_{i=1}^{N} \widetilde{r}_i^{\,p^{f(n-1)}}\left[\widetilde{g}_i^{\,p^{f(n-1)}}\right] = \sum_{i=1}^{N} \widetilde{r}_i^{\,p^{fn}}\left[\widetilde{g}_i^{\,p^{fn}}\right] \in T(RG).$$

We now define the elements $r_i = \widetilde{r}_i^{\,p^{f(n-1)}} \in R$ and $g_i = \widetilde{g}_i^{\,p^{f(n-1)}} \in G$ for $i = 1, \ldots, N$. Since $\widetilde{r}_i^{\,p^{fn}} = \widetilde{r}_i^{\,p^{f(n-1)}} \in R$, in view of Lemma 3.24(ii), the above equation can be rewritten as

$$r_{HS}(E) = \sum_{i=1}^{N} r_i[g_i] = \sum_{i=1}^{N} r_i\left[g_i^{p^f}\right] \in T(RG)$$

and hence the proof is finished. $\square$

3.2.2 The Main Results

Let k be a commutative ring and G a group. If $\rho = \sum_{[g] \in \mathcal{C}(G)} \rho_g [g] \in T(kG)$, where $\rho_g \in k$ for all $g \in G$, then the support $\operatorname{supp} \rho$ of ρ is the finite subset of $\mathcal{C}(G)$ that consists of all conjugacy classes $[g]$, for which ρ_g is non-zero. In this subsection, we state and prove two results of Bass and Linnell that describe certain properties of the support of the Hattori-Stallings rank $r_{HS}(E)$ of an idempotent matrix E with entries in $\mathbf{C}G$.

If Π is a set of prime numbers we denote by $\mathbf{N}_{\Pi}$ the multiplicatively closed subset of $\mathbf{N}$ generated by all primes numbers $p \notin \Pi$; in other words, $\mathbf{N}_{\Pi}$ is the set of products of primes numbers $p \notin \Pi$. We also consider the subring $\mathbf{Q}_{\Pi}$ of $\mathbf{Q}$, which consists of all fractions of the form a/b, where $a \in \mathbf{Z}$ and $b \in \mathbf{N}_{\Pi}$. It is easily seen that $\mathbf{Q}_{\Pi}$ is the intersection $\bigcap_{p \in \Pi} \mathbf{Z}_{(p)}$, where $\mathbf{Z}_{(p)} \subseteq \mathbf{Q}$ is the localization of $\mathbf{Z}$ at the prime ideal $p\mathbf{Z}$ for all $p \in \Pi$. If $\Pi = \emptyset$ then $\mathbf{N}_{\Pi} = \mathbf{N}_+$ and $\mathbf{Q}_{\Pi} = \mathbf{Q}$.

Theorem 3.26 *(Bass) Let G be a group and E an idempotent matrix with entries in the complex group algebra $\mathbf{C}G$. Consider the subset $\mathcal{X} \subseteq G$ that consists of all elements $g \in G$ for which $r_g(E) \neq 0$; in other words, $\mathcal{X} = \{g \in G : [g] \in \operatorname{supp} r_{HS}(E)\}$. Then, there is a finite set Π of prime numbers, such that:*

(i) For any prime number $p \notin \Pi$ the map $[g] \mapsto [g^p]$, $[g] \in \mathcal{C}(G)$, restricts to a bijection of the finite set $\operatorname{supp} r_{HS}(E)$ onto itself. In particular, there is a positive integer u such that the elements g and g^{p^u} are conjugate in G for all prime numbers $p \notin \Pi$ and $g \in \mathcal{X}$.

(ii) If $g \in \mathcal{X}$ is a torsion element then its order $o(g)$ is a product of prime numbers in Π.

(iii) If $g \in \mathcal{X}$ is an element of infinite order then there exists a subgroup $K(g)$ of G containing g, which is isomorphic with the group $(\mathbf{Q}_{\Pi}, +)$.

Proof. Since all claims are trivial if $\mathcal{X}$ is the empty set, we may assume that $\mathcal{X} \neq \emptyset$. As in the proof of Claim 3.8, we consider a typical element $a \in \mathbf{C}G$ and write $a = \sum_{i=1}^{t} a_i g_i$, where t is a non-negative integer, the g_i's are distinct elements of G and the a_i's non-zero complex numbers. We denote by S_a the subset of $\mathbf{C}$ consisting of the a_i's. If $A = (a_{ij})_{i,j}$ is a matrix with entries in $\mathbf{C}G$, we denote by S_A the union $\bigcup_{i,j} S_{a_{ij}}$. We now replace $\mathbf{C}$ by its subring $k = \mathbf{Z}[S_E]$ and view E as a matrix with entries in kG; in particular, $r_g(E) \in k$ for all $g \in G$. For any $g \in \mathcal{X}$ we consider the set Π_g that consists of those prime numbers $p \in \mathbf{Z}$, for which $r_g(E) \in pk$. Since k is a finitely generated integral domain of characteristic 0, the set Π_g is finite (cf. Corollary A.27 of Appendix A). Since $\Pi_g = \Pi_{g'}$ if $[g] = [g'] \in \mathcal{C}(G)$, the finiteness of the set $[\mathcal{X}] = \operatorname{supp} r_{HS}(E)$ implies that the set $\Pi = \bigcup_{g \in \mathcal{X}} \Pi_g$ is also finite.

In order to prove assertion (i), we fix a prime number $p \notin \Pi$ and note that, by the very definition of Π, $r_g(E) \notin pk$ for all $g \in \mathcal{X}$. In particular, the ideal pk is proper in k and hence the quotient ring $R = k/pk$ is non-zero. We consider the idempotent matrix $\overline{E}$ with entries in RG, which is obtained from

E by passage to the quotient. In view of the naturality of the Hattori-Stallings rank with respect to coefficient ring homomorphisms (cf. Proposition 1.46), we have

$$r_x(\overline{E}) = \overline{r_x(E)} \in R$$

for all $x \in G$, where $\overline{r_x(E)}$ denotes the residue class of $r_x(E) \in k$ modulo pk. Since $r_g(E) \notin pk$ for all $g \in \mathcal{X}$, we conclude that

$$\operatorname{supp} r_{HS}(\overline{E}) = \operatorname{supp} r_{HS}(E) = [\mathcal{X}] \subseteq \mathcal{C}(G) .$$

Since the commutative ring R has characteristic p, we may invoke Proposition 3.23 and choose an integer $N \in \mathbf{N}$ and elements $r_1, \ldots, r_N \in R$ and $g_1, \ldots, g_N \in G$, such that

$$r_{HS}(\overline{E}) = \sum_{i=1}^{N} r_i[g_i] = \sum_{i=1}^{N} r_i^p[g_i^p] \in T(RG) . \tag{3.6}$$

Without any loss of generality, we can make the following two assumptions (replacing, if necessary, N by a lower integer):

- If $i \neq j$ then $[g_i] \neq [g_j] \in \mathcal{C}(G)$. Indeed, if g_i is conjugate to g_j for two indices i and j, then g_i^p is conjugate to g_j^p as well. In view of Lemma 3.2(ii), we have $r_i^p + r_j^p = (r_i + r_j)^p \in R$ and hence we may reduce by one the number of summands in both sums of (3.6). Repeating this reduction process finitely many times, we arrive at an equation of the form (3.6) with $[g_i] \neq [g_j]$ if $i \neq j$.
- All coefficients r_i are non-zero. Indeed, if some $r_i = 0$ then $r_i^p = 0$ as well and hence we may remove the corresponding summands from both sums of (3.6).

Under these two assumptions, we have

$$[\mathcal{X}] = \operatorname{supp} r_{HS}(\overline{E}) = \{[g_1], \ldots, [g_N]\} \subseteq \{[g_1^p], \ldots, [g_N^p]\}$$

and

$$\operatorname{card}([\mathcal{X}]) = \operatorname{card}(\operatorname{supp} r_{HS}(\overline{E})) = \operatorname{card}(\{[g_1], \ldots, [g_N]\}) = N .$$

Comparing cardinalities, we conclude that the inclusion above is an equality

$$[\mathcal{X}] = \{[g_1], \ldots, [g_N]\} = \{[g_1^p], \ldots, [g_N^p]\} . \tag{3.7}$$

We now consider the operator

$$\mathcal{F} : \mathcal{C}(G) \longrightarrow \mathcal{C}(G) ,$$

which is given by $[x] \mapsto [x^p]$, $[x] \in \mathcal{C}(G)$; it is clear that this operator is well-defined. Even though $\mathcal{F}$ is, in general, neither injective nor surjective, (3.7) shows that the restriction $\mathcal{F}|_{[\mathcal{X}]}$ is a bijection of the finite set $[\mathcal{X}]$ onto itself.

Since the group of symmetries of $[\mathcal{X}]$ is finite of order $u = N! = \text{card}([\mathcal{X}])!$, we conclude that $\mathcal{F}^u$ is the identity operator on $[\mathcal{X}]$. In particular, we have $[g] = \mathcal{F}^u[g] = [g^{p^u}]$, i.e. the elements g and g^{p^u} are conjugate in G for all $g \in \mathcal{X}$.

Assertion (ii) is an immediate consequence of (i) above, in view of the following lemma.

Lemma 3.27 *Let Π be a set of prime numbers, G a group and $g \in G$ a torsion element with the following property: For any prime number $p \notin \Pi$ there is a positive integer $u(p)$, such that g is conjugate to $g^{p^{u(p)}}$ in G. Then, the order of g is a product of prime numbers in Π.*

Proof. Let $n = o(g)$ be the order of g and suppose that p is a prime number dividing n, with $p \notin \Pi$. If $m = n/p \in \mathbf{Z}$ then, g being conjugate to $g^{p^{u(p)}}$, the element g^m is conjugate to $\left(g^{p^{u(p)}}\right)^m = g^{p^{u(p)}m} = g^{p^{u(p)-1}n} = (g^n)^{p^{u(p)-1}} = 1^{p^{u(p)-1}} = 1$. But then $g^m = 1$, a contradiction as $m < n$. $\square$

Proof of Theorem 3.26 (cont.) The following lemma shows that assertion (iii) is a formal consequence of (i) as well. $\square$

Lemma 3.28 *Let G be a group and $g \in G$ an element of infinite order. Consider a set Π of prime numbers and assume that for any prime number $p \notin \Pi$ there exists a positive integer $u(p)$, such that g and $g^{p^{u(p)}}$ are conjugate in G. Then, there exists a subgroup $K \leq G$, such that:*
(i) K contains g,
(ii) K is isomorphic with the group $(\mathbf{Q}_\Pi, +)$ and
(iii) K is the ascending union of a sequence of cyclic groups that are generated by conjugates of g.

Proof. It suffices to consider the case where Π is a proper subset of the set of all prime numbers. In that case, we may consider a sequence $(p_i)_i$ of prime numbers that are not contained in Π, such that every prime number $p \notin \Pi$ is repeated infinitely often therein. We define $v_i = p_i^{u(p_i)}$ for all i and note that the sequence $(k_n)_n$, where $k_n = \prod_{i=1}^n v_i$ for all $n \geq 0$ (with the assumption that $k_0 = 1$), is such that:
(α) $v_n k_n^{-1} = k_{n-1}^{-1} \in \mathbf{Q}$ for all $n \geq 1$ and
(β) for any $m \in \mathbf{N}_\Pi$ there is an integer $n \geq 0$, such that m divides k_n.[2]
In particular, the sequence $(C_n)_n$ of cyclic groups, where $C_n = \mathbf{Z} \cdot k_n^{-1} \subseteq \mathbf{Q}$ for all $n \geq 0$, is increasing (in view of (α)), whereas $\bigcup_n C_n = \mathbf{Q}_\Pi$ (in view of (β)). The elements g and g^{v_i} are conjugate in G and hence there exists $x_i \in G$ such that

$$x_i^{-1} g x_i = g^{v_i}$$

---

[2] It is precisely at this point where we need the hypothesis that any prime number $p \notin \Pi$ is repeated infinitely often in the sequence $(p_i)_i$.

for all i. We now define

$$h_n = x_1 x_2 \cdots x_n \quad \text{and} \quad z_n = h_n g h_n^{-1}$$

for all $n \geq 0$, with the assumption that $h_0 = 1$. Being conjugate to g, all of the z_n's are elements of infinite order. It follows that if K_n is the cyclic subgroup of G generated by z_n, then there is an isomorphism

$$\varphi_n : K_n \longrightarrow C_n \, ,$$

which maps z_n onto $k_n^{-1} \in C_n$ for all $n \geq 0$. Since $h_n x_n^{-1} = h_{n-1}$, we have

$$
\begin{aligned}
z_n^{v_n} &= \left(h_n g h_n^{-1} \right)^{v_n} \\
&= h_n g^{v_n} h_n^{-1} \\
&= h_n x_n^{-1} g x_n h_n^{-1} \\
&= h_{n-1} g h_{n-1}^{-1} \\
&= z_{n-1}
\end{aligned}
$$

for all $n \geq 1$. In particular, $z_{n-1} \in K_n$ for all $n \geq 1$ and hence the sequence $(K_n)_n$ is increasing. Moreover, the relations $z_n^{v_n} = z_{n-1}$, $n \geq 1$, combined with (α) above, show that there is a ladder of isomorphisms

$$
\begin{array}{ccccccccc}
K_0 & \hookrightarrow & \cdots & \hookrightarrow & K_{n-1} & \hookrightarrow & K_n & \hookrightarrow & \cdots & \subseteq G \\
{\scriptstyle \varphi_0} \downarrow & & & & {\scriptstyle \varphi_{n-1}} \downarrow & & {\scriptstyle \varphi_n} \downarrow & & & \\
C_0 & \hookrightarrow & \cdots & \hookrightarrow & C_{n-1} & \hookrightarrow & C_n & \hookrightarrow & \cdots & \subseteq \mathbf{Q}
\end{array}
$$

It follows that the union $K = \bigcup_n K_n$, a subgroup of G containing $g = z_0$, is isomorphic with the union $\bigcup_n C_n = \mathbf{Q}_\Pi$.   □

We now describe a few consequences of Bass' theorem that concern the idempotent conjecture.

Corollary 3.29 *Let G be a torsion-free group that contains no subgroup isomorphic with the additive group of the ring $\mathbf{Q}_\Pi$, for any finite set Π of prime numbers. Then, G satisfies the idempotent conjecture.*

Proof. If $e \in \mathbf{C}G$ is an idempotent then our assumption on G, combined with Theorem 3.26(iii), implies that $r_g(e) = 0$ for all $g \in G \setminus \{1\}$. Then, e is trivial, in view of Proposition 3.15.   □

A group is called Noetherian if any subgroup of it is finitely generated. Of course, finite groups are Noetherian. The following result can be used in order to obtain less trivial examples of Noetherian groups. A group is called polycyclic-by-finite if it is an iterated extension of cyclic and finite groups.

Proposition 3.30 *(i) A finitely generated abelian group is Noetherian.*
   (ii) An extension of Noetherian groups is Noetherian.
   (iii) A polycyclic-by-finite group is Noetherian.

Proof. (i) It is well-known that any subgroup of a finitely generated abelian group is finitely generated.

(ii) Let G be a group and $N \trianglelefteq G$ a normal subgroup, such that N and the quotient group G/N are Noetherian. In order to show that G is also Noetherian, let us consider a subgroup $H \leq G$. Then, both groups $H \cap N$ and $H/(H \cap N)$ are finitely generated, being subgroups of N and G/N respectively. It follows that H is finitely generated as well.

(iii) This is an immediate consequence of (i) and (ii) above, since finite groups are Noetherian. □

Corollary 3.31 *A torsion-free Noetherian group satisfies the idempotent conjecture.*

Proof. If p is a prime number then the group $(\mathbf{Z}[p^{-1}], +)$ is not finitely generated. On the other hand, $\mathbf{Z}[p^{-1}] \subseteq \mathbf{Q}_\Pi$ for any set Π of prime numbers with $p \notin \Pi$. Therefore, a Noetherian group contains no subgroup isomorphic with the group $(\mathbf{Q}_\Pi, +)$, for any proper subset Π of the set of prime numbers. This finishes the proof, in view of Corollary 3.29. □

Let E be an idempotent matrix with entries in the complex group algebra of a group G. Bass' result (Theorem 3.26) gives no information about the p-th powers of the elements $g \in G$ for which $r_g(E) \neq 0$, in the case where $p \in \Pi$ is one of the exceptional prime numbers therein. If we assume that E has entries in the group algebra kG, where k is a subring of $\mathbf{C}$ with $k \cap \mathbf{Q} = \mathbf{Z}$, then we can obtain information on the p-th power map for all primes numbers p. This is the theme of the following result of Linnell, whose proof is very similar to that of Bass' theorem.

Theorem 3.32 *(Linnell) Let k be a subring of the field $\mathbf{C}$ of complex numbers with $k \cap \mathbf{Q} = \mathbf{Z}$, G a group and E an idempotent matrix with entries in the group algebra kG. Consider the subset $\mathcal{X} \subseteq G$ that consists of all elements $g \in G$, for which $r_g(E) \neq 0$; in other words, $\mathcal{X} = \{g \in G : [g] \in \operatorname{supp} r_{HS}(E)\}$. Then:*

(i) There is a positive integer u such that g and g^{n^u} are conjugate in G for all $n \geq 1$ and all $g \in \mathcal{X}$.

(ii) If $g \in \mathcal{X}$ is a torsion element then $g = 1$.

(iii) If $g \in \mathcal{X}$ is an element of infinite order then there exists a subgroup $K(g)$ of G containing g, which is generated by some conjugates of g, is isomorphic with the additive group $\mathbf{Q}$ of rational numbers and lies in finitely many G-conjugacy classes.

In the special case where $k = \mathbf{Z}$ is the ring of integers, we also have:

(i)' $r_g(E) = r_{g^n}(E)$ for all $g \in \mathcal{X}$ and all $n \geq 1$.

Proof. As in the proof of Theorem 3.26, we may replace k by its subring generated by all complex numbers occurring in the entries of E and hence reduce to the case where the ring k is finitely generated. Let us fix a prime number p. Then, $p \in k$ is not a unit, since $k \cap \mathbf{Q} = \mathbf{Z}$ (cf. Lemma 1.50). Hence, there is a maximal ideal $\mathbf{m} \subseteq k$ with $p \in \mathbf{m}$. Since k is finitely generated as a $\mathbf{Z}$-algebra, the residue field $k/\mathbf{m}$ is finitely generated as an $\mathbf{F}_p$-algebra. Therefore, $k/\mathbf{m}$ is a finite field, say with p^f elements (cf. Corollary A.22(ii) of Appendix A). Of course, the positive integer $f = f(p)$ depends on p (as well as on the chosen maximal ideal containing p). If $k = \mathbf{Z}$ is the ring of integers then the unique maximal ideal of k containing p is the ideal $\mathbf{m} = p\mathbf{Z}$ with corresponding residue field $k/\mathbf{m}$ equal to the field $\mathbf{F}_p$; it follows that, in this case, $f(p) = 1$.

Since $[\mathcal{X}] = \operatorname{supp} r_{HS}(E)$ is a finite set, we may apply the Krull intersection theorem (cf. Corollary A.35 of Appendix A), in order to conclude that there exists $n_0 \in \mathbf{N}$ such that $r_g(E) \notin \mathbf{m}^n$ for all integers $n \geq n_0$ and $g \in \mathcal{X}$. We fix such an integer n and consider the quotient ring $R = k/\mathbf{m}^n$ and the idempotent matrix $\overline{E}$ with entries in RG, which is obtained from E by passage to the quotient. In view of the naturality of the Hattori-Stallings rank with respect to coefficient ring homomorphisms (cf. Proposition 1.46), we have

$$r_x(\overline{E}) = \overline{r_x(E)} \in R \tag{3.8}$$

for all $x \in G$, where $\overline{r_x(E)}$ denotes the residue class of $r_x(E) \in k$ modulo $\mathbf{m}^n$. Since $r_g(E) \notin \mathbf{m}^n$ for all $g \in \mathcal{X}$, we conclude that

$$\operatorname{supp} r_{HS}(\overline{E}) = \operatorname{supp} r_{HS}(E) = [\mathcal{X}] \subseteq \mathcal{C}(G) .$$

We now invoke Corollary 3.25 and choose an integer $N \in \mathbf{N}$ and suitable elements $r_1, \ldots, r_N \in R$ and $g_1, \ldots, g_N \in G$, such that

$$r_{HS}(\overline{E}) = \sum_{i=1}^{N} r_i[g_i] = \sum_{i=1}^{N} r_i\left[g_i^{p^f}\right] \in T(RG) . \tag{3.9}$$

As in the proof of Theorem 3.26, we can assume (replacing, if necessary, N by a lower integer) that $[g_i] \neq [g_j] \in \mathcal{C}(G)$ for all $i \neq j$ and $r_i \neq 0$ for all i. Then,

$$[\mathcal{X}] = \operatorname{supp} r_{HS}(\overline{E}) = \{[g_1], \ldots, [g_N]\} \subseteq \left\{\left[g_1^{p^f}\right], \ldots, \left[g_N^{p^f}\right]\right\}$$

and

$$\operatorname{card}([\mathcal{X}]) = \operatorname{card}(\operatorname{supp} r_{HS}(\overline{E})) = \operatorname{card}(\{[g_1], \ldots, [g_N]\}) = N .$$

Comparing cardinalities, we conclude that the set $\left\{\left[g_1^{p^f}\right], \ldots, \left[g_N^{p^f}\right]\right\}$ has exactly N elements, i.e. $\left[g_i^{p^f}\right] \neq \left[g_j^{p^f}\right] \in \mathcal{C}(G)$ if $i \neq j$, whereas the inclusion above is an equality

$$[\mathcal{X}] = \{[g_1], \ldots, [g_N]\} = \left\{\left[g_1^{p^f}\right], \ldots, \left[g_N^{p^f}\right]\right\} . \tag{3.10}$$

We consider the operator

$$\mathcal{F} : \mathcal{C}(G) \longrightarrow \mathcal{C}(G) \,,$$

which is given by $[x] \mapsto [x^p]$, $[x] \in \mathcal{C}(G)$, and note that (3.10) shows that the restriction $\mathcal{F}^f|_{[\mathcal{X}]}$ is a bijection of the finite set $[\mathcal{X}]$ onto itself. Since the order of the group of symmetries of $[\mathcal{X}]$ is $N! = \mathrm{card}([\mathcal{X}])!$, we conclude that $(\mathcal{F}^f)^{N!} = \mathcal{F}^{f \cdot N!}$ is the identity operator on $[\mathcal{X}]$ and hence $[g] = \mathcal{F}^{f \cdot N!}[g] = \left[g^{p^{f \cdot N!}} \right]$ for all $g \in \mathcal{X}$. For the given prime number p, we denote the positive integer $f \cdot N! = f(p) \cdot N!$ by $u(p)$. Therefore, we have shown that the elements g and $g^{p^{u(p)}}$ are conjugate in G for all $g \in \mathcal{X}$.

According to Theorem 3.26(i), there is a finite set Π of prime numbers and a positive integer u_0, such that for any element $g \in \mathcal{X}$ and any prime number $p \notin \Pi$ we have $g^p \in \mathcal{X}$ and $[g] = \left[g^{p^{u_0}} \right] \in \mathcal{C}(G)$.[3] For any prime number $p \in \Pi$ we consider the positive integer $u(p)$ constructed above and define u to be the least common multiple of u_0 and the $u(p)$'s, for $p \in \Pi$. For any element $x \in G$ the subset of $\mathbf{N}$ consisting of those integers t for which x and x^t are conjugate in G is easily seen to be multiplicatively closed. It follows that g and g^{p^u} are conjugate in G for all prime numbers p and all $g \in \mathcal{X}$. Since any positive integer is a product of prime numbers, we conclude that g and g^{n^u} are conjugate in G for all $n \geq 1$ and all $g \in \mathcal{X}$, thereby proving (i).

For later use, we make at this point the following claim.

Claim 3.33 *For any $g \in \mathcal{X}$ the cyclic subgroup generated by g lies in finitely many G-conjugacy classes.*

Proof. We consider the finite set Π of prime numbers provided to us by Theorem 3.26 and the function $p \mapsto f(p)$, that was defined in the beginning of the proof of Theorem 3.32 on the set of all prime numbers. Then, $g^p \in \mathcal{X}$ for any $g \in \mathcal{X}$ and any prime number p with $p \notin \Pi$. On the other hand, (3.10) shows that $g^{p^{f(p)}} \in \mathcal{X}$ for all prime numbers p and all $g \in \mathcal{X}$. We note that any positive integer n can be written uniquely as the product of three integers a, b and c, where:

- a is a product of prime numbers $p \notin \Pi$,
- b is a product of powers of the form $p^{f(p)}$, for prime numbers $p \in \Pi$, and
- c is a product of prime numbers $p \in \Pi$, such that the power p^i divides c only if $i < f(p)$.

Let C be the set consisting of all c's as above. Since Π is a finite set, it is clear that the set C is finite as well. If a, b and c are positive integers of the above form, $n = abc$ and $g \in \mathcal{X}$, then $g^{ab} \in \mathcal{X}$ and hence $[g^n] = [(g^{ab})^c]$ is an element of the set

---

[3] In fact, it follows from the proof of Theorem 3.26(i) that u_0 can be chosen to be equal to $N!$.

$$\Phi = \{[x] \in \mathcal{C}(G) : \text{there exists } [y] \in [\mathcal{X}] \text{ and } c \in C, \text{ such that } [x] = [y^c]\} .$$

Both sets $[\mathcal{X}] = \operatorname{supp} r_{HS}(E)$ and C being finite, it follows that the set Φ is also finite. For any subset $\Psi \subseteq \mathcal{C}(G)$, let us denote by Ψ^{-1} the set consisting of all $[x] \in \mathcal{C}(G)$ for which $[x^{-1}] \in \Psi$. It is clear that if Ψ is a finite set then the same is true for Ψ^{-1}; in fact, $\operatorname{card} \Psi = \operatorname{card} \Psi^{-1}$. Since

$$\{[g^n] : n \in \mathbf{Z}\} = \{[g^n] : n \geq 1\} \cup \{[1]\} \cup \{[g^{-n}] : n \geq 1\} \subseteq \Phi \cup \{[1]\} \cup \Phi^{-1},$$

we conclude that the cyclic group generated by g lies in finitely many G-conjugacy classes, as needed. ∎

Proof of Theorem 3.32 (cont.) We fix a conjugacy class $[g] \in [\mathcal{X}]$; then, $[g]$ is one of the $[g_i]$'s, say $[g] = [g_1]$. Since the $[g_i]$'s are distinct, (3.9) shows that $r_g(\overline{E}) = r_{g_1}(\overline{E}) = r_1$. But the $\left[g_i^{p^f}\right]$'s are distinct as well and hence we may use (3.9) once again, in order to conclude that $r_{g^{p^f}}(\overline{E}) = r_{g_1^{p^f}}(\overline{E}) = r_1$. In particular, it follows that $r_g(\overline{E}) = r_{g^{p^f}}(\overline{E}) \in R$; in view of (3.8), this means that $r_g(E) \equiv r_{g^{p^f}}(E)$ modulo $\mathbf{m}^n$. This being the case for all $n \geq n_0$, Corollary A.35 of Appendix A implies that $r_g(E) = r_{g^{p^f}}(E) \in k$.

In the special case where $k = \mathbf{Z}$, we have $f = 1$ for all prime numbers p and hence $r_g(E) = r_{g^p}(E)$ for all such p's and all $g \in \mathcal{X}$. If p is a prime number and $g \in \mathcal{X}$, then $g^p \in \mathcal{X}$ as well (since $r_{g^p}(E) = r_g(E) \neq 0$); therefore, $r_{g^p}(E) = r_{g^{pq}}(E)$ for all prime numbers q and hence $r_g(E) = r_{g^{pq}}(E)$ for all prime numbers p, q. Since any positive integer is a product of prime numbers, we may continue in this way and conclude that $r_g(E) = r_{g^n}(E)$ for all $g \in \mathcal{X}$ and all $n \geq 1$. Hence, we have proved assertion (i)$'$.

Assertion (ii) is an immediate consequence of (i), in view of Lemma 3.27 (applied in the special case where the set Π of prime numbers considered therein is empty).

In order to prove assertion (iii), we fix an element $g \in \mathcal{X}$ of infinite order and note that Lemma 3.28, combined with (i) above, shows that there exists a subgroup $K \leq G$ containing g, which is isomorphic with the additive group $\mathbf{Q}$ of rational numbers and is equal to the ascending union of a sequence of cyclic groups that are generated by conjugates of g. Hence, for any $x \in K$ there exists a conjugate z of g, such that $x = z^n$ for some $n \in \mathbf{Z}$; in particular, $[x] = [z^n] = [g^n] \in \mathcal{C}(G)$. It follows that if $K_0 \leq K$ is the cyclic group generated by g, then $[K] = [K_0] \subseteq \mathcal{C}(G)$. Since the set $[K_0]$ is finite, in view of Claim 3.33, we conclude that the subgroup K lies indeed in finitely many G-conjugacy classes. ∎

Addendum 3.34 Let G be a group and E a non-zero idempotent matrix with entries in the integral group ring $\mathbf{Z}G$. Consider the Hattori-Stallings rank $r_{HS}(E) = \sum_{[g] \in \mathcal{C}(G)} r_g(E)[g]$ and let $\mathcal{X}$ be the set consisting of all elements $g \in G$, for which $r_g(E) \neq 0$. Then, Corollary 3.14(ii) implies that $1 \in \mathcal{X}$ and hence the finite set $[\mathcal{X}] = \operatorname{supp} r_{HS}(E)$ is non-empty; let $N = \operatorname{card}([\mathcal{X}])$.

If $g \in \mathcal{X}$ is an element of infinite order then the subgroup $K = K(g)$ of G that was constructed in the proof of assertion (iii) of Theorem 3.32 lies in the union of $2N - 1$ G-conjugacy classes. Indeed, assertion (i)$'$ of loc.cit. implies that the set $\{[g^n] : n \geq 1\}$ is contained in the $(N - 1)$-element set $[\mathcal{X}] \setminus \{[1]\}$. Therefore, the argument used in the final part of the proof of Claim 3.33 shows that the cyclic subgroup K_0 of G generated by g lies in the union of $(N-1)+1+(N-1) = 2N-1$ G-conjugacy classes. Since $[K] = [K_0] \subseteq \mathcal{C}(G)$, as noted in the proof of Theorem 3.32(iii), we conclude that $\mathrm{card}([K]) \leq 2N - 1$.

We now describe a few consequences of Linnell's theorem that concern Bass' conjecture.

Corollary 3.35 *Let k be a subring of the field $\mathbf{C}$ of complex numbers with $k \cap \mathbf{Q} = \mathbf{Z}$, G a group, E an idempotent matrix with entries in the group algebra kG and $g \in G \setminus \{1\}$ an element of finite order. Then, $r_g(E) = 0$.*

Proof. This is immediate from Theorem 3.32(ii). □

Corollary 3.36 *Let G be a group containing no subgroup isomorphic with the additive group of rational numbers. Then, G satisfies Bass' conjecture.*

Proof. Assume on the contrary that there exists a subring k of the field $\mathbf{C}$ of complex numbers satisfying the condition $k \cap \mathbf{Q} = \mathbf{Z}$, an idempotent matrix E with entries in kG and an element $g \in G \setminus \{1\}$, for which $r_g(E) \neq 0$. In view of Corollary 3.35, g is an element of infinite order. Then, Theorem 3.32(iii) shows that G has a subgroup $K(g)$, which is isomorphic with the additive group $\mathbf{Q}$ of rational numbers. This contradicts our hypothesis on G and therefore finishes the proof. □

Corollary 3.37 *Noetherian groups satisfy Bass' conjecture.*

Proof. A Noetherian group contains no subgroup isomorphic with the group $(\mathbf{Q}, +)$; hence, the result follows from Corollary 3.36. □

Corollary 3.38 *Residually finite groups and torsion groups satisfy Bass' conjecture.*

Proof. It is clear that the only element of a finite group which is divisible by the order of the group is the identity. It follows that a residually finite group contains no (infinitely) divisible non-identity elements. In particular, a residually finite group contains no subgroup isomorphic with the additive group of rational numbers. Of course, this latter statement is also true for torsion groups. This finishes the proof, in view of Corollary 3.36. □

Proposition 3.39 *Let G be a group and $H \leq G$ a subgroup of finite index. If H satisfies Bass' conjecture then so does G.*

Proof. Assume on the contrary that there exists a subring k of the field $\mathbf{C}$ of complex numbers satisfying the condition $k \cap \mathbf{Q} = \mathbf{Z}$, an idempotent matrix E with entries in kG and an element $g \in G \setminus \{1\}$, for which $r_g(E) \neq 0$. In view of Corollary 3.35, g is an element of infinite order. Since $H \leq G$ is a subgroup of finite index, there is an integer $n \geq 1$ such that $g^n \in H$. Using Theorem 3.32(i), we can find a positive integer $m \in n\mathbf{Z}$, such that $[g] = [g^m] \in \mathcal{C}(G)$; for example, we may take $m = n^u$, with $u > 0$ as in loc.cit. In particular, we have

$$ r_{g^m}(E) = r_g(E) \neq 0 \ . $$

Since m is a multiple of n, g^m is a power of g^n and hence $g^m \in H$. Moreover, g being an element of infinite order, we have $g^m \neq 1$. On the other hand, we may use our assumption that H satisfies Bass' conjecture and invoke Remark 2.37, in order to conclude that $r_h(E) = 0$ for all $h \in H \setminus \{1\}$. This is a contradiction, since $g^m \in H \setminus \{1\}$, whereas $r_{g^m}(E) \neq 0$. $\qquad\square$

Remark 3.40 Since abelian groups satisfy Bass' conjecture (cf. Theorem 2.26), Proposition 3.39 implies that groups having an abelian subgroup of finite index satisfy Bass' conjecture as well. In this way, Proposition 3.39 generalizes Proposition 2.35.

3.2.3 An Application: the Case of Solvable Groups

In this subsection, we examine Bass' conjecture in the special case of solvable groups and show how Linnell's theorem (Theorem 3.32) can be used in order to prove that solvable groups of finite Hirsch number satisfy the conjecture.[4]

Let G be a group and $N, H \trianglelefteq G$ normal subgroups with $N \leq H$. Then, $\overline{H} = H/N$ is a normal subgroup of the quotient group $\overline{G} = G/N$; we refer to it as a normal subquotient of G. We may pull the conjugation action of $\overline{G}$ on $\overline{H}$ back to G and regard $\overline{H}$ as a group with a G-action. By an obvious abuse of language, we refer to this action as action by conjugation.

Proposition 3.41 *Let G be a finitely generated solvable group which does not satisfy Bass' conjecture. Then, there is an abelian normal subquotient V of G with the following properties:*

(i) V is a $\mathbf{Q}$-vector space on which G acts (by conjugation) $\mathbf{Q}$-linearly,

(ii) the $\mathbf{Q}$-vector space V is infinite dimensional,

(iii) V has no proper non-trivial G-invariant subgroup and

(iv) there is a one-dimensional $\mathbf{Q}$-linear subspace $V_0 \subseteq V$ which lies in finitely many G-orbits.

Proof. Since G does not satisfy Bass' conjecture, there is a subring k of the field $\mathbf{C}$ of complex numbers with $k \cap \mathbf{Q} = \mathbf{Z}$, an idempotent matrix E with entries

---

[4] In fact, F. Farrell and P. Linnell have shown in [28], using techniques that are beyond the scope of this book, that all solvable groups satisfy Bass' conjecture.

in the group algebra kG and an element $g \in G \setminus \{1\}$, for which $r_g(E) \neq 0$. Then, g is an element of infinite order (Corollary 3.35) and hence Linnell's theorem (Theorem 3.32) implies the existence of a subgroup $K \leq G$, such that the triple (G, K, g) has the following properties:

(L1) There exists a positive integer u such that g and g^{n^u} are conjugate in G for all $n \geq 1$.

(L2) The group K contains g, is generated by some conjugates of g, is isomorphic with the additive group $\mathbf{Q}$ of rational numbers and lies in finitely many G-conjugacy classes.

The following lemma asserts that the above properties are preserved under group homomorphisms.

Lemma 3.42 *Consider a group G, a subgroup $K \leq G$ and an element $g \in K$, such that the triple (G, K, g) satisfies conditions (L1) and (L2). Let $\varphi : G \longrightarrow G'$ be a group homomorphism and define $K' = \varphi(K) \leq G'$. If $g' = \varphi(g) \neq 1$ then the triple (G', K', g') satisfies conditions (L1) and (L2) as well.*

Proof. It is immediate that the triple (G', K', g') has property (L1). Since $g' \neq 1$, it follows from Lemma 3.27 (with Π equal to the empty set) that g' is an element of infinite order, i.e. $< g > \cap \ker \varphi = \{1\}$. It follows that the subgroup $K \cap \ker \varphi$ of K intersects trivially the infinite cyclic group $< g >$. Since $K \simeq (\mathbf{Q}, +)$, in view of property (L2), this can happen only if the group $K \cap \ker \varphi$ is itself trivial. But then φ restricts to an isomorphism $K \simeq K'$ and hence the triple (G', K', g') satisfies property (L2) as well.   □

Proof of Proposition 3.41 (cont.) We consider the set

$$\mathcal{A} = \{N : N \trianglelefteq G \text{ and } g \notin N\},$$

which, ordered by inclusion, is easily seen to satisfy the hypotheses of Zorn's lemma. Let $N \in \mathcal{A}$ be a maximal element and consider the triple $(\overline{G}, \overline{K}, \overline{g})$, where $\overline{G} = G/N$, $\overline{K} = KN/N$ and $\overline{g} = gN$. The maximality of N in $\mathcal{A}$ implies that $\overline{g}$ is contained in any non-trivial normal subgroup of $\overline{G}$. In view of Lemma 3.42, we may change notation $(\overline{G} \rightsquigarrow G, \overline{K} \rightsquigarrow K$ and $\overline{g} \rightsquigarrow g)$ and assume that the triple (G, K, g) satisfies conditions (L1) and (L2)[5], whereas g is contained in any non-trivial normal subgroup of G. In this case, G has a unique minimal non-trivial normal subgroup V, namely the one generated by the conjugates of g. Since the group K is generated by some of these conjugates, in view of property (L2), it follows that $K \leq V$. Let

$$\varrho : G \longrightarrow \text{Aut}(V)$$

be the group homomorphism associated with the conjugation action of G on V, i.e. let $\varrho(x)(v) = xvx^{-1}$ for all $x \in G$ and $v \in V$. We shall prove that the pair (V, ϱ) has all of the properties in the statement of the proposition.

---

[5] In particular, g is still divisible in K and has infinite order.

(i) Being a minimal non-trivial normal subgroup of G, V can have no proper non-trivial characteristic subgroup. Since V is solvable, the derived group DV is a proper characteristic subgroup of V; it follows that DV is trivial and hence V is abelian. We now consider the torsion subgroup $V_t \subseteq V$. Of course, V_t is a characteristic subgroup of V. On the other hand, $g \in V$ is an element of infinite order and hence $g \notin V_t$; therefore, V_t is proper in V. It follows that V_t is the trivial group and hence V is torsion-free. We also note that the abelian group V is divisible, since it is generated by the elements $\varrho(x)(g)$, $x \in G$, which are divisible in V. Being a torsion-free divisible abelian group, V is a $\mathbf{Q}$-vector space. Since $\mathrm{Aut}_{\mathbf{Z}}(V) = \mathrm{Aut}_{\mathbf{Q}}(V)$, the additive (i.e. $\mathbf{Z}$-linear) action ϱ of G on V is also $\mathbf{Q}$-linear.

(ii) We claim that V is infinite dimensional as a $\mathbf{Q}$-vector space. Assume on the contrary that $\dim_{\mathbf{Q}} V = m < \infty$. In this case, we can choose a $\mathbf{Q}$-basis $\{v_1, \ldots, v_m\}$ of V with $v_1 = g$ and consider the associated matrix representation

$$\varrho : G \longrightarrow GL_m(\mathbf{Q}) \ .$$

Since G is finitely generated, ϱ factors through $GL_m\left(\mathbf{Z}\left[\frac{1}{N}\right]\right)$ for some N. Indeed, if $x_1, \ldots, x_s$ are generators of G, we may choose N to be the least common multiple of all denominators occurring in the entries of the matrices $\varrho(x_1), \varrho(x_1^{-1}), \ldots, \varrho(x_s), \varrho(x_s^{-1}) \in GL_m(\mathbf{Q})$. It follows that the G-orbit $G \cdot v_1$ is contained in the subgroup $\bigoplus_{i=1}^m \mathbf{Z}\left[\frac{1}{N}\right]v_i$ of V. In particular, if p is a prime number not dividing N, then this G-orbit cannot contain $\frac{1}{p^u}v_1 \in V$ for any positive integer u. On the other hand, property (L1) implies that there exists a positive integer u and an element $x \in G$, such that $g = xg^{p^u}x^{-1}$. We may write this equation additively and recall that $g = v_1$, in order to conclude that $v_1 = \varrho(x)(p^u v_1) \in V$. But then $v_1 = p^u\varrho(x)(v_1)$ and hence $\frac{1}{p^u}v_1 = \varrho(x)(v_1)$ is contained in the G-orbit of v_1. This contradiction shows that $\dim_{\mathbf{Q}} V$ must be infinite.

(iii) If V' is a proper G-invariant subgroup of V, then V' is normal in G and hence $V' = 1$, in view of the minimality of V.

(iv) The group $K \simeq \mathbf{Q}$ is a one-dimensional subspace of the $\mathbf{Q}$-vector space V lying in finitely many G-orbits, in view of property (L2). $\square$

In order to give a concrete application of the result above, let us consider a solvable group G and its derived series $(D^n G)_n$. Then, the quotient group $D^n G/D^{n+1}G$ is abelian and has rank $r_n = r_n(G)$, which is equal to the dimension of the $\mathbf{Q}$-vector space $(D^n G/D^{n+1}G) \otimes \mathbf{Q}$ for all n. Since G is solvable, $D^n G$ is the trivial group for $n \gg 0$ and hence $r_n = 0$ for $n \gg 0$. The sum $r = \sum_n r_n$ is the Hirsch number $h(G)$ of the solvable group G. It follows easily from the definition that

$$h(G) = h(D^i G) + h(G/D^i G) \tag{3.11}$$

for all $i \geq 1$. If the group G is abelian then $h(G)$ is the rank of G. The following result can be viewed as a generalization of some well-known properties of the notion of rank for abelian groups to the concept of Hirsch number for solvable ones.

Lemma 3.43 *Let G be a solvable group.*
(i) If $N \trianglelefteq G$ is a normal subgroup then $h(G/N) \leq h(G)$.
(ii) If $A \leq G$ is an abelian subgroup then $h(A) \leq h(G)$.
(iii) If A is an abelian subquotient of G then $h(A) \leq h(G)$.[6]

Proof. (i) The derived series $(D^n(G/N))_n$ of the quotient group G/N is given by $D^n(G/N) = (D^nG)N/N$ for all n. In particular, the abelian group $D^n(G/N)/D^{n+1}(G/N) \simeq (D^nG)N/(D^{n+1}G)N$ is a quotient of the abelian group $D^nG/D^{n+1}G$ and hence $r_n(G/N) \leq r_n(G)$ for all n. Summing these inequalities up for all n, we conclude that $h(G/N) \leq h(G)$.

(ii) We use induction on the degree of solvability n of G. The result is clear if $n = 1$, i.e. if G is abelian. If $n > 1$ we consider the abelian subgroup $D^{n-1}G \leq G$ and the quotient group $G/D^{n-1}G$, which is solvable of degree $n - 1$. The group $A' = A \cap D^{n-1}G$ is contained in $D^{n-1}G$ and hence $h(A') \leq h(D^{n-1}G)$, whereas the quotient A/A' can be embedded as a subgroup of $G/D^{n-1}G$ and hence $h(A/A') \leq h(G/D^{n-1}G)$, in view of the induction hypothesis. Therefore, we have

$$h(A) = h(A') + h(A/A') \leq h(D^{n-1}G) + h(G/D^{n-1}G) = h(G) \,,$$

where the first equality describes a well-known property of the rank of abelian groups and the last one is a special case of (3.11).

(iii) The subquotient A of G is a subgroup of the quotient G/N, for a suitable normal subgroup $N \trianglelefteq G$. Then, $h(A) \leq h(G/N) \leq h(G)$, where the first (resp. second) inequality follows from (ii) (resp. from (i)) above. $\square$

Corollary 3.44 *If G is a finitely generated solvable group with finite Hirsch number, then G satisfies Bass' conjecture.*

Proof. Since $h(G)$ is finite, Lemma 3.43(iii) implies that G does not admit an infinite dimensional **Q**-vector space as a subquotient. Hence, the result follows invoking Proposition 3.41. $\square$

We conclude this subsection by describing an example of Hall of a finitely generated solvable group $\mathcal{H}$, which contains an infinite dimensional **Q**-vector space V as a normal subgroup, in such a way that the conjugation action of $\mathcal{H}$ on V satisfies the conditions in the statement of Proposition 3.41. One can show that Hall's group $\mathcal{H}$ satisfies Bass' conjecture, by using the homological techniques of Chap. 4 (cf. Exercise 4.3.5).

Let V be a vector space over the field **Q** of rational numbers with a countable basis $\{v_n : n \in \mathbf{Z}\}$ and fix an enumeration $(p_n)_n$ of all prime numbers,

---

[6] In fact, this inequality holds without assuming that A is abelian, but we do not need this more general result.

which is indexed by the set $\mathbf{Z}$. Consider the group $\mathrm{Aut}(V)$ of $\mathbf{Q}$-linear automorphisms of V and its subgroup Λ generated by the automorphisms x and y, where $x(v_n) = v_{n+1}$ and $y(v_n) = p_n v_n$ for all $n \in \mathbf{Z}$. We list a few properties of the pair (V, Λ):

(i) The normal subgroup Λ' of Λ generated by y is free abelian with basis the elements $x^{-n} y x^n$, $n \in \mathbf{Z}$. The orbit of each basis vector $v_i \in V$ under the action of Λ' is the set consisting of all scalar multiples $q v_i$, where $q \in \mathbf{Q}_+$ is a positive rational number, whereas $\lambda' \in \Lambda'$ stabilizes v_i only if $\lambda' = 1$.

Proof. For any $n \in \mathbf{Z}$ the automorphism $x^{-n} y x^n \in \mathrm{Aut}(V)$ maps $v_i \in V$ onto $p_{n+i} v_i$ for all i; therefore, $x^{-n} y x^n$ commutes with $x^{-m} y x^m$ for all $n, m \in \mathbf{Z}$. In particular, $x^{-n} y x^n$ commutes with y for all n and hence the subgroup Λ' is easily seen to be generated by the set $\{x^{-n} y x^n : n \in \mathbf{Z}\}$. It follows that Λ' is abelian. The claim that Λ' is freely generated by the above set, as well as the statements made about the orbit and the stabilizer of a basis vector v_i under the Λ'-action, follow from the fundamental theorem of arithmetic, which asserts that the multiplicative group of positive rational numbers is free abelian with basis the set of prime numbers.

(ii) The group Λ/Λ' is infinite cyclic generated by the class of x.

Proof. It is clear that the group Λ/Λ' is generated by the class of x. Hence, it suffices to verify that no non-trivial power of x can lie in Λ'. This follows since the group Λ' stabilizes the subspace $V_0 = \mathbf{Q} v_0$, whereas x^n maps V_0 onto $V_n = \mathbf{Q} v_n$ for all n.

(iii) The group Λ is metabelian.

Proof. This is an immediate consequence of (i) and (ii).

(iv) The Λ-orbit of v_0 consists of all scalar multiples $q v_n$, where $n \in \mathbf{Z}$ and $q \in \mathbf{Q}_+$, i.e. $\Lambda \cdot v_0 = \bigcup_n (\mathbf{Q}_+) v_n$.

Proof. For any integer n the Λ'-orbit of v_n consists of the scalar multiples $q v_n$, $q \in \mathbf{Q}_+$, whereas the orbit of v_0 under the action of the cyclic group generated by x is the set $\{v_n : n \in \mathbf{Z}\}$. The claim follows from this, since any element of Λ is the product of an element of Λ' and a power of x.

(v) The one-dimensional subspace $V_0 = \mathbf{Q} v_0 \subseteq V$ is contained in the union of the Λ-orbits $\Lambda \cdot 0 = \{0\}$, $\Lambda \cdot v_0$ and $\Lambda \cdot (-v_0)$.

Proof. This follows from (iv) above.

(vi) V has no proper non-trivial Λ-invariant subgroup.

Proof. Let $V' \neq 0$ be a Λ-invariant subgroup of V and choose a non-zero vector $v = \sum_{i=1}^n q_i v_{a(i)} \in V'$, where the q_i's are non-zero rational numbers, the $a(i)$'s distinct integers and $n > 0$ is minimal such. We compute

$$p_{a(1)}v - y(v) = \sum_{i=1}^{n} p_{a(1)} q_i v_{a(i)} - \sum_{i=1}^{n} q_i y(v_{a(i)})$$
$$= \sum_{i=1}^{n} p_{a(1)} q_i v_{a(i)} - \sum_{i=1}^{n} p_{a(i)} q_i v_{a(i)}$$
$$= \sum_{i=1}^{n} (p_{a(1)} - p_{a(i)}) q_i v_{a(i)}$$
$$= (p_{a(1)} - p_{a(2)}) q_2 v_{a(2)} + \cdots + (p_{a(1)} - p_{a(n)}) q_n v_{a(n)} \ .$$

Since $p_{a(1)}v - y(v) \in V'$, the minimality of n implies that $p_{a(1)}v - y(v)$ must be the zero vector. It follows that $n = 1$ and hence $v = q_1 v_{a(1)}$. Replacing, if necessary, v by $-v$, we may assume that $q_1 > 0$. In this case, the Λ-orbit $\Lambda \cdot v$ of v consists of all vectors of the form qv_n, where $n \in \mathbf{Z}$ and $q \in \mathbf{Q}_+$ (cf. (iv) above). Since these elements generate V as an abelian group, the Λ-invariant subgroup V' must be actually equal to V.

(vii) For any $v \in V \setminus \{0\}$ the stabilizer $\Lambda_v = \{\lambda \in \Lambda : \lambda(v) = v\}$ is trivial.

Proof. Let $v = \sum_{i=1}^{n} q_i v_{a(i)}$, where the q_i's are non-zero rational numbers and $a(1) < \cdots < a(n)$. Any $\lambda \in \Lambda$ can be written as a product $x^m \lambda'$, for a suitable $\lambda' \in \Lambda'$ and some $m \in \mathbf{Z}$. Since λ' restricts to an automorphism of the subspace $\mathbf{Q}v_i$ for all i, we conclude that

$$\lambda(v) = \sum_{i=1}^{n} q_i \lambda(v_{a(i)})$$
$$= \sum_{i=1}^{n} q_i (x^m \lambda')(v_{a(i)})$$
$$= \sum_{i=1}^{n} q_i' v_{m+a(i)} \ ,$$

for suitable non-zero rational numbers q_i', $i = 1, \ldots, n$. In fact, q_i' is such that the restriction of λ' on $\mathbf{Q}v_{a(i)}$ is multiplication by $q_i' q_i^{-1}$ for all $i = 1, \ldots, n$. It follows that $\lambda \in \Lambda_v$ if and only if $m = 0$ and $q_i = q_i'$ for all i. As noted in (i) above, $q_1 = q_1'$ (i.e. λ' acts as the identity on the one-dimensional subspace $\mathbf{Q}v_{a(1)}$) if and only if $\lambda' = 1$. Therefore, $\lambda \in \Lambda_v$ if and only if $\lambda = 1$, as needed.

We define Hall's group $\mathcal{H}$ as the semi-direct product of Λ by V and list below some of its properties:

(α) The group $\mathcal{H}$ is generated by x, y and v_0.

Proof. In order to verify this, it suffices to show that any element of V can be expressed in terms of x, y and v_0. But this is clear since the Λ-orbit of v_0 generates V as an abelian group, in view of (iv) above.

(β) Hall's group $\mathcal{H}$ is solvable of (solvability) degree 3.

Proof. This follows from (iii).

(γ) If $N \trianglelefteq \mathcal{H}$ is a normal subgroup then $V \subseteq N$ or $N \cap V = \{1\}$.

Proof. Since V has no proper non-trivial Λ-invariant subgroup, in view of (vi) above, it is a minimal non-trivial normal subgroup of $\mathcal{H}$. Therefore, being contained in V, the normal subgroup $N \cap V \trianglelefteq \mathcal{H}$ is either trivial or else equal to the whole of V.

(δ) For any $v \in V \setminus \{0\}$ the centralizer C_v of v in $\mathcal{H}$ is equal to V.

Proof. It is easily seen that the centralizer C_v is the semi-direct product of the stabilizer $\Lambda_v = \{\lambda \in \Lambda : \lambda(v) = v\}$ by V. Hence, the claim follows from (vii).

3.3 Exercises

1. Let G be a group. The goal of this Exercise is to show that the property of Lemma 3.9(i) characterizes the operators in the von Neumann algebra $\mathcal{N}G$. To that end, let us fix an operator $a \in \mathcal{B}(\ell^2 G)$, for which $<a(\delta_g), \delta_{hg}> = <a(\delta_1), \delta_h>$ for all $g, h \in G$.
   (i) Show that for any operator $b \in \mathcal{B}(\ell^2 G)$ and any elements $g, h \in G$ the families of complex numbers $(< a(\delta_1), \delta_x > \cdot < b(\delta_g), \delta_{x^{-1}h} >)_x$ and $(<a(\delta_1), \delta_x > \cdot <b(\delta_{xg}), \delta_h >)_x$ are summable with sum $<ab(\delta_g), \delta_h>$ and $<ba(\delta_g), \delta_h>$ respectively.
   (ii) Assume that $b \in \mathcal{B}(\ell^2 G)$ is an operator in the commutant $L(\mathbf{C}G)'$ of the subalgebra $L(\mathbf{C}G) \subseteq \mathcal{B}(\ell^2 G)$. Then, show that $ab = ba$. In particular, conclude that $a \in L(\mathbf{C}G)'' = \mathcal{N}G$.
2. Let G be a finite group of order n.
   (i) Consider a positive integer t and an idempotent $t \times t$ matrix E with entries in the complex group algebra $\mathbf{C}G$. Let

$$L : \mathbf{C}G \longrightarrow \mathrm{End}_{\mathbf{C}}(\mathbf{C}G) = \mathbf{M}_n(\mathbf{C})$$

   be the left regular representation and

$$L_t : \mathbf{M}_t(\mathbf{C}G) \longrightarrow \mathbf{M}_t(\mathbf{M}_n(\mathbf{C})) = \mathbf{M}_{tn}(\mathbf{C})$$

   the induced algebra homomorphism. Show that $\mathrm{tr}(L_t E) = n\, r_1(E)$ and hence obtain an elementary proof of the theorems of Kaplansky and Zaleskii, in the special case of a finite group.
   (ii) Let k be a subring of the field $\mathbf{C}$ of complex numbers and consider the trace functional $r_1 : kG \longrightarrow k$. Show that the image of the induced additive map $r_{1*} : K_0(kG) \longrightarrow k$ is contained in $k \cap \mathbf{Z} \cdot \frac{1}{n}$.
3. Let G be a group and E an idempotent matrix with entries in the complex group algebra $\mathbf{C}G$. Consider a normal subgroup $N \trianglelefteq G$ and let $[N] \subseteq \mathcal{C}(G)$ be the set of all G-conjugacy classes of elements of N. Show that the sum $\sum_{[g] \in [N]} r_g(E)$ is a non-negative rational number.
4. Let R be a commutative ring. The goal of this Exercise is to show that the matrix group $GL_m(R)$ satisfies Bass' conjecture for all m.
   (i) Assume that the ring R is Noetherian and let $\mathbf{m} \subseteq R$ be a maximal ideal. Show that $\mathbf{m}^n/\mathbf{m}^{n+1}$ is a finite dimensional $R/\mathbf{m}$-vector space for all $n \geq 0$.

(ii) Assume that the ring R is finitely generated and let $\mathbf{m} \subseteq R$ be a maximal ideal. Show that the ring $R/\mathbf{m}^n$ is finite for all $n \geq 1$.
(*Hint:* Using Corollaries A.27 and A.22(ii) of Appendix A, show that the field $R/\mathbf{m}$ is finite.)
(iii) Assume that the ring R is finitely generated and let $r \in R \setminus \{0\}$. Show that there exists an ideal $I \subseteq R$, such that the ring R/I is finite and $r \notin I$.
(*Hint:* Let $\mathbf{m} \subseteq R$ be a maximal ideal containing the annihilator $\mathrm{ann}_R(r)$ of r in R. Then, Proposition A.33(ii) of Appendix A implies that $r \notin \mathbf{m}^n$ for some $n \gg 0$.)
(iv) Assume that the ring R is finitely generated and consider two distinct $m \times m$ matrices A, B with entries in R. Show that there exists a finite ring R' and a ring homomorphism $\varphi : R \longrightarrow R'$, such that $\varphi_m(A) \neq \varphi_m(B)$, where $\varphi_m : \mathbf{M}_m(R) \longrightarrow \mathbf{M}_m(R')$ is the homomorphism induced by φ.
(v) Show that the linear group $GL_m(R)$ is locally residually finite, i.e. any finitely generated subgroup G of $GL_m(R)$ is residually finite.[7] In particular, conclude that the group $GL_m(R)$ satisfies Bass' conjecture.

5. Let G be a group and k a subring of the field $\mathbf{C}$ of complex numbers. We consider an idempotent matrix E with entries in kG and let $r_{HS}(E) = \sum_{[g] \in \mathcal{C}(G)} r_g(E)[g]$ be its Hattori-Stallings rank.
(i) Show that there is a positive integer u having the following property: For any prime number p with $p^{-1} \notin k$ and any element $g \in G$ with $r_g(E) \neq 0$, we have $[g] = [g^{p^u}] \in \mathcal{C}(G)$.
(*Hint:* Review the proof of Theorem 3.32.)
(ii) Let $g \in G$ be an element of finite order with $r_g(E) \neq 0$. Show that the order $o(g)$ of g is invertible in k.
(iii) Consider the subring $\Lambda_G \subseteq \mathbf{Q}$ defined in Exercise 1.3.10 and assume that $k \cap \Lambda_G = \mathbf{Z}$. Then, show that $r_g(E) = 0$ for any element $g \in G$ of finite order with $g \neq 1$.
(iv) Assume that G is a Noetherian group, such that $k \cap \Lambda_G = \mathbf{Z}$. Show that the pair (k, G) satisfies Bass' conjecture, whereas the group ring kG has no non-trivial idempotents.

6. Let G be a group, $H \leq G$ a subgroup of finite index and k a subring of the field $\mathbf{C}$ of complex numbers, such that the pair (k, H) satisfies Bass' conjecture. We consider the subring $\Lambda_G \subseteq \mathbf{Q}$ defined in Exercise 1.3.10 and assume that $k \cap \Lambda_G = \mathbf{Z} = k \cap \mathbf{Z} \cdot \frac{1}{n!}$, where $n = [G : H]$. Then, show that the pair (k, G) satisfies Bass' conjecture, whereas the group ring kG has no non-trivial idempotents.
(*Hint:* Follow the argument in the proof of Proposition 3.39, using the result of Exercise 5(i),(iii) above.)

---

[7] In the special case where R is assumed to be a field this result is due to Malcev [49]; the simple proof outlined in this Exercise is due to R. Coleman.

7. A group G is said to satisfy the generalized Bass' conjecture if the map $r_g : K_0(kG) \longrightarrow k$ is identically zero for any pair (k, g), where k is a subring of the field $\mathbf{C}$ of complex numbers and $g \in G$ a group element whose order is not invertible in k.[8] In this way, Exercise 5(ii) implies that torsion groups satisfy the generalized Bass' conjecture.

(i) Show that a group G that satisfies the generalized Bass' conjecture satisfies (the ordinary) Bass' conjecture as well.

(ii) Show that a torsion-free group G that satisfies the generalized Bass' conjecture satisfies the idempotent conjecture.

---

[8] By convention, ∞ is not invertible in any commutative ring.

Notes and Comments on Chap. 3. The study of idempotent matrices with entries in a complex group algebra by reduction to positive characteristic was initiated by A. Zaleskii [75], in order to prove the rationality of r_1 (complementing I. Kaplansky's theorem on the positivity of r_1; cf. [38]). This technique was subsequently used by E. Formanek, who showed in [29] that torsion-free Noetherian groups satisfy the idempotent conjecture (cf. Corollary 3.31), and H. Bass, who proved Theorem 3.26 in [4]. Bass has also proved in [loc.cit.] that the coefficients of the Hattori-Stallings rank of an idempotent matrix with entries in a complex group algebra are algebraic numbers and generate an abelian extension of $\mathbf{Q}$; in that direction, see also [2, 24] and [54]. The relation between the integrality properties of these coefficients and the torsion of a polycyclic-by-finite group has been studied by G. Cliff, S. Sehgal and A. Weiss in [13, 14] and [74]. P. Linnell's result is proved in [43], by using Lemma 3.22, which is itself due to Cliff [12]. In fact, Linnell proved Theorem 3.32 in the case where $k = \mathbf{Z}$ is the ring of integers. The case where k is a ring of algebraic integers is due to J. Schafer [62], whereas the general case was proved by J. Moody [53]. Using the homological techniques that will be developed in Chap. 4, B. Eckmann proved in [19] that solvable groups with finite Hirsch number satisfy Bass' conjecture (cf. Corollary 3.44). The validity of Bass' conjecture for all solvable groups has been obtained by F. Farrell and P. Linnell [28], by means of K-theoretic techniques that are beyond the scope of this book. The construction of the solvable group $\mathcal{H}$, given at the end of the chapter, is due to P. Hall [30].

4

A Homological Approach

4.1 Cyclic Homology of Algebras

In this chapter, we present another method that can be used in the study of the idempotent conjectures, based on cyclic homology. Our goal in this section is to define the cyclic homology groups of an algebra and then compute these groups in the special case of a group algebra. This computation will be used in order to obtain some results on the idempotent and Bass' conjectures in the next section.

For any algebra A the 0-th cyclic homology group $HC_0(A)$ coincides with the abelianization $T(A) = A/[A, A]$. There is an endomorphism S of cyclic homology of degree -2, which plays an important role in the applications to the study of idempotents. A fundamental property of cyclic homology is its relation to K-theory; there are additive group homomorphisms

$$\mathrm{ch}_n : K_0(A) \longrightarrow HC_{2n}(A)$$

for all $n \geq 0$, which are compatible with the operator S and coincide with the Hattori-Stallings rank in degree 0. In this way, the ch_n's provide a factorization of the Hattori-Stallings rank through the higher cyclic homology groups. In the special case where A is the group algebra of a group G, we compute these cyclic homology groups in terms of the homology of certain subquotients of G. Hence, in this special case, the Hattori-Stallings rank factors through abelian groups that can be analyzed using the techniques of homological algebra.

In §4.1.1 we define the Hochschild and cyclic homology groups of an algebra and establish the relationship between these groups and the corresponding ones of a matrix algebra with entries therein. In the following subsection, we construct the K-theory characters ch_n, $n \geq 0$, and establish some of their basic properties. Finally, in §4.1.3 we compute the cyclic homology groups of a group algebra in terms of group homology.

Throughout this section, we work over a fixed field k. Unless otherwise specified, all tensor products will be over k.

4.1.1 Basic Definitions and Results

We consider a k-algebra A and the tensor powers $A^{\otimes n} = A \otimes \cdots \otimes A$, $n \geq 1$. We define for all $n \geq 1$ and all $i \in \{0, \ldots, n\}$ the k-linear operator

$$d_i^n : A^{\otimes n+1} \longrightarrow A^{\otimes n} ,$$

by letting

$$d_i^n(a_0 \otimes \cdots \otimes a_n) = \begin{cases} a_0 \otimes \cdots \otimes a_i a_{i+1} \otimes \cdots \otimes a_n & \text{if } i < n \\ a_n a_0 \otimes a_1 \otimes \cdots \otimes a_{n-1} & \text{if } i = n \end{cases}$$

for any elementary tensor $a_0 \otimes \cdots \otimes a_n \in A^{\otimes n+1}$.

Lemma 4.1 *Let A be a k-algebra and d_i^n the operators defined above. Then, $d_i^{n-1} d_j^n = d_{j-1}^{n-1} d_i^n$ whenever $n \geq 2$ and $0 \leq i < j \leq n$.*

Proof. We have to verify that both operators agree on any elementary tensor $\mathbf{a} = a_0 \otimes \cdots \otimes a_n \in A^{\otimes n+1}$. For example, if $i+1 < j < n$ then both operators map $\mathbf{a}$ onto $a_0 \otimes \cdots \otimes a_i a_{i+1} \otimes \cdots \otimes a_j a_{j+1} \otimes \cdots \otimes a_n \in A^{\otimes n-1}$. The remaining cases are left as an exercise to the reader. □

Using the d_i^n's, we now define the operators

$$b_n', b_n : A^{\otimes n+1} \longrightarrow A^{\otimes n} ,$$

by letting $b_n' = \sum_{i=0}^{n-1}(-1)^i d_i^n$ and $b_n = \sum_{i=0}^{n}(-1)^i d_i^n$ for all $n \geq 1$.

Proposition 4.2 *Let A be a k-algebra and consider the operators b_n' and b_n, $n \geq 1$, that were defined above.*
  (i) The compositions $b_{n-1}' b_n'$ and $b_{n-1} b_n$ vanish for all $n \geq 2$.
  (ii) The complex $(C(A), b')$, which is defined by letting $C_n(A) = A^{\otimes n+1}$ for all $n \geq 0$ and whose differential b' is equal to $(b_n')_n$, is contractible.

Proof. (i) Using Lemma 4.1, we compute

$$\begin{aligned}
b_{n-1}' b_n' &= \sum_{i=0}^{n-2} \sum_{j=0}^{n-1} (-1)^{i+j} d_i^{n-1} d_j^n \\
&= \sum_{i<j\leq n-1} (-1)^{i+j} d_i^{n-1} d_j^n + \sum_{n-2\geq i\geq j} (-1)^{i+j} d_i^{n-1} d_j^n \\
&= \sum_{i<j\leq n-1} (-1)^{i+j} d_{j-1}^{n-1} d_i^n + \sum_{n-2\geq i\geq j} (-1)^{i+j} d_i^{n-1} d_j^n \\
&= \sum_{\alpha\leq\beta\leq n-2} (-1)^{\alpha+\beta+1} d_\beta^{n-1} d_\alpha^n + \sum_{n-2\geq i\geq j} (-1)^{i+j} d_i^{n-1} d_j^n \\
&= 0
\end{aligned}$$

and

$$\begin{aligned}
b_{n-1} b_n &= \sum_{i=0}^{n-1} \sum_{j=0}^{n} (-1)^{i+j} d_i^{n-1} d_j^n \\
&= \sum_{i<j\leq n} (-1)^{i+j} d_i^{n-1} d_j^n + \sum_{n-1\geq i\geq j} (-1)^{i+j} d_i^{n-1} d_j^n \\
&= \sum_{i<j\leq n} (-1)^{i+j} d_{j-1}^{n-1} d_i^n + \sum_{n-1\geq i\geq j} (-1)^{i+j} d_i^{n-1} d_j^n \\
&= \sum_{\alpha\leq\beta\leq n-1} (-1)^{\alpha+\beta+1} d_\beta^{n-1} d_\alpha^n + \sum_{n-1\geq i\geq j} (-1)^{i+j} d_i^{n-1} d_j^n \\
&= 0 .
\end{aligned}$$

(ii) For all $n \geq 0$ we consider the k-linear operator

$$s_n : A^{\otimes n+1} \longrightarrow A^{\otimes n+2} ,$$

which is defined by letting $s_n(a_0 \otimes \cdots \otimes a_n) = 1 \otimes a_0 \otimes \cdots \otimes a_n$ for any elementary tensor $a_0 \otimes \cdots \otimes a_n \in A^{\otimes n+1}$. It is easily seen that

$$d_i^{n+1} s_n = \begin{cases} s_{n-1} d_{i-1}^n & \text{if } 0 < i \leq n \\ \text{id} & \text{if } i = 0 \end{cases}$$

for all $n \geq 0$. It follows that

$$
\begin{aligned}
b'_{n+1} s_n + s_{n-1} b'_n &= \sum_{i=0}^{n} (-1)^i d_i^{n+1} s_n + \sum_{i=0}^{n-1} (-1)^i s_{n-1} d_i^n \\
&= \text{id} + \sum_{i=1}^{n} (-1)^i s_{n-1} d_{i-1}^n + \sum_{i=0}^{n-1} (-1)^i s_{n-1} d_i^n \\
&= \text{id} + \sum_{\alpha=0}^{n-1} (-1)^{\alpha+1} s_{n-1} d_\alpha^n + \sum_{i=0}^{n-1} (-1)^i s_{n-1} d_i^n \\
&= \text{id}
\end{aligned}
$$

is the identity operator on $A^{\otimes n+1}$ for all $n \geq 0$. Hence, the sequence $(s_n)_n$ is a contracting homotopy for the chain complex $(C(A), b')$.   □

Definition 4.3 *The Hochschild homology of a k-algebra A is the homology of the complex $(C(A), b)$, which is itself defined by letting $C_n(A) = A^{\otimes n+1}$ for all $n \geq 0$ and whose differential b is equal to $(b_n)_n$. We denote the n-th Hochschild homology group of A by $HH_n(A)$ for all $n \geq 0$.*

Examples 4.4 (i) Let A be a k-algebra. Then, the differential b_1 maps any elementary tensor $a \otimes a' \in A \otimes A = C_1(A)$ onto $aa' - a'a \in A = C_0(A)$. Hence, $HH_0(A) = A / \text{im } b_1$ is the abelianization $T(A) = A / [A, A]$.

(ii) If $A = k$ then $C_n(A) = A^{\otimes n+1} = k$ for all $n \geq 0$. The operator $d_i^n : k \longrightarrow k$ is the identity map and hence the differential $b_n : k \longrightarrow k$ is the zero (resp. the identity) map if n is odd (resp. even). It follows that $HH_0(k) = k$ and $HH_n(k) = 0$ if $n > 0$.

Remark 4.5 Let A, B be two k-algebras and $f : A \longrightarrow B$ a morphism of non-unital algebras, i.e. a k-linear map which preserves multiplication. Taking into account the form of the differentials $(b_n)_n$, it follows that

$$f_* = \left(f^{\otimes n+1} \right)_n : (C(A), b) \longrightarrow (C(B), b)$$

is a chain map. Therefore, there are induced k-linear maps

$$f_n : HH_n(A) \longrightarrow HH_n(B), \ n \geq 0 .$$

Let A be a k-algebra. We wish to describe the Hochschild homology groups of A in terms of the Tor functors of homological algebra (cf. §D.1.3 of Appendix D). To that end, we consider the algebra $A^e = A \otimes A^{op}$ and view the tensor powers $A^{\otimes n}$, $n \geq 1$, as (left) A^e-modules by letting $a \otimes a'^{op} \in A^e$ act on $A^{\otimes n}$ as the operator

$$a_1 \otimes a_2 \otimes \cdots \otimes a_{n-1} \otimes a_n \mapsto aa_1 \otimes a_2 \otimes \cdots \otimes a_{n-1} \otimes a_n a' \, ,$$

$a_1 \otimes a_2 \otimes \cdots \otimes a_{n-1} \otimes a_n \in A^{\otimes n}$. For all $n \geq 2$ we consider the A^e-module $A^e \otimes A^{\otimes n-2}$, which is obtained from the k-module $A^{\otimes n-2}$ by extension of scalars, and note that there is an isomorphism of A^e-modules

$$\phi_n : A^{\otimes n} \longrightarrow A^e \otimes A^{\otimes n-2} \, ,$$

which identifies an elementary tensor $a_1 \otimes a_2 \otimes \cdots \otimes a_{n-1} \otimes a_n \in A^{\otimes n}$ with $(a_1 \otimes a_n^{op}) \otimes a_2 \otimes \cdots \otimes a_{n-1} \in A^e \otimes A^{\otimes n-2}$. We may also view A as a right A^e-module by letting $a \otimes a'^{op} \in A^e$ act on A as the operator $a_1 \mapsto a'a_1a$, $a_1 \in A$. Tensoring ϕ_n with the identity operator of the right A^e-module A, we obtain an isomorphism of k-modules

$$\varphi_n : A \otimes_{A^e} A^{\otimes n} \longrightarrow A \otimes_{A^e} (A^e \otimes A^{\otimes n-2}) \simeq A^{\otimes n-1} \, ,$$

which identifies a tensor $a \otimes (a_1 \otimes a_2 \otimes \cdots \otimes a_{n-1} \otimes a_n) \in A \otimes_{A^e} A^{\otimes n}$ with $a_n a a_1 \otimes a_2 \otimes \cdots \otimes a_{n-1} \in A^{\otimes n-1}$ for all $n \geq 2$.

Proposition 4.6 *For any k-algebra A and any non-negative integer n there is a natural isomorphism $HH_n(A) \simeq Tor_n^{A^e}(A, A)$.*

Proof. We consider the left A^e-modules $A^{\otimes n}$, $n \geq 2$. Since $A^{\otimes n}$ is isomorphic with the extended A^e-module $A^e \otimes A^{\otimes n-2}$, it is free and hence projective. It is clear that the operators $d_i^n : A^{\otimes n+1} \longrightarrow A^{\otimes n}$ are A^e-linear for all $0 \leq i < n$; in particular, the operator

$$b_n' : A^{\otimes n+1} \longrightarrow A^{\otimes n}$$

is A^e-linear for all $n \geq 1$. We now consider the complex

$$\mathcal{F} : A^{\otimes 2} \xleftarrow{b_2'} A^{\otimes 3} \xleftarrow{b_3'} \cdots \xleftarrow{b_n'} A^{\otimes n+1} \xleftarrow{b_{n+1}'} A^{\otimes n+2} \xleftarrow{b_{n+2}'} \cdots$$

In view of Lemma 4.2(ii), the augmentation

$$\varepsilon = b_1' : A^{\otimes 2} \longrightarrow A$$

makes

$$0 \longleftarrow A \xleftarrow{\varepsilon} \mathcal{F}$$

an A^e-projective resolution of A. Therefore, the groups $Tor_*^{A^e}(A, A)$ can be computed as the homology groups of the complex $A \otimes_{A^e} \mathcal{F}$. It is an immediate consequence of the definitions that the following diagram is commutative for all $n \geq 1$

$$A \otimes_{A^e} A^{\otimes n+2} \xrightarrow{\varphi_{n+2}} A^{\otimes n+1}$$
$$1 \otimes b'_{n+1} \downarrow \qquad\qquad \downarrow b_n$$
$$A \otimes_{A^e} A^{\otimes n+1} \xrightarrow{\varphi_{n+1}} A^{\otimes n}$$

Therefore, the complex $A \otimes_{A^e} \mathcal{F}$ is identified with the Hochschild complex $(C(A), b)$

$$A \xleftarrow{b_1} A^{\otimes 2} \xleftarrow{b_2} \cdots \xleftarrow{b_{n-1}} A^{\otimes n} \xleftarrow{b_n} A^{\otimes n+1} \xleftarrow{b_{n+1}} \cdots .$$

It follows that

$$\operatorname{Tor}_n^{A^e}(A, A) = H_n(A \otimes_{A^e} \mathcal{F}) \simeq H_n(C(A), b) = HH_n(A)$$

for all n and hence the proof is finished. $\qquad\qquad\qquad\qquad\qquad\square$

The following result is an immediate consequence of the description of Hochschild homology in terms of Tor groups. The resulting computation in Example 4.8 will turn out to be very useful in the sequel.

Corollary 4.7 *Let A_1, A_2 be two k-algebras and $A = A_1 \times A_2$ the corresponding direct product. Then, the projection maps from A to the A_i's induce an isomorphism $HH_n(A) \simeq HH_n(A_1) \oplus HH_n(A_2)$ for all n.*

Proof. Since A^e decomposes into the direct product of the rings A_1^e, A_2^e, $A_1 \otimes A_2^{op}$ and $A_2 \otimes A_1^{op}$, the result follows from Proposition D.4(ii) of Appendix D, in view of Proposition 4.6. $\qquad\qquad\qquad\qquad\qquad\square$

Example 4.8 Let $A = k[\sigma, \tau]/(\sigma^2 - \sigma, \tau^2 - \tau, \sigma\tau)$. Then, there is a k-algebra isomorphism

$$f : A \longrightarrow k \times k \times k$$

which maps $\overline{\sigma}$ onto $(0, 1, 0)$ and $\overline{\tau}$ onto $(0, 0, 1)$. Using the obvious extension of Corollary 4.7 to the case of a direct product of finitely many k-algebras, we conclude that $HH_n(A) \simeq HH_n(k) \oplus HH_n(k) \oplus HH_n(k)$ for all n. In particular, the group $HH_n(A)$ vanishes if $n > 0$ (cf. Example 4.4(ii)).

Let A be a k-algebra. We fix a positive integer t and consider the algebra $\mathbf{M}_t(A)$ of $t \times t$ matrices with entries in A. As usual, we denote by E_{ij} the matrix units in $\mathbf{M}_t(A)$. In order to relate the Hochschild homology groups of A to those of $\mathbf{M}_t(A)$, we consider the morphism of non-unital algebras

$$\iota : A \longrightarrow \mathbf{M}_t(A) ,$$

which maps any element $a \in A$ onto the matrix $aE_{11} \in \mathbf{M}_t(A)$, and the induced k-linear maps

$$\iota_n : HH_n(A) \longrightarrow HH_n(\mathbf{M}_t(A)), \; n \geq 0$$

(cf. Remark 4.5). Our goal is to prove that the ι_n's are isomorphisms. To that end, we extend the ordinary trace $\operatorname{tr} : \mathbf{M}_t(A) \longrightarrow A$ and define for any $n \geq 0$ the k-linear map

$$\mathrm{tr}_n : \mathbf{M}_t(A)^{\otimes n+1} \longrightarrow A^{\otimes n+1} \, ,$$

as follows: For any elementary tensor $\mathbf{a} = a_0 E_{i_0 j_0} \otimes a_1 E_{i_1 j_1} \otimes \cdots \otimes a_n E_{i_n j_n} \in \mathbf{M}_t(A)^{\otimes n+1}$, where $a_0, a_1, \ldots, a_n$ are elements of A, we define

$$\mathrm{tr}_n(\mathbf{a}) = \begin{cases} a_0 \otimes a_1 \otimes \cdots \otimes a_n \text{ if } j_0 = i_1,\ j_1 = i_2,\ \ldots,\ j_{n-1} = i_n \text{ and } j_n = i_0 \\ \\ 0 \qquad\qquad \text{otherwise} \end{cases}$$

The behavior of these generalized traces with respect to the operators defining the Hochschild complex is described in the following result.

Lemma 4.9 *Let A be a k-algebra, t a positive integer and $\mathbf{M}_t(A)$ the algebra of $t \times t$ matrices with entries in A. Then, $\mathrm{tr}_{n-1} d_i^n = d_i^n \mathrm{tr}_n$ for all $n \geq 1$ and all $i \in \{0, 1, \ldots, n\}$, where we have used the same notation for the d-operators that correspond to both algebras A and $\mathbf{M}_t(A)$. In particular, $\mathrm{tr}_{n-1} b_n = b_n \mathrm{tr}_n$ for all n and hence $Tr = (\mathrm{tr}_n)_n$ is a chain map between the Hochschild complexes of $\mathbf{M}_t(A)$ and A.*

Proof. It suffices to check the validity of the equality $\mathrm{tr}_{n-1} d_i^n = d_i^n \mathrm{tr}_n$ on the elementary tensors of the form $a_0 E_{i_0 j_0} \otimes a_1 E_{i_1 j_1} \otimes \cdots \otimes a_n E_{i_n j_n}$, where the a's are elements of A and the E's the matrix units in $\mathbf{M}_t(A)$. In that case, the result follows immediately from the definitions, in view of the way that the matrix units get multiplied in $\mathbf{M}_t(A)$. $\qquad\square$

Theorem 4.10 *(Morita invariance of Hochschild homology) Let A be a k-algebra, t a positive integer and $\mathbf{M}_t(A)$ the algebra of $t \times t$ matrices with entries in A. Then, for all n the map*

$$\iota_n : HH_n(A) \longrightarrow HH_n(\mathbf{M}_t(A))$$

is an isomorphism with inverse the k-linear map

$$Tr_n : HH_n(\mathbf{M}_t(A)) \longrightarrow HH_n(A) \, ,$$

which is induced by the chain map Tr of Lemma 4.9.

Proof. It follows immediately from the definitions that the composition $\mathrm{tr}_n \circ \iota^{\otimes n+1}$ is the identity map on $A^{\otimes n+1}$ for all $n \geq 0$. Therefore, the composition

$$HH_n(A) \xrightarrow{\iota_n} HH_n(\mathbf{M}_t(A)) \xrightarrow{Tr_n} HH_n(A)$$

is the identity map on $HH_n(A)$ for all $n \geq 0$. In order to show that $\iota_n \circ Tr_n$ is the identity map on $HH_n(\mathbf{M}_t(A))$, we shall construct a chain homotopy between the composition $(\iota^{\otimes n+1})_n \circ (\mathrm{tr}_n)_n$ and the identity map of the chain complex $(C(\mathbf{M}_t(A)), b)$. To that end, we consider for all $n \geq 0$ and all $s \in \{0, 1, \ldots, n\}$ the k-linear operator

$$\phi_s^n : \mathbf{M}_t(A)^{\otimes n+1} \longrightarrow \mathbf{M}_t(A)^{\otimes n+2} ,$$

which is defined as follows: For any elementary tensor

$$\mathbf{a} = a_0 E_{i_0 j_0} \otimes a_1 E_{i_1 j_1} \otimes \cdots \otimes a_n E_{i_n j_n} \in \mathbf{M}_t(A)^{\otimes n+1} ,$$

where $a_0, a_1, \ldots, a_n \in A$, we define $\phi_s^n(\mathbf{a})$ to be the elementary tensor

$$a_0 E_{i_0 1} \otimes a_1 E_{11} \otimes \cdots \otimes a_s E_{11} \otimes E_{1 j_s} \otimes a_{s+1} E_{i_{s+1} j_{s+1}} \otimes \cdots \otimes a_n E_{i_n j_n}$$

if $j_0 = i_1$, $j_1 = i_2$, $\ldots$, $j_{s-1} = i_s$ and 0 otherwise. It is easily verified that

$$d_i^{n+1} \phi_s^n = \begin{cases} \phi_{s-1}^{n-1} d_i^n & \text{if } i < s \\ d_i^{n+1} \phi_{s-1}^n & \text{if } i = s > 0 \\ d_i^{n+1} \phi_{s+1}^n & \text{if } i = s+1 < n+1 \\ \phi_s^{n-1} d_{i-1}^n & \text{if } i > s+1 \\ \mathrm{id} & \text{if } i = s = 0 \\ \iota^{\otimes n+1} \mathrm{tr}_n & \text{if } i = s+1 = n+1 \end{cases} \tag{4.1}$$

We now consider for all $n \geq 0$ the k-linear operator

$$\Phi_n = \sum_{s=0}^{n} (-1)^s \phi_s^n : \mathbf{M}_t(A)^{\otimes n+1} \longrightarrow \mathbf{M}_t(A)^{\otimes n+2}$$

and compute

$$b_{n+1} \Phi_n = \sum_{i=0}^{n+1} \sum_{s=0}^{n} (-1)^{i+s} d_i^{n+1} \phi_s^n$$

$$= \sum_1 + \sum_2 + \sum_3 + \sum_4 + d_0^{n+1} \phi_0^n - d_{n+1}^{n+1} \phi_n^n,$$

where $\sum_1$ denotes summation over the set $\{i < s\}$, $\sum_2$ summation over the set $\{i > s+1\}$, $\sum_3$ summation over the set $\{i = s > 0\}$ and $\sum_4$ summation over the set $\{i = s+1 < n+1\}$. In view of (4.1), we have

$$\sum_1 + \sum_2 = \sum_{i<s} (-1)^{i+s} d_i^{n+1} \phi_s^n + \sum_{i>s+1} (-1)^{i+s} d_i^{n+1} \phi_s^n$$

$$= \sum_{i<s} (-1)^{i+s} \phi_{s-1}^{n-1} d_i^n + \sum_{i>s+1} (-1)^{i+s} \phi_s^{n-1} d_{i-1}^n$$

$$= -\sum_{\alpha \leq \beta} (-1)^{\alpha+\beta} \phi_\beta^{n-1} d_\alpha^n - \sum_{\alpha > \beta} (-1)^{\alpha+\beta} \phi_\beta^{n-1} d_\alpha^n$$

$$= -\sum_{\beta=0}^{n-1} \sum_{\alpha=0}^{n} (-1)^{\alpha+\beta} \phi_\beta^{n-1} d_\alpha^n$$

$$= -\Phi_{n-1} b_n .$$

Similarly, we have

$$\sum_3 + \sum_4 = \sum_{i=1}^{n} d_i^{n+1} \phi_i^n - \sum_{i=1}^{n} d_i^{n+1} \phi_{i-1}^n$$

$$= \sum_{i=1}^{n} d_i^{n+1} \phi_{i-1}^n - \sum_{i=1}^{n} d_i^{n+1} \phi_{i-1}^n$$

$$= 0 .$$

Since $d_0^{n+1}\phi_0^n = \text{id}$ and $d_{n+1}^{n+1}\phi_n^n = \iota^{\otimes n+1}\text{tr}_n$, we conclude that

$$b_{n+1}\Phi_n = -\Phi_{n-1}b_n + \text{id} - \iota^{\otimes n+1}\text{tr}_n .$$

This is the case for all n and hence $\Phi = (\Phi_n)_n$ is a chain homotopy between the identity and $(\iota^{\otimes n+1})_n \circ (\text{tr}_n)_n$, as needed. $\qquad\square$

Remark 4.11 Let A be a k-algebra. For any positive integer t and any $i \in \{1,\dots,t\}$ we may consider the morphism of non-unital algebras

$$\iota' : A \longrightarrow \mathbf{M}_t(A) ,$$

which maps any element $a \in A$ onto the matrix $aE_{ii} \in \mathbf{M}_t(A)$. Then, the induced k-linear map between the Hochschild homology groups

$$\iota'_n : HH_n(A) \longrightarrow HH_n(\mathbf{M}_t(A))$$

is an isomorphism for all n; in fact, $\iota'_n = \iota_n$, where ι_n is the isomorphism of Theorem 4.10. Indeed, the composition

$$A^{\otimes n+1} \xrightarrow{\iota'^{\otimes n+1}} \mathbf{M}_t(A)^{\otimes n+1} \xrightarrow{\text{tr}_n} A^{\otimes n+1}$$

is the identity map on $A^{\otimes n+1}$ for all $n \geq 0$ and hence the composition

$$HH_n(A) \xrightarrow{\iota'_n} HH_n(\mathbf{M}_t(A)) \xrightarrow{Tr_n} HH_n(A)$$

is the identity map on $HH_n(A)$ for all $n \geq 0$. The result follows, since Tr_n is an isomorphism with inverse ι_n.

It is clear that an inner automorphism of a k-algebra A induces the identity on the abelianization $T(A) = A/[A, A]$. We generalize this observation, by proving that such an automorphism induces the identity on all Hochschild homology groups of A.

Proposition 4.12 *Let A be a k-algebra. Then, the conjugation action of the group $U(A)$ on A induces the trivial action on the Hochschild homology groups $HH_n(A)$, $n \geq 0$.*

Proof. Let $g \in U(A)$ be a unit of A and $I_g : A \longrightarrow A$ the associated inner automorphism. We consider the invertible 2×2 matrix

$$G = \begin{pmatrix} g & 0 \\ 0 & 1 \end{pmatrix} \in GL_2(A)$$

and the commutative diagram

$$
\begin{array}{ccccc}
A & \xrightarrow{\iota} & \mathbf{M}_2(A) & \xleftarrow{\iota'} & A \\
{\scriptstyle I_g}\downarrow & & {\scriptstyle I_G}\downarrow & & {\scriptstyle id}\downarrow \\
A & \xrightarrow{\iota} & \mathbf{M}_2(A) & \xleftarrow{\iota'} & A
\end{array}
$$

Here, I_G denotes the inner automorphism of $\mathbf{M}_2(A)$ associated with G and ι (resp. ι') the morphism of non-unital algebras given by $a \mapsto aE_{11}$, $a \in A$ (resp. $a \mapsto aE_{22}$, $a \in A$). There is a corresponding commutative diagram between the Hochschild homology groups

$$
\begin{array}{ccccc}
HH_n(A) & \xrightarrow{\iota_n} & HH_n(\mathbf{M}_2(A)) & \xleftarrow{\iota'_n} & HH_n(A) \\
{\scriptstyle (I_g)_n}\downarrow & & {\scriptstyle (I_G)_n}\downarrow & & {\scriptstyle id}\downarrow \\
HH_n(A) & \xrightarrow{\iota_n} & HH_n(\mathbf{M}_2(A)) & \xleftarrow{\iota'_n} & HH_n(A)
\end{array}
$$

where all arrows are isomorphisms, and hence a commutative diagram

$$
\begin{array}{ccc}
HH_n(A) & \xrightarrow{\iota'^{-1}_n \iota_n} & HH_n(A) \\
{\scriptstyle (I_g)_n}\downarrow & & {\scriptstyle id}\downarrow \\
HH_n(A) & \xrightarrow{\iota'^{-1}_n \iota_n} & HH_n(A)
\end{array}
$$

It follows readily from this that $(I_g)_n$ is the identity on $HH_n(A)$ for all $n \geq 0$, as needed. $\qquad\square$

In order to define the cyclic homology groups of a k-algebra A, we consider the operator

$$
t_n : A^{\otimes n+1} \longrightarrow A^{\otimes n+1} ,
$$

which is defined by letting $t_n(a_0 \otimes a_1 \otimes \cdots \otimes a_n) = (-1)^n a_n \otimes a_0 \otimes \cdots \otimes a_{n-1}$ for any elementary tensor $a_0 \otimes a_1 \otimes \cdots \otimes a_n \in A^{\otimes n+1}$. It is clear that t_n^{n+1} is the identity operator on $A^{\otimes n+1}$; therefore, t_n defines an action of the cyclic group of order $n+1$ on $A^{\otimes n+1}$. The homology of $\mathbf{Z}/(n+1)\mathbf{Z}$ with coefficients in $A^{\otimes n+1}$ is the homology of the complex

$$
A^{\otimes n+1} \xleftarrow{1-t_n} A^{\otimes n+1} \xleftarrow{N_n} A^{\otimes n+1} \xleftarrow{1-t_n} A^{\otimes n+1} \xleftarrow{N_n} \cdots , \tag{4.2}
$$

where $N_n = \sum_{i=0}^{n} t_n^i$ (cf. Example D.8(iii) of Appendix D). The following result describes the behavior of the t's with respect to the d-operators that were defined at the beginning of this subsection.

Lemma 4.13 *Let A be a k-algebra. Then:*
(i) $d_0^m t_n = (-1)^n d_n^m$,
(ii) $d_i^m t_n = -t_{n-1} d_{i-1}^m$ *for all $i > 0$ and*
(iii) $d_i^m t_n^j = \begin{cases} (-1)^j t_{n-1}^j d_{i-j}^m & \text{if } i \geq j \\ (-1)^{n+j-1} t_{n-1}^{j-1} d_{n+1+i-j}^m & \text{if } i < j \end{cases}$

Proof. (i) This is clear, since both operators map an elementary tensor $a_0 \otimes a_1 \otimes \cdots \otimes a_n \in A^{\otimes n+1}$ onto $(-1)^n a_n a_0 \otimes a_1 \otimes \cdots \otimes a_{n-1}$.

(ii) This is clear as well, since both operators map an elementary tensor $a_0 \otimes a_1 \otimes \cdots \otimes a_n \in A^{\otimes n+1}$ onto $(-1)^n a_n \otimes a_0 \otimes \cdots \otimes a_{i-1}a_i \otimes \ldots \otimes a_{n-1}$ (resp. onto $(-1)^n a_{n-1}a_n \otimes a_0 \otimes \cdots \otimes a_{n-2}$) if $i < n$ (resp. if $i = n$).

(iii) The case where $i \geq j$ follows by induction on j, in view of (ii) above. In particular, we have $d_i^n t_n^i = (-1)^i t_{n-1}^i d_0^n$ and hence

$$d_i^n t_n^{i+1} = (-1)^i t_{n-1}^i d_0^n t_n = (-1)^{n+i} t_{n-1}^i d_n^n \ ,$$

in view of (i) above. Therefore, if $i < j$ then

$$d_i^n t_n^j = (-1)^{n+i} t_{n-1}^i d_n^n t_n^{j-i-1} = (-1)^{n+j-1} t_{n-1}^{j-1} d_{n+1+i-j}^n$$

and this finishes the proof. □

Corollary 4.14 *Let A be a k-algebra. Then, $b_n(1 - t_n) = (1 - t_{n-1})b_n'$ and $b_n' N_n = N_{n-1} b_n$ for all $n \geq 1$, where $N_n = \sum_{i=0}^n t_n^i$.*

Proof. We compute

$$
\begin{aligned}
b_n(1 - t_n) &= b_n - b_n t_n \\
&= b_n - \sum_{i=0}^n (-1)^i d_i^n t_n \\
&= b_n - (-1)^n d_n^n - \sum_{i=1}^n (-1)^{i-1} t_{n-1} d_{i-1}^n \\
&= b_n' - \sum_{\alpha=0}^{n-1} (-1)^\alpha t_{n-1} d_\alpha^n \\
&= b_n' - t_{n-1} b_n' \\
&= (1 - t_{n-1}) b_n',
\end{aligned}
$$

where the third equality follows from Lemma 4.13(i),(ii). We also have

$$
\begin{aligned}
b_n' N_n &= \sum_{i=0}^{n-1} \sum_{j=0}^n (-1)^i d_i^n t_n^j \\
&= \sum_{0 \leq i < j \leq n} (-1)^i d_i^n t_n^j + \sum_{0 \leq j \leq i < n} (-1)^i d_i^n t_n^j \\
&= \sum_{0 \leq i < j \leq n} (-1)^{n+1+i-j} t_{n-1}^{j-1} d_{n+1+i-j}^n + \\
&\quad \sum_{0 \leq j \leq i < n} (-1)^{i-j} t_{n-1}^j d_{i-j}^n \\
&= \sum_{\alpha+\beta \geq n} (-1)^\beta t_{n-1}^\alpha d_\beta^n + \sum_{\alpha+\beta < n} (-1)^\beta t_{n-1}^\alpha d_\beta^n \\
&= \sum_{\alpha=0}^{n-1} \sum_{\beta=0}^n (-1)^\beta t_{n-1}^\alpha d_\beta^n \\
&= N_{n-1} b_n,
\end{aligned}
$$

where the third equality follows from Lemma 4.13(iii). □

In view of Corollary 4.14, we may consider for any k-algebra A the double complex $C_{**}(A)$ pictured below

$$
\begin{array}{ccccccccc}
\vdots & & \vdots & & \vdots & & \vdots & & \\
b_{n+1}\downarrow & & -b'_{n+1}\downarrow & & b_{n+1}\downarrow & & -b'_{n+1}\downarrow & & \\
A^{\otimes n+1} & \xleftarrow{1-t_n} & A^{\otimes n+1} & \xleftarrow{N_n} & A^{\otimes n+1} & \xleftarrow{1-t_n} & A^{\otimes n+1} & \xleftarrow{N_n} & \cdots \\
b_n\downarrow & & -b'_n\downarrow & & b_n\downarrow & & -b'_n\downarrow & & \\
A^{\otimes n} & \xleftarrow{1-t_{n-1}} & A^{\otimes n} & \xleftarrow{N_{n-1}} & A^{\otimes n} & \xleftarrow{1-t_{n-1}} & A^{\otimes n} & \xleftarrow{N_{n-1}} & \cdots \\
b_{n-1}\downarrow & & -b'_{n-1}\downarrow & & b_{n-1}\downarrow & & -b'_{n-1}\downarrow & & \\
\vdots & & \vdots & & \vdots & & \vdots & & \\
b_2\downarrow & & -b'_2\downarrow & & b_2\downarrow & & -b'_2\downarrow & & \\
A^{\otimes 2} & \xleftarrow{1-t_1} & A^{\otimes 2} & \xleftarrow{N_1} & A^{\otimes 2} & \xleftarrow{1-t_1} & A^{\otimes 2} & \xleftarrow{N_1} & \cdots \\
b_1\downarrow & & -b'_1\downarrow & & b_1\downarrow & & -b'_1\downarrow & & \\
A & \xleftarrow{1-t_0} & A & \xleftarrow{N_0} & A & \xleftarrow{1-t_0} & A & \xleftarrow{N_0} & \cdots
\end{array}
$$

The n-th row of the double complex consists of the chain complex (4.2), computing the homology of the cyclic group of order $n+1$ with coefficients in $A^{\otimes n+1}$. On the other hand, the even-numbered (resp. odd-numbered) columns of $C_{**}(A)$ coincide with the Hochschild complex $(C(A),b)$ (resp. the acyclic complex $(C(A),-b')$).

Definition 4.15 *The cyclic homology of a k-algebra A is the homology of the total complex $\mathrm{Tot}\,C_{**}(A)$. We denote the n-th cyclic homology group of A by $HC_n(A)$ for all $n \geq 0$.*

Remark 4.16 Let A, B be two k-algebras and $f : A \longrightarrow B$ a morphism of non-unital algebras. It is clear that

$$
f_{**} = (f^{\otimes n+1})_{n,m} : C_{**}(A) \longrightarrow C_{**}(B)
$$

is a chain bicomplex map. Therefore, there are induced k-linear maps

$$
f_n : HC_n(A) \longrightarrow HC_n(B), \ n \geq 0 .
$$

Since Hochschild homology is, in principle, easier to compute than cyclic homology, the following result turns out to be very useful for computation purposes.

Proposition 4.17 *For any k-algebra A there is a graded endomorphism S of degree -2 of the complex $\mathrm{Tot}\,C_{**}(A)$, such that the associated operator in homology fits into a natural long exact sequence (Connes' exact sequence)*

$$
\cdots \xrightarrow{B} HH_n(A) \xrightarrow{I} HC_n(A) \xrightarrow{S} HC_{n-2}(A) \xrightarrow{B} HH_{n-1}(A) \xrightarrow{I} \cdots .
$$

By an abuse of language, the operator S will be referred to as the periodicity operator in cyclic homology.

Proof. Let S be the endomorphism of the bicomplex $C_{**}(A)$ which vanishes on the first two columns and maps identically the i-th column onto the $(i-2)$-th one for all $i \geq 2$. Then, S induces a surjective endomorphism of the complex $\operatorname{Tot} C_{**}(A)$ of degree -2, with kernel K equal to the total complex associated with the double complex formed by the first two columns of $C_{**}(A)$. Since the complex $(C(A), -b')$ is acyclic (cf. Proposition 4.2(ii)), Corollary D.2(ii) of Appendix D implies that $H_n(K) = H_n(C(A), b) = HH_n(A)$ for all $n \geq 0$. Then, the long exact sequence in the statement is the one induced in homology by the short exact sequence of chain complexes

$$ 0 \longrightarrow K \longrightarrow \operatorname{Tot} C_{**}(A) \overset{S}{\longrightarrow} \operatorname{Tot} C_{**}(A)[2] \longrightarrow 0 \ , $$

where $\operatorname{Tot} C_{**}(A)[2]$ denotes the chain complex $\operatorname{Tot} C_{**}(A)$ with degrees shifted 2 units to the left. $\qquad\square$

Corollary 4.18 *Let A be a k-algebra.*
(i) The natural map $I : HH_n(A) \longrightarrow HC_n(A)$ is bijective for $n = 0$ and surjective for $n = 1$. In particular, the group $HC_0(A)$ is the abelianization $T(A) = A/[A, A]$.
(ii) Assume that $HH_n(A) = 0$ for all $n \geq 1$. Then, the periodicity operator $S : HC_n(A) \longrightarrow HC_{n-2}(A)$ is bijective for all $n \geq 2$.

Proof. (i) This follows from the exact sequence

$$ HH_1(A) \overset{I}{\to} HC_1(A) \overset{S}{\to} HC_{-1}(A) \overset{B}{\to} HH_0(A) \overset{I}{\to} HC_0(A) \overset{S}{\to} HC_{-2}(A) $$

and the vanishing of the group $HC_n(A)$ for $n < 0$.
  (ii) This follows from the exact sequence

$$ HH_n(A) \overset{I}{\longrightarrow} HC_n(A) \overset{S}{\longrightarrow} HC_{n-2}(A) \overset{B}{\longrightarrow} HH_{n-1}(A) $$

and our assumption about the vanishing of the Hochschild homology groups in positive degrees. $\qquad\square$

Example 4.19 Since $HH_0(k) = k$ and $HH_n(k) = 0$ if n is positive (cf. Example 4.4(ii)), Corollary 4.18 implies that $HC_n(k) = 0$ (resp. k) if n is odd (resp. even and ≥ 0).

Let A be a k-algebra, t a positive integer and $\mathbf{M}_t(A)$ the algebra of $t \times t$ matrices with entries in A. We conclude this subsection by relating the cyclic homology groups of A to those of $\mathbf{M}_t(A)$. This relationship will be used in the following subsection, in order to define the character of an element of $K_0(A)$ in the even cyclic homology groups of A. We consider the morphism of non-unital algebras

$$ \iota : A \longrightarrow \mathbf{M}_t(A) \ , $$

which maps any element $a \in A$ onto the matrix $aE_{11} \in \mathbf{M}_t(A)$, and the induced k-linear maps

$$\iota_n : HC_n(A) \longrightarrow HC_n(\mathbf{M}_t(A)), \ n \geq 0$$

(cf. Remark 4.16). We also consider the generalized traces

$$\mathrm{tr}_n : \mathbf{M}_t(A)^{\otimes n+1} \longrightarrow A^{\otimes n+1}, \ n \geq 0 \, ,$$

that were defined earlier. The behavior of the tr_n's with respect to the operators defining the cyclic bicomplex is described in the following result.

Lemma 4.20 *Let A be a k-algebra, t a positive integer and $\mathbf{M}_t(A)$ the algebra of $t \times t$ matrices with entries in A. Then, $\mathrm{tr}_n t_n = t_n \mathrm{tr}_n$ for all $n \geq 0$, where we have used the same notation for the t-operators that correspond to both algebras A and $\mathbf{M}_t(A)$. In particular, $\mathrm{tr}_n(1 - t_n) = (1 - t_n)\mathrm{tr}_n$ and $\mathrm{tr}_n N_n = N_n \mathrm{tr}_n$ for all $n \geq 0$ and hence the tr_n's define a chain map Tr between the cyclic bicomplexes of $\mathbf{M}_t(A)$ and A.*

Proof. It suffices to check the validity of the equality $\mathrm{tr}_n t_n = t_n \mathrm{tr}_n$ on the elementary tensors of the form $a_0 E_{i_0 j_0} \otimes a_1 E_{i_1 j_1} \otimes \cdots \otimes a_n E_{i_n j_n}$, where the a's are elements of A and the E's are the matrix units in $\mathbf{M}_t(A)$. In that case, the result follows immediately from the definitions. It is clear that we also have $\mathrm{tr}_n(1 - t_n) = (1 - t_n)\mathrm{tr}_n$ and $\mathrm{tr}_n N_n = N_n \mathrm{tr}_n$. Since the tr_n's are compatible with the d_i^n's (cf. Lemma 4.9), it follows that they are compatible with the b- and b'-operators as well. Therefore, they induce a chain map between the cyclic bicomplexes, as needed. $\square$

Theorem 4.21 *(Morita invariance of cyclic homology) Let us consider a k-algebra A, a positive integer t and the algebra $\mathbf{M}_t(A)$ of $t \times t$ matrices with entries in A. Then, for all n the map*

$$\iota_n : HC_n(A) \longrightarrow HC_n(\mathbf{M}_t(A))$$

is an isomorphism with inverse the k-linear map

$$Tr_n : HC_n(\mathbf{M}_t(A)) \longrightarrow HC_n(A) \, ,$$

which is induced by the chain bicomplex map Tr of Lemma 4.20.

Proof. Since the composition $Tr_n \circ \iota_n$ is obviously the identity map on the group $HC_n(A)$, it suffices to prove that ι_n is an isomorphism for all n. In view of the naturality of the long exact sequence of Proposition 4.17, the morphism of non-unital algebras

$$\iota : A \longrightarrow \mathbf{M}_t(A)$$

induces a morphism of long exact sequences

$$
\begin{array}{ccccccccc}
\cdots & \xrightarrow{B} & HH_n(A) & \xrightarrow{I} & HC_n(A) & \xrightarrow{S} & HC_{n-2}(A) & \xrightarrow{B} & \cdots \\
& & \iota_n \downarrow & & \iota_n \downarrow & & \iota_{n-2} \downarrow & & \\
\cdots & \xrightarrow{B} & HH_n(\mathbf{M}_t(A)) & \xrightarrow{I} & HC_n(\mathbf{M}_t(A)) & \xrightarrow{S} & HC_{n-2}(\mathbf{M}_t(A)) & \xrightarrow{B} & \cdots
\end{array}
$$

We note that the ι_n's are isomorphisms between the Hochschild homology groups, in view of Theorem 4.10. Therefore, we may prove by induction on n that the ι_n's are isomorphisms between the cyclic homology groups as well, using the 5-lemma. $\square$

Remark 4.22 Let A be a k-algebra. For any positive integer t and any $i \in \{1, \ldots, t\}$ we may consider the morphism of non-unital algebras

$$\iota' : A \longrightarrow \mathbf{M}_t(A) ,$$

which maps an element $a \in A$ onto the matrix $aE_{ii} \in \mathbf{M}_t(A)$. Then, the induced k-linear map between the cyclic homology groups

$$\iota'_n : HC_n(A) \longrightarrow HC_n(\mathbf{M}_t(A))$$

is an isomorphism for all n; in fact, $\iota'_n = \iota_n$, where ι_n is the isomorphism of Theorem 4.21. Indeed, the composition

$$HC_n(A) \xrightarrow{\iota'_n} HC_n(\mathbf{M}_t(A)) \xrightarrow{Tr_n} HC_n(A)$$

is clearly the identity map on $HC_n(A)$ for all n and hence the result follows, Tr_n being an isomorphism with inverse ι_n.

As in the case of Hochschild homology, we can now show that inner automorphisms of an algebra induce the identity in cyclic homology.

Proposition 4.23 *Let A be a k-algebra. Then, the conjugation action of the group $U(A)$ on A induces the trivial action on the cyclic homology groups $HC_n(A)$, $n \geq 0$.*

Proof. This follows by the same argument as in the proof of Proposition 4.12, where we use Remark 4.22 instead of Remark 4.11. $\square$

Let A be a k-algebra. We consider two positive integers t, t' with $t < t'$ and let

$$\jmath = \jmath_{t,t'} : \mathbf{M}_t(A) \longrightarrow \mathbf{M}_{t'}(A)$$

be the morphism of non-unital algebras, which maps a $t \times t$ matrix X with entries in A onto the $t' \times t'$ matrix $\jmath(X) = \begin{pmatrix} X & 0 \\ 0 & 0 \end{pmatrix}$. Then, $\jmath$ induces k-linear maps between the cyclic homology groups

$$\jmath_n : HC_n(\mathbf{M}_t(A)) \longrightarrow HC_n(\mathbf{M}_{t'}(A)), \ n \geq 0$$

(cf. Remark 4.16). The following result shows that these maps are compatible with the Morita isomorphisms.

Lemma 4.24 *Let A be a k-algebra. For any pair (t, t') of positive integers with $t < t'$ and any integer $n \geq 0$, there is a commutative diagram*

$$
\begin{array}{ccc}
HC_n(\mathbf{M}_t(A)) & \xrightarrow{Tr_n} & HC_n(A) \\
{\scriptstyle \jmath_n} \downarrow & & \| \\
HC_n(\mathbf{M}_{t'}(A)) & \xrightarrow{Tr_n} & HC_n(A)
\end{array}
$$

Proof. In view of the definition of the generalized traces, the following diagram is commutative for all $n \geq 0$

$$
\begin{array}{ccc}
\mathbf{M}_t(A)^{\otimes n+1} & \xrightarrow{tr_n} & A^{\otimes n+1} \\
{\scriptstyle \jmath^{\otimes n+1}} \downarrow & & \| \\
\mathbf{M}_{t'}(A)^{\otimes n+1} & \xrightarrow{tr_n} & A^{\otimes n+1}
\end{array}
$$

This clearly finishes the proof. □

4.1.2 The Relation to K-theory

Our goal in the present subsection is to describe the relationship between the K_0-group of a k-algebra A and the cyclic homology groups of A. More precisely, we construct additive group homomorphisms

$$
\mathrm{ch}_n : K_0(A) \longrightarrow HC_{2n}(A), \ n \geq 0 \,,
$$

that have the following properties:

(i) $\mathrm{ch}_0 : K_0(A) \longrightarrow HC_0(A) = T(A)$ is the Hattori-Stallings rank and
(ii) $S \circ \mathrm{ch}_n = \mathrm{ch}_{n-1}$ for all $n \geq 1$.

Hence, the ch_n's provide factorizations of the Hattori-Stallings rank $\mathrm{ch}_0 = r_{HS}$ through the higher even cyclic homology groups of the algebra.

If $e \in A$ is an idempotent, we consider the element $e^{\otimes n+1} \in A^{\otimes n+1}$ and note that $d_i^n(e^{\otimes n+1}) = e^{\otimes n}$ for all $0 \leq i \leq n$. It is equally clear that the endomorphism t_n of $A^{\otimes n+1}$ acts on $e^{\otimes n+1}$ as multiplication by $(-1)^n$. The following result describes the behavior of the $e^{\otimes n+1}$'s with respect to the differentials of the cyclic bicomplex.

Lemma 4.25 *Let A be a k-algebra and $e \in A$ an idempotent. Then:*
(i) $b_{2n}(e^{\otimes 2n+1}) = e^{\otimes 2n}$ *for all* $n \geq 1$,
(ii) $b'_{2n-1}(e^{\otimes 2n}) = e^{\otimes 2n-1}$ *for all* $n \geq 1$,
(iii) $(1 - t_{2n-1})(e^{\otimes 2n}) = 2e^{\otimes 2n}$ *for all* $n \geq 1$ *and*
(iv) $N_{2n}(e^{\otimes 2n+1}) = (2n+1)e^{\otimes 2n+1}$ *for all* $n \geq 0$.

Proof. Assertion (i) is clear since $b_{2n}(e^{\otimes 2n+1})$ is the alternating sum of $2n+1$ copies of $e^{\otimes 2n}$, whereas (ii) follows since $b'_{2n-1}(e^{\otimes 2n})$ is the alternating sum of $2n-1$ copies of $e^{\otimes 2n-1}$. Finally, assertions (iii) and (iv) are consequences of the equalities $t_{2n-1}(e^{\otimes 2n}) = -e^{\otimes 2n}$ and $t_{2n}(e^{\otimes 2n+1}) = e^{\otimes 2n+1}$. □

Proposition 4.26 *Let A be a k-algebra and $e \in A$ an idempotent.*
   (i) The element

$$\xi_n(e) = \left(\frac{(-1)^n (2n)!}{n!} e^{\otimes 2n+1}, \frac{(-1)^n (2n)!}{2(n!)} e^{\otimes 2n}, \frac{(-1)^{n-1}(2n-2)!}{(n-1)!} e^{\otimes 2n-1}, \dots \right)$$

is a $2n$-cycle of the complex $\operatorname{Tot} C_{**}(A)$ for all $n \geq 0$ and
   (ii) If S is the endomorphism of the complex $\operatorname{Tot} C_{**}(A)$ defined in Proposition 4.17, then $S(\xi_n(e)) = \xi_{n-1}(e)$ for all $n \geq 1$.

Proof. (i) In view of Lemma 4.25, we have

$$b_{2n}\left(\frac{(-1)^n (2n)!}{n!} e^{\otimes 2n+1} \right) = (1 - t_{2n-1})\left(\frac{(-1)^n (2n)!}{2(n!)} e^{\otimes 2n} \right)$$

and

$$-b'_{2n-1}\left(\frac{(-1)^n (2n)!}{2(n!)} e^{\otimes 2n} \right) = N_{2n-2}\left(\frac{(-1)^{n-1}(2n-2)!}{(n-1)!} e^{\otimes 2n-1} \right)$$

for all $n \geq 1$. It follows that the chain $\xi_n(e) \in (\operatorname{Tot} C_{**}(A))_{2n}$ is indeed a cycle for all $n \geq 0$.
   (ii) This is an immediate consequence of the definition of S, in view of the form of the $\xi_n(e)$'s.   □

Remarks 4.27 (i) Let A be a k-algebra and $e \in A$ an idempotent. Then, $\xi_0(e) = e \in A = (\operatorname{Tot} C_{**}(A))_0$ and hence the cyclic homology class $[\xi_0(e)]$ is the residue class of $e \in A$ in the quotient $A/[A, A] = HC_0(A)$.
   (ii) Let A, B be two k-algebras and $f : A \longrightarrow B$ a morphism of non-unital algebras. If $e \in A$ is an idempotent then $f(e) \in B$ is also an idempotent and $[\xi_n(f(e))] = f_{2n}[\xi_n(e)] \in HC_{2n}(B)$ for all n, where

$$f_{2n} : HC_{2n}(A) \longrightarrow HC_{2n}(B)$$

is the k-linear map induced by f (cf. Remark 4.16).

We now establish a key additivity property of the cyclic homology classes that are associated with idempotent elements as above.

Lemma 4.28 *Let A be a k-algebra and $e, f \in A$ two orthogonal idempotents. Then, $e + f \in A$ is an idempotent and $[\xi_n(e+f)] = [\xi_n(e)] + [\xi_n(f)] \in HC_{2n}(A)$ for all $n \geq 0$.*

Proof. We consider the universal k-algebra $R = k[\sigma, \tau]/(\sigma^2 - \sigma, \tau^2 - \tau, \sigma\tau)$ on two orthogonal idempotents $s = \bar{\sigma}$ and $t = \bar{\tau}$ and the homomorphism of k-algebras $\lambda : R \longrightarrow A$, which is given by $s \mapsto e$ and $t \mapsto f$. Then, the induced k-linear map

$$\lambda_{2n} : HC_{2n}(R) \longrightarrow HC_{2n}(A)$$

is such that $\lambda_{2n}[\xi_n(s)] = [\xi_n(\lambda(s))] = [\xi_n(e)]$ (cf. Remark 4.27(ii)) and, similarly, $\lambda_{2n}[\xi_n(t)] = [\xi_n(f)]$ and $\lambda_{2n}[\xi_n(s+t)] = [\xi_n(e+f)]$ for all $n \geq 0$. Therefore, in view of the additivity of λ_{2n}, it suffices to prove that

$$[\xi_n(s+t)] = [\xi_n(s)] + [\xi_n(t)] \in HC_{2n}(R) \qquad (4.3)$$

for all n. We note that (4.3) is valid for $n = 0$, since $[\xi_0(s+t)] = s+t$, $[\xi_0(s)] = s$ and $[\xi_0(t)] = t$ in the group $HC_0(R) = R$ (cf. Remark 4.27(i)). Since the Hochschild homology groups of R vanish in positive degrees (cf. Example 4.8), we may invoke Corollary 4.18(ii) in order to conclude that

$$S^n : HC_{2n}(R) \longrightarrow HC_0(R) = R$$

is an isomorphism for all $n \geq 0$. Then, (4.3) follows since

$$
\begin{aligned}
S^n[\xi_n(s+t)] &= [\xi_0(s+t)] \\
&= [\xi_0(s)] + [\xi_0(t)] \\
&= S^n[\xi_n(s)] + S^n[\xi_n(t)] \\
&= S^n([\xi_n(s)] + [\xi_n(t)])
\end{aligned}
$$

for all $n \geq 0$. In the above chain of equalities, the first and the third ones follow from Proposition 4.26(ii).   □

Let A be a k-algebra, t a positive integer and E an idempotent $t \times t$ matrix with entries in A. For any integer $n \geq 0$ we define the cyclic homology class $\mathrm{ch}_n(E) \in HC_{2n}(A)$ as the image of the class $[\xi_n(E)] \in HC_{2n}(\mathbf{M}_t(A))$ under the map

$$\mathrm{Tr}_{2n} : HC_{2n}(\mathbf{M}_t(A)) \longrightarrow HC_{2n}(A)$$

defined in §4.1.1; in other words, we define $\mathrm{ch}_n(E) = \mathrm{Tr}_{2n}[\xi_n(E)]$.

Proposition 4.29 *Let A be a k-algebra.*

(i) We consider two positive integers t, t' with $t < t'$ and the morphism of non-unital algebras

$$\jmath = \jmath_{t,t'} : \mathbf{M}_t(A) \longrightarrow \mathbf{M}_{t'}(A)$$

which maps a $t \times t$ matrix X with entries in A onto the matrix $\begin{pmatrix} X & 0 \\ 0 & 0 \end{pmatrix}$. If $E \in \mathbf{M}_t(A)$ is an idempotent matrix and $E' = \jmath(E) \in \mathbf{M}_{t'}(A)$, then $\mathrm{ch}_n(E) = \mathrm{ch}_n(E') \in HC_{2n}(A)$ for all n.

(ii) If t is a positive integer, $E \in \mathbf{M}_t(A)$ an idempotent and $G \in GL_t(A)$ an invertible matrix, then $\mathrm{ch}_n(E) = \mathrm{ch}_n(GEG^{-1}) \in HC_{2n}(A)$ for all n.

(iii) Let t_1, t_2 be positive integers and E_1, E_2 idempotent matrices with entries in A of sizes t_1 and t_2 respectively. Then, the idempotent matrix $E = \begin{pmatrix} E_1 & 0 \\ 0 & E_2 \end{pmatrix}$ is such that $\mathrm{ch}_n(E) = \mathrm{ch}_n(E_1) + \mathrm{ch}_n(E_2)$ for all n.

(iv) Let t be a positive integer and $E \in \mathbf{M}_t(A)$ an idempotent matrix. Then, $\mathrm{ch}_0(E)$ is the Hattori-Stallings rank $r_{HS}(E) \in A/[A, A] = HC_0(A)$. Moreover, $S(\mathrm{ch}_n(E)) = \mathrm{ch}_{n-1}(E)$ for all $n \geq 1$.

Proof. (i) We fix an integer $n \geq 0$ and consider the commutative diagram

$$
\begin{array}{ccc}
HC_{2n}(\mathbf{M}_t(A)) & \xrightarrow{Tr_{2n}} & HC_{2n}(A) \\
{\scriptstyle J_{2n}} \downarrow & & \| \\
HC_{2n}(\mathbf{M}_{t'}(A)) & \xrightarrow{Tr_{2n}} & HC_{2n}(A)
\end{array}
$$

of Lemma 4.24. In view of Remark 4.27(ii), we have $[\xi_n(E')] = [\xi_n(\jmath(E))] = \jmath_{2n}[\xi_n(E)]$ and hence

$$\mathrm{ch}_n(E') = \mathrm{Tr}_{2n}[\xi_n(E')] = (\mathrm{Tr}_{2n} \circ \jmath_{2n})[\xi_n(E)] = \mathrm{Tr}_{2n}[\xi_n(E)] = \mathrm{ch}_n(E) \ .$$

(ii) The cycle $\xi_n(GEG^{-1})$ is obtained from the cycle $\xi_n(E)$ by letting the inner automorphism $I_G \in \mathrm{Aut}(\mathbf{M}_t(A))$ associated with G act on the cyclic bicomplex of $\mathbf{M}_t(A)$. Since I_G induces the identity operator on the cyclic homology of $\mathbf{M}_t(A)$ (cf. Proposition 4.23), we have $[\xi_n(GEG^{-1})] = [\xi_n(E)] \in HC_{2n}(\mathbf{M}_t(A))$ and hence

$$\mathrm{ch}_n(GEG^{-1}) = \mathrm{Tr}_{2n}[\xi_n(GEG^{-1})] = \mathrm{Tr}_{2n}[\xi_n(E)] = \mathrm{ch}_n(E) \in HC_{2n}(A)$$

for all n.

(iii) We fix a non-negative integer n and consider the idempotent matrices $E_1' = \begin{pmatrix} E_1 & 0 \\ 0 & 0 \end{pmatrix}$, $E_2' = \begin{pmatrix} 0 & 0 \\ 0 & E_2 \end{pmatrix}$ and $E_2'' = \begin{pmatrix} E_2 & 0 \\ 0 & 0 \end{pmatrix}$ in $\mathbf{M}_{t_1+t_2}(A)$. In view of (i) above, we have $\mathrm{ch}_n(E_1) = \mathrm{ch}_n(E_1')$ and $\mathrm{ch}_n(E_2) = \mathrm{ch}_n(E_2'')$. Since the matrices E_2' and E_2'' are conjugate, it follows from (ii) that $\mathrm{ch}_n(E_2'') = \mathrm{ch}_n(E_2')$ and hence $\mathrm{ch}_n(E_2) = \mathrm{ch}_n(E_2')$. Applying Lemma 4.28 to the orthogonal idempotent matrices E_1' and E_2', we conclude that

$$[\xi_n(E)] = [\xi_n(E_1')] + [\xi_n(E_2')] \in HC_{2n}(\mathbf{M}_{t_1+t_2}(A))$$

and hence

$$
\begin{aligned}
\mathrm{ch}_n(E) &= \mathrm{Tr}_{2n}[\xi_n(E)] \\
&= \mathrm{Tr}_{2n}[\xi_n(E_1')] + \mathrm{Tr}_{2n}[\xi_n(E_2')] \\
&= \mathrm{ch}_n(E_1') + \mathrm{ch}_n(E_2') \\
&= \mathrm{ch}_n(E_1) + \mathrm{ch}_n(E_2) \ .
\end{aligned}
$$

(iv) First of all, we note that $\xi_0(E) = E$ (cf. Remark 4.27(i)). Since $\mathrm{tr}_0 : \mathbf{M}_t(A) \longrightarrow A$ is the usual trace map, $\mathrm{ch}_0(E) = \mathrm{Tr}_0[\xi_0(E)] = \mathrm{Tr}_0[E] = [\mathrm{tr}_0(E)] = [\mathrm{tr}(E)]$ is indeed the residue class of the trace of E in the quotient group $A/[A, A] = HC_0(A)$. In order to prove the final claim, we compute

$$
\begin{aligned}
S(\mathrm{ch}_n(E)) &= (S \circ \mathrm{Tr}_{2n})[\xi_n(E)] \\
&= (\mathrm{Tr}_{2n-2} \circ S)[\xi_n(E)] \\
&= \mathrm{Tr}_{2n-2}[S(\xi_n(E))] \\
&= \mathrm{Tr}_{2n-2}[\xi_{n-1}(E)] \\
&= \mathrm{ch}_{n-1}(E)
\end{aligned}
$$

for all $n \geq 1$. In the above chain of equalities, the second one follows since the Tr_n's are compatible with the periodicity operators[1], whereas the fourth one results from Proposition 4.26(ii). □

In the following theorem, we reformulate the results obtained above in K-theoretic terms.

Theorem 4.30 *Let A be a k-algebra. Then, for any non-negative integer n there is an additive group homomorphism*

$$ch_n : K_0(A) \longrightarrow HC_{2n}(A) ,$$

such that the sequence $(ch_n)_n$ has the following properties:
(i) ch_0 is the Hattori-Stallings rank and
(ii) $S \circ ch_n = ch_{n-1}$ for all $n \geq 1$.
The ch_n's will be referred to as the Connes-Karoubi character maps.

Proof. We fix a positive integer n and consider the cyclic homology classes $ch_n(E) \in HC_{2n}(A)$ that are associated with idempotent matrices E with entries in A as above. In view of Remark 1.32, the assertions proved in Proposition 4.29(i),(ii),(iii) imply the existence of a group homomorphism

$$\mathrm{ch}_n : K_0(A) \longrightarrow HC_{2n}(A) ,$$

which maps the K-theory class of an idempotent matrix E as above onto $\mathrm{ch}_n(E)$. Then, Proposition 4.29(iv) shows that the ch_n's have properties (i) and (ii) in the statement and hence the proof is finished. □

Remark 4.31 In the special case where k is a field of characteristic 0 and A the algebra of regular functions on a smooth affine algebraic variety X over k, the cyclic homology groups of A can be expressed as a certain direct sum of de Rham cohomology groups of X. Moreover, a vector bundle V over X determines an element $[V]$ in the K_0-group of A and the Connes-Karoubi characters $\mathrm{ch}_n[V]$ defined in Theorem 4.30 recover the classical Chern characters of V. The reader can find a detailed exposition of these results in Loday's book [44].

4.1.3 The Cyclic Homology of Group Algebras

In the previous subsections, we defined the cyclic homology groups of a k-algebra A and established the relationship between these groups and the K-theory group $K_0(A)$. With an eye to the applications we have in mind, we now assume that the field k has characteristic 0 and specialize the previous discussion to the case where $A = kG$ is the group algebra of a group G, which will remain fixed throughout this subsection. Our goal is to study the structure of the inverse systems

---

[1] Note that the Tr_n's are induced by a chain bicomplex map; cf. Exercise 4.3.1.

$$\cdots \xrightarrow{S} HC_{2n}(kG) \xrightarrow{S} HC_{2n-2}(kG) \xrightarrow{S} \cdots \xrightarrow{S} HC_0(kG)$$

and

$$\cdots \xrightarrow{S} HC_{2n+1}(kG) \xrightarrow{S} HC_{2n-1}(kG) \xrightarrow{S} \cdots \xrightarrow{S} HC_1(kG) .$$

In analogy with the decomposition

$$HC_0(kG) = T(kG) = \bigoplus_{[g]\in\mathcal{C}(G)} k \cdot [g], \qquad (4.4)$$

we shall obtain a decomposition of the inverse systems above into a direct sum of inverse systems, indexed by the set $\mathcal{C}(G)$ of G-conjugacy classes.

For any element $g \in G$ we consider its centralizer C_g in G and the quotient group $N_g = C_g/{<}g{>}$. Using group homology in even degrees, we define the inverse system $\mathcal{X}^{even}(g, k) = (X_n^{even}(g,k))_n$ as follows:

- If g has finite order then

$$X_n^{even}(g, k) = \bigoplus_{i=0}^{n} H_{2i}(N_g, k)$$

for all $n \geq 0$. In particular, $X_0^{even}(g, k) = H_0(N_g, k) = k$. We note that $X_n^{even}(g, k) = X_{n-1}^{even}(g, k) \oplus H_{2n}(N_g, k)$ and define the structural map

$$X_n^{even}(g, k) \longrightarrow X_{n-1}^{even}(g, k)$$

to be the natural projection with kernel $H_{2n}(N_g, k)$ for all n.

- If g has infinite order we consider the central extension

$$1 \longrightarrow \mathbf{Z} \xrightarrow{g} C_g \longrightarrow N_g \longrightarrow 1$$

and let $\alpha_g \in H^2(N_g, \mathbf{Z})$ be the corresponding cohomology class (cf. §D.2.2 of Appendix D). We now define

$$X_n^{even}(g, k) = H_{2n}(N_g, k)$$

for all n, whereas the structural map

$$X_n^{even}(g, k) = H_{2n}(N_g, k) \longrightarrow H_{2n-2}(N_g, k) = X_{n-1}^{even}(g, k)$$

is the cap-product with the cohomology class α_g (cf. §D.2.3 of Appendix D). In this case too, we have $X_0^{even}(g, k) = H_0(N_g, k) = k$.

In exactly the same way, using group homology in odd degrees, we can associate with any element $g \in G$ an inverse system $\mathcal{X}^{odd}(g, k)$.

Remark 4.32 For any $g \in G$ the isomorphism type of the centralizer C_g depends only upon the conjugacy class $[g] \in \mathcal{C}(G)$ and the same is true for the quotient $N_g = C_g/{<}g{>}$. In fact, it is clear that the isomorphism types of the inverse systems $\mathcal{X}^{even}(g, k)$ and $\mathcal{X}^{odd}(g, k)$ depend only upon the conjugacy class $[g]$ of g.

We can now state the main result of this subsection on the structure of the inverse system $(HC_{\omega+2n}(kG), S)_n$, $\omega = 0, 1$, as follows:

Theorem 4.33 *There is a natural decomposition of inverse systems*

$$(HC_{\omega+2n}(kG), S)_n \simeq \bigoplus_{[g]\in\mathcal{C}(G)} \mathcal{X}^\omega(g,k) \,,$$

where the parity ω can be even$(=0)$ or odd$(=1)$. In the even case, this decomposition generalizes the decomposition of (4.4).

Proof. The strategy of the proof (which will occupy the remainder of this subsection) will be to replace the double complex $C_{**}(kG)$ with another double complex $\Gamma_{**}(G,k)$, which admits itself a natural decomposition into the direct sum of subcomplexes indexed by the set $\mathcal{C}(G)$ of G-conjugacy classes, and then compute the homology of the total complexes that are associated with each one of these subcomplexes.

Let M be the right kG-module which is equal to kG as a k-vector space with right G-action given by the rule $m \cdot g = g^{-1}mg$ for all $g \in G$ and all $m \in M$; here, $g^{-1}mg$ denotes the usual product of elements in $M = kG$. We also consider the G-set G^{n+1} with diagonal action and the associated kG-module $S_n(G,k) = k[G^{n+1}]$ for all n. We now let

$$\Gamma_n(G,k) = M \otimes_{kG} S_n(G,k)$$

and prove the following result, relating the $\Gamma_n(G,k)$'s to the tensor powers $C_n(kG) = kG^{\otimes n+1}$, $n \geq 0$.

Lemma 4.34 *(i) There is a natural isomorphism of k-vector spaces*

$$\theta_n : C_n(kG) \longrightarrow \Gamma_n(G,k) \,,$$

which maps an elementary tensor $g_0 \otimes g_1 \otimes \cdots \otimes g_n \in C_n(kG)$ onto the tensor $g_0g_1\cdots g_n \otimes (g_0, g_0g_1, \ldots, g_0g_1\cdots g_n) \in \Gamma_n(G,k)$ for any elements $g_0, g_1, \ldots, g_n \in G$ and any $n \geq 0$.

(ii) There is a natural decomposition of k-vector spaces

$$\Gamma_n(G,k) \simeq \bigoplus_{[g]\in\mathcal{C}(G)} \Gamma_n(g,k) \,,$$

where $\Gamma_n(g,k) = k \otimes_{kC_g} S_n(G,k)$ for all $g \in G$ and all n. The embedding $\Gamma_n(g,k) \hookrightarrow \Gamma_n(G,k)$ identifies the element $1 \otimes (g_0, \ldots, g_n) \in \Gamma_n(g,k)$ with $g \otimes (g_0, \ldots, g_n) \in \Gamma_n(G,k)$ for all $g, g_0, \ldots, g_n \in G$ and all $n \geq 0$.

Proof. (i) Let us consider the k-linear map

$$\eta_n : \Gamma_n(G,k) \longrightarrow C_n(kG) \,,$$

which maps an elementary tensor $m \otimes (g_0, \ldots, g_n) \in \Gamma_n(G,k)$ onto the tensor $g_n^{-1}mg_0 \otimes g_0^{-1}g_1 \otimes \cdots \otimes g_{n-1}^{-1}g_n \in C_n(kG)$ for all $m \in M$, $g_0, \ldots, g_n \in G$ and $n \geq 0$; here, $g_n^{-1}mg_0$ denotes the usual product of elements in $M = kG$. It is easily seen that η_n is well-defined; in fact, it is the inverse of θ_n for all n.

(ii) For any $g \in G$ we consider the kG-submodule $M_g = \bigoplus_{x \in [g]} k \cdot x$ of M. Then, there is a natural isomorphism of right kG-modules

$$M_g \simeq k \otimes_{kC_g} kG \ ,$$

which identifies $g \in M_g$ with $1 \otimes 1 \in k \otimes_{kC_g} kG$. Since $M = \bigoplus_{[g] \in \mathcal{C}(G)} M_g$ (as right kG-modules), we have an associated decomposition

$$
\begin{aligned}
\Gamma_n(G, k) &= M \otimes_{kG} S_n(G, k) \\
&= \bigoplus_{[g] \in \mathcal{C}(G)} M_g \otimes_{kG} S_n(G, k) \\
&\simeq \bigoplus_{[g] \in \mathcal{C}(G)} (k \otimes_{kC_g} kG) \otimes_{kG} S_n(G, k) \\
&\simeq \bigoplus_{[g] \in \mathcal{C}(G)} k \otimes_{kC_g} S_n(G, k) \\
&= \bigoplus_{[g] \in \mathcal{C}(G)} \Gamma_n(g, k)
\end{aligned}
$$

for all n, as needed.   □

By means of the isomorphisms $(\theta_n)_n$, the differentials of the double complex $C_{**}(kG)$ induce certain differentials on the bigraded k-vector space $\Gamma_{**}(G, k)$, which is defined by letting $\Gamma_{ij}(G, k) = \Gamma_j(G, k)$ for all $i, j \geq 0$. In order to identify these differentials, we consider for all $n \geq 1$ and all $i \in \{0, 1, \ldots, n\}$ the k-linear map

$$\delta_i^n : S_n(G, k) \longrightarrow S_{n-1}(G, k) \ ,$$

which is defined by letting $\delta_i^n(g_0, \ldots, g_n) = (g_0, \ldots, \widehat{g_i}, \ldots, g_n)$ for any element $(g_0, \ldots, g_n) \in G^{n+1}$; here, the symbol $\widehat{}$ over an element denotes omission of that element. It is clear that the k-linear maps δ_i^n are, in fact, kG-linear for all n, i. We also consider the kG-linear maps

$$\delta_n : S_n(G, k) \longrightarrow S_{n-1}(G, k) \ ,$$

which are defined by letting $\delta_n = \sum_{i=0}^{n}(-1)^i \delta_i^n$ for all $n \geq 1$. Then, as noted in Proposition D.6(i) of Appendix D, the δ_n's define a kG-free resolution of the trivial kG-module k

$$k \xleftarrow{\varepsilon} S_0(G, k) \xleftarrow{\delta_1} S_1(G, k) \xleftarrow{\delta_2} \cdots \xleftarrow{\delta_n} S_n(G, k) \xleftarrow{\delta_{n+1}} \cdots \qquad (4.5)$$

Moreover, if we define the kG-linear maps

$$\delta_n' : S_n(G, k) \longrightarrow S_{n-1}(G, k) \ ,$$

by letting $\delta_n' = \sum_{i=0}^{n-1}(-1)^i \delta_i^n$ for all $n \geq 1$, then the δ_n''s induce a contractible chain complex

$$S_0(G,k) \xleftarrow{\delta_1'} S_1(G,k) \xleftarrow{\delta_2'} \cdots \xleftarrow{\delta_n'} S_n(G,k) \xleftarrow{\delta_{n+1}'} \cdots$$

(cf. Proposition D.6(ii) of Appendix D). We now define the k-linear operators

$$\widetilde{\delta}_i^n : \Gamma_n(G,k) \longrightarrow \Gamma_{n-1}(G,k)$$

by letting $\widetilde{\delta}_i^n = 1 \otimes \delta_i^n$ for all $n \geq 1$ and $i \in \{0, 1, \ldots, n\}$. Similarly, we define the k-linear operators

$$\widetilde{\delta}_i^n(g) : \Gamma_n(g,k) \longrightarrow \Gamma_{n-1}(g,k)$$

by letting $\widetilde{\delta}_i^n(g) = 1 \otimes \delta_i^n$ for all $n \geq 1$, $i \in \{0, 1, \ldots, n\}$ and $g \in G$. The operators just defined are related to the operators d_i^n that were defined in §4.1.1, as described in the following result.

Lemma 4.35 *(i) The diagram*

$$
\begin{array}{ccc}
C_n(kG) & \xrightarrow{\theta_n} & \Gamma_n(G,k) \\
d_i^n \downarrow & & \downarrow \widetilde{\delta}_i^n \\
C_{n-1}(kG) & \xrightarrow{\theta_{n-1}} & \Gamma_{n-1}(G,k)
\end{array}
$$

is commutative for all $n \geq 1$ and $i \in \{0, 1, \ldots, n\}$.
  (ii) The diagram

$$
\begin{array}{ccc}
\Gamma_n(g,k) & \hookrightarrow & \Gamma_n(G,k) \\
\widetilde{\delta}_i^n(g) \downarrow & & \downarrow \widetilde{\delta}_i^n \\
\Gamma_{n-1}(g,k) & \hookrightarrow & \Gamma_{n-1}(G,k)
\end{array}
$$

whose horizontal arrows are the embeddings of Lemma 4.34(ii), is commutative for all $n \geq 1$, $i \in \{0, 1, \ldots, n\}$ and $g \in G$.

Proof. Both assertions are immediate consequences of the definitions. □

Using the $\widetilde{\delta}_i^n$'s, we define the operators

$$\beta_n : \Gamma_n(G,k) \longrightarrow \Gamma_{n-1}(G,k) \quad \text{and} \quad \beta_n' : \Gamma_n(G,k) \longrightarrow \Gamma_{n-1}(G,k) \,,$$

by letting $\beta_n = \sum_{i=0}^n (-1)^i \widetilde{\delta}_i^n = 1 \otimes \delta_n$ and $\beta_n' = \sum_{i=0}^{n-1}(-1)^i \widetilde{\delta}_i^n = 1 \otimes \delta_n'$ for all $n \geq 1$. Similarly, for all $n \geq 1$ and all $g \in G$ we define the operators

$$\beta_n(g) : \Gamma_n(g,k) \longrightarrow \Gamma_{n-1}(g,k) \quad \text{and} \quad \beta_n'(g) : \Gamma_n(g,k) \longrightarrow \Gamma_{n-1}(g,k) \,,$$

by letting $\beta_n(g) = \sum_{i=0}^n (-1)^i \widetilde{\delta}_i^n(g)$ and $\beta_n'(g) = \sum_{i=0}^{n-1}(-1)^i \widetilde{\delta}_i^n(g)$; we note that $\beta_n(g) = 1 \otimes \delta_n$ and $\beta_n'(g) = 1 \otimes \delta_n'$.

Corollary 4.36 *(i) The compositions $\beta_{n-1}\beta_n$ and $\beta_{n-1}'\beta_n'$ vanish for all $n \geq 1$.*
  (ii) The θ_n's induce an isomorphism of chain complexes

$$\theta : (C(kG), b) \longrightarrow (\Gamma(G, k), \beta) ,$$

where $\Gamma(G, k) = (\Gamma_n(G, k))_n$ and $\beta = (\beta_n)_n$. Hence, the Hochschild homology groups of kG are naturally identified with the homology groups of the complex $(\Gamma(G, k), \beta)$.

(iii) The θ_n's induce an isomorphism of chain complexes

$$\theta : (C(kG), b') \longrightarrow (\Gamma(G, k), \beta') ,$$

where $\beta' = (\beta'_n)_n$. In particular, the chain complex $(\Gamma(G, k), \beta')$ is contractible and hence acyclic.

(iv) There are natural decompositions of chain complexes

$$(\Gamma(G, k), \beta) = \bigoplus_{[g] \in \mathcal{C}(G)} (\Gamma(g, k), \beta(g))$$

and

$$(\Gamma(G, k), \beta') = \bigoplus_{[g] \in \mathcal{C}(G)} (\Gamma(g, k), \beta'(g)) ,$$

where $\Gamma(g, k) = (\Gamma_n(g, k))_n$, $\beta(g) = (\beta_n(g))_n$ and $\beta'(g) = (\beta'_n(g))_n$ for all $g \in G$.

(v) There is a natural isomorphism

$$HH_n(kG) \simeq \bigoplus_{[g] \in \mathcal{C}(G)} H_n(C_g, k)$$

for all $n \geq 0$.

Proof. Assertions (i), (ii), (iii) and (iv) follow from Lemma 4.35, in view of Lemma 4.34. Viewing the acyclic complex (4.5) as a kC_g-free resolution of the trivial kC_g-module k, we conclude that the chain complex $(\Gamma(g, k), \beta(g))$ computes the homology groups of C_g with coefficients in k for all $g \in G$. Therefore, assertion (v) follows from (ii) and (iv). $\square$

The action of the cyclic group of order $n + 1$ on $C_n(kG)$ induces, by means of θ_n, an action on $\Gamma_n(G, k)$ for all $n \geq 0$. In order to find a formula for this latter action, we consider the k-linear operator

$$\tau_n : \Gamma_n(G, k) \longrightarrow \Gamma_n(G, k) ,$$

which is defined by letting $g \otimes (g_0, \ldots, g_n) \mapsto (-1)^n g \otimes (g^{-1} g_n, g_0, \ldots, g_{n-1})$ for any elements $g, g_0, \ldots, g_n \in G$. It is easily seen that τ_n is a well-defined operator, such that τ_n^{n+1} is the identity on $\Gamma_n(G, k)$. Similarly, for any $g \in G$ we consider the k-linear operator

$$\tau_n(g) : \Gamma_n(g, k) \longrightarrow \Gamma_n(g, k) ,$$

which is defined by letting $1 \otimes (g_0, \ldots, g_n) \mapsto (-1)^n \otimes (g^{-1} g_n, g_0, \ldots, g_{n-1})$ for any elements $g_0, \ldots, g_n \in G$. In this case too, one can easily verify that $\tau_n(g)$ is well-defined and $\tau_n(g)^{n+1} = \mathrm{id}$.

Lemma 4.37 *(i) The diagram*

$$\begin{array}{ccc} C_n(kG) & \xrightarrow{\ \theta_n\ } & \Gamma_n(G,k) \\ t_n \downarrow & & \downarrow \tau_n \\ C_n(kG) & \xrightarrow{\ \theta_n\ } & \Gamma_n(G,k) \end{array}$$

is commutative for all $n \geq 0$.
    (ii) The diagram

$$\begin{array}{ccc} \Gamma_n(g,k) & \hookrightarrow & \Gamma_n(G,k) \\ \tau_n(g) \downarrow & & \downarrow \tau_n \\ \Gamma_n(g,k) & \hookrightarrow & \Gamma_n(G,k) \end{array}$$

whose horizontal arrows are the embeddings of Lemma 4.34(ii), is commutative for all $n \geq 0$ and all $g \in G$.

Proof. Both assertions are immediate consequences of the definitions. $\square$

We now consider the double complex $\Gamma_{**}(G,k)$, which is defined by letting $\Gamma_{ij}(G,k) = \Gamma_j(G,k)$ for all $i,j \geq 0$ and whose differentials are given in analogy with those of the double complex $C_{**}(kG)$, by replacing the b_n's (resp. the b'_n's, resp. the t_n's) with the β_n's (resp. the β'_n's, resp. the τ_n's). The appropriate commutation rules for the differentials of $\Gamma_{**}(G,k)$ follow from the corresponding ones for $C_{**}(kG)$, in view of Corollary 4.36(ii),(iii) and Lemma 4.37(i). Moreover, the 2-periodicity of the double complex $\Gamma_{**}(G,k)$ in the horizontal direction enables us to define a surjective endomorphism S of it, in analogy with the corresponding operator defined on the cyclic bicomplex of kG. In particular, there is an induced chain map

$$S : \operatorname{Tot}\Gamma_{**}(G,k) \longrightarrow \operatorname{Tot}\Gamma_{**}(G,k)[2] . \tag{4.6}$$

Similarly, for any element $g \in G$ we consider the double complex $\Gamma_{**}(g,k)$, which is defined by letting $\Gamma_{ij}(g,k) = \Gamma_j(g,k)$ for all $i,j \geq 0$ and whose differentials are given in analogy with those of the double complex $C_{**}(kG)$, by replacing the b_n's (resp. the b'_n's, resp. the t_n's) with the $\beta_n(g)$'s (resp. the $\beta'_n(g)$'s, resp. the $\tau_n(g)$'s). We also consider the chain map

$$S : \operatorname{Tot}\Gamma_{**}(g,k) \longrightarrow \operatorname{Tot}\Gamma_{**}(g,k)[2] , \tag{4.7}$$

which is induced from the endomorphism S of the double complex $\Gamma_{**}(g,k)$ that vanishes on the first two columns and maps identically the i-th column onto the $(i-2)$-th one for all $i \geq 2$.

Corollary 4.38 *(i) The θ_n's induce an isomorphism of double complexes*

$$\theta : C_{**}(kG) \longrightarrow \Gamma_{**}(G,k) .$$

In particular, the homology groups of the chain complex $\operatorname{Tot}\Gamma_{**}(G,k)$ are naturally identified with the cyclic homology groups of kG.

(ii) There is a natural decomposition of double complexes

$$\Gamma_{**}(G,k) \simeq \bigoplus\nolimits_{[g]\in\mathcal{C}(G)} \Gamma_{**}(g,k) \ .$$

In particular, the cyclic homology groups of kG decompose into the direct sum of the homology groups of the complexes $Tot\,\Gamma_{**}(g,k)$, $[g] \in \mathcal{C}(G)$.

(iii) The periodicity operator S associated with kG is induced by the chain map (4.6), in view of the identification of (i) above, and decomposes into the direct sum of the operators induced by the chain maps (4.7), in view of the decomposition of (ii) above.

Proof. Assertion (i) follows from Corollary 4.36(ii),(iii) and Lemma 4.37(i), whereas (ii) is a consequence of Lemma 4.34(ii), in view of Corollary 4.36(iv) and Lemma 4.37(ii). Finally, assertion (iii) is an immediate consequence of the definitions. □

We complete the proof of Theorem 4.33 by computing the homology groups of the complexes $Tot\,\Gamma_{**}(g,k)$ for all $g \in G$. In fact, we shall prove that there is a natural isomorphism of inverse systems

$$(H_{\omega+2n}(Tot\,\Gamma_{**}(g,k)), S)_n \simeq \mathcal{X}^\omega(g,k) \tag{4.8}$$

for all $g \in G$, where the parity ω can be even $(= 0)$ or odd $(= 1)$ and S is the periodicity operator induced by the chain map (4.7). To that end, we consider an element $g \in G$ and let $C = C_g$ be its centralizer in G. We define $\widetilde{\Gamma}_n(g,k) = k \otimes_{kC} S_n(C,k)$ and consider the map

$$\gamma_n : \widetilde{\Gamma}_n(g,k) \longrightarrow \Gamma_n(g,k) \ ,$$

which is induced by the inclusion $S_n(C,k) \hookrightarrow S_n(G,k)$ for all n. This latter inclusion being a split monomorphism of kC-modules (cf. Exercise 4.3.2), we conclude that the k-linear map γ_n is injective for all n. It follows that the γ_n's define a monomorphism of double complexes

$$\gamma : \widetilde{\Gamma}_{**}(g,k) \longrightarrow \Gamma_{**}(g,k) \ ,$$

where $\widetilde{\Gamma}_{**}(g,k)$ is the double complex which is defined by letting $\widetilde{\Gamma}_{ij}(g,k) = \widetilde{\Gamma}_j(g,k)$ for all $i,j \geq 0$ and whose differentials are given by the same formulae as the corresponding ones for $\Gamma_{**}(g,k)$.[2] Furthermore, the operator S on $\Gamma_{**}(g,k)$ restricts along γ to an operator S on $\widetilde{\Gamma}_{**}(g,k)$.

Lemma 4.39 *For any element $g \in G$ the monomorphism of double complexes $\gamma : \widetilde{\Gamma}_{**}(g,k) \longrightarrow \Gamma_{**}(g,k)$ defined above is a quasi-isomorphism. In particular, there is a natural isomorphism of inverse systems*

---

[2] More formally, $\widetilde{\Gamma}_{**}(g,k)$ is the double complex associated with the pair (C,g), in the same way that the double complex $\Gamma_{**}(g,k)$ is associated with the pair (G,g).

$$\left(H_{\omega+2n}\left(Tot\widetilde{\Gamma}_{**}(g,k)\right),S\right)_n \simeq (H_{\omega+2n}(Tot\Gamma_{**}(g,k)),S)_n \ ,$$

where the parity ω can be even $(= 0)$ or odd $(= 1)$.

Proof. In view of Proposition D.1(i) of Appendix D, it suffices to show that γ induces isomorphisms between the homology groups of the columns of the double complexes. As far as the even-numbered columns are concerned, we note that the inclusion

$$(S(C,k),\delta) \hookrightarrow (S(G,k),\delta)$$

is a morphism of kC-projective resolutions of k lifting the identity. Hence, the chain map

$$\gamma : \left(\widetilde{\Gamma}(g,k),\widetilde{\beta}(g)\right) \longrightarrow (\Gamma(g,k),\beta(g))$$

is a quasi-isomorphism, identifying the homology groups of both sides with $H_*(C,k)$. Concerning the odd-numbered columns, we note that the complex $(\Gamma(g,k),-\beta'(g))$ is acyclic, in view of Corollary 4.36(iii),(iv). The same result, applied to the group C and its element g, shows that the complex $\left(\widetilde{\Gamma}(g,k),-\widetilde{\beta}'(g)\right)$ is acyclic as well. □

In view of Lemma 4.39, the existence of a natural isomorphism as in (4.8) for an element $g \in G$ is equivalent to the existence of a natural isomorphism

$$\left(H_{\omega+2n}\left(Tot\,\widetilde{\Gamma}_{**}(g,k)\right),S\right)_n \simeq \mathcal{X}^\omega(g,k) \ ,$$

where the parity ω can be even $(= 0)$ or odd $(= 1)$. In other words, we may replace the pair (G,g) with the pair (C,g). Changing notation, it suffices to prove the existence of a natural isomorphism as in (4.8), under the assumption that g *is central in G.*

   Therefore, we fix a central element $g \in G$ and consider the double complex $\Gamma_{**} = \Gamma_{**}(g,k)$. We recall that

- $\Gamma_{ij} = \Gamma_j = k \otimes_{kG} S_j(G,k)$ for all $i,j \geq 0$.
- The even-numbered columns consist of the complex $(\Gamma, \beta(g))$ that computes the homology groups of G with coefficients in k.
- The odd-numbered columns consist of the complex $(\Gamma, -\beta'(g))$, which is acyclic.
- The n-th row of Γ_{**} is the complex

$$\Gamma_n \xleftarrow{1-\tau_n(g)} \Gamma_n \xleftarrow{N_n(g)} \Gamma_n \xleftarrow{1-\tau_n(g)} \cdots$$

that computes the homology of the cyclic group of order $n+1$ with coefficients in Γ_n, where the action is given by means of the operator $\tau_n(g)$, which maps $1 \otimes (g_0,\ldots,g_n)$ onto $(-1)^n \otimes (g^{-1}g_n,g_0,\ldots,g_{n-1})$ for all $g_0,\ldots,g_n \in G$.

In order to identify the homology of the complex $\operatorname{Tot}\Gamma_{**}$, we distinguish two cases:

Case I: Assume that $g \in G$ is a central element of finite order. We consider the quotient group $\overline{G} = G/{<}g{>}$ and the identity element $\overline{1} \in \overline{G}$. In the same way that the double complex Γ_{**} is associated with the pair (G, g), we may consider the double complex $\overline{\Gamma}_{**}$ associated with the pair $(\overline{G}, \overline{1})$. We note that the quotient map $\pi : G \longrightarrow \overline{G}$ induces a surjective map of double complexes

$$\pi : \Gamma_{**} \longrightarrow \overline{\Gamma}_{**} \ .$$

Lemma 4.40 *The chain bicomplex map* $\pi : \Gamma_{**} \longrightarrow \overline{\Gamma}_{**}$ *defined above is a quasi-isomorphism. In particular, there is a natural isomorphism of inverse systems*

$$\left(H_{\omega+2n}(\operatorname{Tot}\Gamma_{**}), S \right)_n \simeq \left(H_{\omega+2n}(\operatorname{Tot}\overline{\Gamma}_{**}), S \right)_n$$

for any parity ω.

Proof. In view of Proposition D.1(i) of Appendix D, it suffices to show that the maps induced by π between the homology groups of the columns of the double complexes are isomorphisms. This is clear for the odd-numbered columns, since the relevant homology groups vanish (cf. Corollary 4.36(iii),(iv)). As far as the even-numbered columns are concerned, we have to show that the chain map

$$\pi : (\Gamma, \beta(g)) \longrightarrow (\overline{\Gamma}, \beta(\overline{1}))$$

is a quasi-isomorphism. In other words, we have to show that

$$\pi_* : H_n(G, k) \longrightarrow H_n(\overline{G}, k)$$

is an isomorphism for all n. This latter assertion follows from Proposition D.21 of Appendix D, since k is (by assumption) a field of characteristic 0, whereas the kernel of the surjective group homomorphism $\pi : G \longrightarrow \overline{G}$ is finite. Finally, being a chain bicomplex map, π induces a map between the homology groups of the corresponding total complexes that commutes with the relevant periodicity operators S (cf. Exercise 4.3.1). This proves the last assertion in the statement and finishes the proof of the lemma.   □

In order to compute the homology of the complex $\operatorname{Tot}\overline{\Gamma}_{**}$, we consider for any $n \geq 0$ the k-linear endomorphism ϱ_n of $S_n(\overline{G}, k) = k\left[\overline{G}^{n+1}\right]$, which is given by letting

$$\varrho_n(x_0, \ldots, x_n) = \frac{1}{(n+1)!} \sum_{\sigma \in \Lambda_{n+1}} (-1)^{\tilde{\sigma}} (x_{\sigma 0}, \ldots, x_{\sigma n})$$

for any elements $x_0, \ldots, x_n \in \overline{G}$. Here, Λ_{n+1} is the group of permutations of the set $\{0, 1, \ldots, n\}$, whereas for any $\sigma \in \Lambda_{n+1}$ we define $\tilde{\sigma}$ to be the cardinality of the set of crossings $\{(i, j) : 0 \leq i < j \leq n \text{ and } \sigma i > \sigma j\}$;

then, $(-1)^{\tilde{\sigma}}$ is the sign of σ. We also consider the k-linear endomorphism $\overline{\tau}_n$ of $S_n(\overline{G}, k)$, which is given by letting

$$\overline{\tau}_n(x_0, \ldots, x_n) = (-1)^n (x_n, x_0, \ldots, x_{n-1})$$

for any elements $x_0, \ldots, x_n \in \overline{G}$. It is clear that the k-linear maps ϱ_n and $\overline{\tau}_n$ are, in fact, $k\overline{G}$-linear.

Lemma 4.41 (i) We have $\varrho_n = \varrho_n \circ \overline{\tau}_n$ for all n.
  (ii) The ϱ_n's induce a chain endomorphism of the resolution (4.5), that corresponds to the group $\overline{G}$, lifting the identity map of k.

Proof. (i) This is an immediate consequence of the definitions, since the permutation $(0, 1, \ldots, n) \mapsto (n, 0, \ldots, n-1)$ of $\{0, 1, \ldots, n\}$ has n crossings.
  (ii) Since the endomorphism ϱ_0 is the identity operator on $S_0(\overline{G}, k)$, the following diagram commutes

$$
\begin{array}{ccc}
k & \xleftarrow{\varepsilon} & S_0(\overline{G}, k) \\
\| & & \downarrow \varrho_0 \\
k & \xleftarrow{\varepsilon} & S_0(\overline{G}, k)
\end{array}
$$

It remains to prove that $(\varrho_n)_n$ is a chain map, i.e. that the following diagram commutes for all $n \geq 1$

$$
\begin{array}{ccc}
S_{n-1}(\overline{G}, k) & \xleftarrow{\delta_n} & S_n(\overline{G}, k) \\
\varrho_{n-1} \downarrow & & \downarrow \varrho_n \\
S_{n-1}(\overline{G}, k) & \xleftarrow{\delta_n} & S_n(\overline{G}, k)
\end{array}
$$

To that end, we fix an $(n+1)$-tuple $\mathbf{x} = (x_0, \ldots, x_n) \in \overline{G}^{n+1}$ and compute

$$
\begin{aligned}
\delta_n \varrho_n(\mathbf{x}) &= \delta_n \left[\frac{1}{(n+1)!} \sum_{\sigma \in \Lambda_{n+1}} (-1)^{\tilde{\sigma}} (x_{\sigma 0}, \ldots, x_{\sigma n}) \right] \\
&= \frac{1}{(n+1)!} \sum_{\sigma \in \Lambda_{n+1}} (-1)^{\tilde{\sigma}} \delta_n(x_{\sigma 0}, \ldots, x_{\sigma n}) \\
&= \frac{1}{(n+1)!} \sum_{\sigma \in \Lambda_{n+1}} \sum_{i=0}^{n} (-1)^{\tilde{\sigma}+i} (x_{\sigma 0}, \ldots, \widehat{x_{\sigma i}}, \ldots, x_{\sigma n})
\end{aligned}
$$

and

$$
\begin{aligned}
\varrho_{n-1} \delta_n(\mathbf{x}) &= \varrho_{n-1} \left[\sum_{j=0}^{n} (-1)^j (x_0, \ldots, \widehat{x_j}, \ldots, x_n) \right] \\
&= \sum_{j=0}^{n} (-1)^j \varrho_{n-1}(x_0, \ldots, \widehat{x_j}, \ldots, x_n) \\
&= \frac{1}{n!} \sum_{j=0}^{n} \sum_{\tau \in \Lambda_n} (-1)^{j+\tilde{\tau}} (x_{\tau 0}, \ldots, x_{\tau(j-1)}, x_{\tau(j+1)}, \ldots, x_{\tau n}) .
\end{aligned}
$$

We note that in the very last summation the summand corresponding to the pair (j, τ) is interpreted by viewing $\tau \in \Lambda_n$ as a permutation of the set $\{0, \ldots, j-1, j+1, \ldots, n\}$. The equality $\delta_n \varrho_n = \varrho_{n-1} \delta_n$ is therefore a consequence of the following result. $\square$

Lemma 4.42 *Let n be a positive integer, j a non-negative integer with $j \leq n$ and τ a permutation of the set $\{0, \ldots, j-1, j+1, \ldots, n\}$. Then:*

(i) There are precisely $n+1$ pairs (i, σ), where i is a non-negative integer with $i \leq n$ and σ a permutation of the set $\{0, \ldots, n\}$, such that the sequences

$$(\tau 0, \ldots, \tau(j-1), \tau(j+1), \ldots, \tau n) \text{ and } (\sigma 0, \ldots, \sigma(i-1), \sigma(i+1), \ldots, \sigma n)$$

coincide.

(ii) For any pair (i, σ) as in (i), we have $j + \tilde{\tau} \equiv i + \tilde{\sigma} \pmod{2}$.

Proof. (i) Having chosen an arbitrary element $i \in \{0, \ldots, n\}$, there is a unique permutation $\sigma \in \Lambda_{n+1}$ with the required property; namely, the one which is defined by letting

$$(\sigma 0, \ldots, \sigma(i-1), \sigma(i+1), \ldots, \sigma n) = (\tau 0, \ldots, \tau(j-1), \tau(j+1), \ldots, \tau n)$$

and $\sigma i = j$.

(ii) Let σ be defined as above, corresponding to the choice of an element $i \in \{0, \ldots, n\}$. In order to compute $\tilde{\sigma}$, we have to count the cardinality of the set of crossings

$$A = \{(x, y) : 0 \leq x < y \leq n \text{ and } \sigma x > \sigma y\}.$$

Let A_1 be the set consisting of those pairs $(x, y) \in A$ for which $x \neq i \neq y$; it is clear that

$$\operatorname{card} A_1 = \tilde{\tau}. \tag{4.9}$$

We now consider the subsets A_2 and A_3 of A, that consist of the pairs of the form (i, y) and (x, i) respectively. Since $\sigma i = j$, it follows that there are exactly $\operatorname{card} A_2$ elements of the set $\{i+1, \ldots, n\}$ and $i - \operatorname{card} A_3$ elements of the set $\{0, \ldots, i-1\}$ that are mapped under σ into the set $\{0, \ldots, j-1\}$. Therefore, it follows that

$$\operatorname{card} A_2 + i - \operatorname{card} A_3 = j. \tag{4.10}$$

Since A is the disjoint union of A_1, A_2 and A_3, we conclude from (4.9) and (4.10) that

$$\tilde{\sigma} \equiv \operatorname{card} A \equiv \operatorname{card} A_1 + \operatorname{card} A_2 + \operatorname{card} A_3 \equiv \tilde{\tau} + i + j \pmod{2}$$

and hence the proof is complete. □

The $k\overline{G}$-linear endomorphism ϱ_n of $S_n(\overline{G}, k)$ induces a k-linear endomorphism $\rho_n = 1 \otimes \varrho_n$ of $\overline{\Gamma}_n = k \otimes_{k\overline{G}} S_n(\overline{G}, k)$ for all n. We recall that the cyclic group of order $n+1$ acts on $\overline{\Gamma}_n$ by means of the operator $\tau_n(\overline{1})$, which maps $1 \otimes (x_0, \ldots, x_n)$ onto $(-1)^n \otimes (x_n, x_0, \ldots, x_{n-1})$ for any elements $x_0, \ldots, x_n \in \overline{G}$; therefore, $\tau_n(\overline{1}) = 1 \otimes \overline{\tau}_n$.

Corollary 4.43 *(i) We have $\rho_n = \rho_n \circ \tau_n(\bar{1})$ for all n.*
(ii) The ρ_n's induce a quasi-isomorphism

$$\rho : (\bar{\Gamma}, \beta(\bar{1})) \longrightarrow (\bar{\Gamma}, \beta(\bar{1})) \ .$$

Proof. Both assertions are immediate consequences of Lemma 4.41.   □

We now consider the double complex $\bar{\Gamma}'_{**}$, whose even-numbered columns consist of the complex $(\bar{\Gamma}, \beta(\bar{1}))$ and whose odd-numbered ones vanish. Then, Corollary 4.43(i) implies that there is a chain bicomplex map

$$\rho : \bar{\Gamma}_{**} \longrightarrow \bar{\Gamma}'_{**} \ ,$$

whose components are the ρ_n's between the even-numbered columns and the zero maps between the odd-numbered ones. Since the maps induced by ρ between the homology groups of the columns of the double complexes are isomorphisms (this is a consequence of Corollary 4.43(ii) for the even-numbered columns and Corollary 4.36(iii),(iv) for the odd-numbered ones), Proposition D.1(i) of Appendix D implies that ρ is a quasi-isomorphism. Being a chain bicomplex map, ρ induces a map between the homology groups of the corresponding total complexes that commutes with the relevant periodicity operators S (cf. Exercise 4.3.1). Hence, there is a natural isomorphism of inverse systems

$$\left(H_{\omega+2n}\left(\text{Tot}\,\bar{\Gamma}_{**}\right), S\right)_n \simeq \left(H_{\omega+2n}\left(\text{Tot}\,\bar{\Gamma}'_{**}\right), S\right)_n \tag{4.11}$$

for any parity ω. On the other hand, it is clear that

$$H_n\left(\text{Tot}\,\bar{\Gamma}'_{**}\right) = \bigoplus_{i\geq 0} H_{n-2i}(\bar{\Gamma}, \beta(\bar{1})) = \bigoplus_{i\geq 0} H_{n-2i}(\overline{G}, k)$$

for all $n \geq 0$. Moreover, the periodicity operator S of the double complex $\bar{\Gamma}'_{**}$ induces for any $n \geq 0$ a k-linear map

$$S : H_n\left(\text{Tot}\,\bar{\Gamma}'_{**}\right) \longrightarrow H_{n-2}\left(\text{Tot}\,\bar{\Gamma}'_{**}\right) \ ,$$

which is identified with the natural projection of $\bigoplus_{i\geq 0} H_{n-2i}(\overline{G}, k)$ onto $\bigoplus_{j\geq 0} H_{n-2-2j}(\overline{G}, k) = \bigoplus_{i\geq 1} H_{n-2i}(\overline{G}, k)$ with kernel $H_n(\overline{G}, k)$. It follows that for any parity ω there is an equality of inverse systems

$$\left(H_{\omega+2n}\left(\text{Tot}\,\bar{\Gamma}'_{**}\right), S\right)_n = \mathcal{X}^\omega(g, k) \ .$$

In view of Lemma 4.40 and (4.11) above, this establishes the existence of a natural isomorphism as in (4.8), in the case where g is a central element of finite order.

Case II: We now consider the case where $g \in G$ is a central element of infinite order. For any integer $n \geq 0$ we let $\sigma_n(g)$ be the k-linear endomorphism of $S_n(G, k)$, which is defined by mapping $(g_0, \ldots, g_n) \in S_n(G, k)$ onto $(-1)^n(g^{-1}g_n, g_0, \ldots, g_{n-1})$ for any $(n+1)$-tuple $(g_0, \ldots, g_n)$ of elements of G. Since g is central in G, the k-linear map $\sigma_n(g)$ is, in fact, kG-linear.

Lemma 4.44 *(i) The operator* $1 - \sigma_n(g) \in End_{kG}S_n(G, k)$ *is injective for all* $n \geq 0$.

(ii) $\delta_0^n \sigma_n(g) = (-1)^n \delta_n^n$ *for all* $n \geq 0$.

(iii) $\delta_i^n \sigma_n(g) = -\sigma_{n-1}(g)\delta_{i-1}^n$ *for all* $i > 0$.

(iv) $\delta_n(1 - \sigma_n(g)) = (1 - \sigma_{n-1}(g))\delta'_n$ *for all* $n \geq 0$.

In particular, there is an injective chain map

$$1 - \sigma(g) : (S(G, k), \delta') \longrightarrow (S(G, k), \delta) ,$$

where $S(G, k) = (S_n(G, k))_n$, $\delta' = (\delta'_n)_n$ *and* $\delta = (\delta_n)_n$.

Proof. (i) If $x \in S_n(G, k)$ is annihilated by $1 - \sigma_n(g)$, then $x = \sigma_n(g)x$ and hence $x = \sigma_n(g)^{n+1}x$. Since the iterate $\sigma_n(g)^{n+1}$ is multiplication by g^{-1}, we conclude that $(1 - g^{-1})x = 0$. We now consider the cyclic subgroup H of G generated by g and note that the group algebra kH is isomorphic with the k-algebra of Laurent polynomials in one variable; in particular, kH is an integral domain. Since $S_n(G, k)$ is free as a kG-module, it is free as a kH-module as well (cf. Lemma 1.1(ii) and Remark 1.6(i)). It follows that left multiplication by $1 - g^{-1}$ is injective on $S_n(G, k)$ and hence $x = 0$.

Assertions (ii) and (iii) are immediate consequences of the definitions (see also Lemma 4.13(i),(ii)).

(iv) This can be proved in exactly the same way as the first equality of Lemma 4.14 was proved, using (ii) and (iii) above, instead of Lemma 4.13(i),(ii). □

We are interested in the cokernel $T(g)$ of the chain map $1 - \sigma(g)$ of Lemma 4.44; it is the chain complex of kG-modules

$$T_0(g) \xleftarrow{\overline{\delta_1}} T_1(g) \xleftarrow{\overline{\delta_2}} \cdots \xleftarrow{\overline{\delta_n}} T_n(g) \xleftarrow{\overline{\delta_{n+1}}} \cdots ,$$

where $T_n(g) = S_n(G, k)/\mathrm{im}(1 - \sigma_n(g))$ for all n and the $\overline{\delta}$'s are induced from the δ's by passage to the quotient. We claim that g acts trivially on $T_n(g)$ for all n. Indeed, for any $x \in S_n(G, k)$ we have

$$\begin{aligned}
x - g^{-1}x &= x - \sigma_n(g)^{n+1}x \\
&= (1 - \sigma_n(g)^{n+1})x \\
&= (1 - \sigma_n(g))\left(\sum_{i=0}^{n} \sigma_n(g)^i x\right)
\end{aligned}$$

and hence $x - g^{-1}x \in \mathrm{im}(1 - \sigma_n(g))$. Being a kG-module, $\mathrm{im}(1 - \sigma_n(g))$ is invariant under the action of g and hence $gx - x = g(x - g^{-1}x) \in \mathrm{im}(1 - \sigma_n(g))$, as needed. Therefore, the action of G on $T_n(g)$ defines by passage to the quotient an action of the quotient group $\overline{G} = G/<g>$ on $T_n(g)$. In other words, $T_n(g)$ can be endowed with the structure of a $k\overline{G}$-module, in such a way that the restriction of that structure along the natural k-algebra map $kG \longrightarrow k\overline{G}$ is the given kG-module structure.

Lemma 4.45 *The $k\overline{G}$-module $T_n(g)$ defined above is projective for all n.*

Proof. Let us consider the kG-module

$$T_n'(g) = S_n(G,k)/\mathrm{im}\big(1 - \sigma_n(g)^{n+1}\big) = S_n(G,k)/\big(1 - g^{-1}\big)S_n(G,k) \ .$$

If $H = <g>$ is the cyclic subgroup of G generated by g, then

$$T_n'(g) = k \otimes_{kH} S_n(G,k) = k \otimes_{kH} kG \otimes_{kG} S_n(G,k) \simeq k\overline{G} \otimes_{kG} S_n(G,k)$$

(cf. Remark 1.6(ii)). Since $S_n(G,k)$ is a free kG-module, it follows that $T_n'(g)$ is a free $k\overline{G}$-module. Therefore, in order to prove that $T_n(g)$ is a projective $k\overline{G}$-module, it suffices to show that it is a direct summand of $T_n'(g)$. We note that the image of the endomorphism $1 - \sigma_n(g)^{n+1}$ of $S_n(G,k)$ is contained in that of $1 - \sigma_n(g)$; hence, there is a $k\overline{G}$-linear map

$$\pi : T_n'(g) \longrightarrow T_n(g) \ ,$$

which is induced from the identity of $S_n(G,k)$ by passage to the quotient. We also consider the kG-linear endomorphism

$$f = \tfrac{1}{n+1}\sum_{i=0}^{n} \sigma_n(g)^i : S_n(G,k) \longrightarrow S_n(G,k) \ .$$

We note that $f(1 - \sigma_n(g)) = \tfrac{1}{n+1}\big(1 - \sigma_n(g)^{n+1}\big)$ and hence f maps the kG-submodule $\mathrm{im}(1 - \sigma_n(g))$ into $\mathrm{im}\big(1 - \sigma_n(g)^{n+1}\big)$. It follows that f induces by passage to the quotient a $k\overline{G}$-linear map

$$\phi : T_n(g) \longrightarrow T_n'(g) \ .$$

Moreover, $1 - f = (1 - \sigma_n(g))\psi$ for some endomorphism ψ of $S_n(G,k)$ and hence $x - f(x) \in \mathrm{im}(1 - \sigma_n(g))$ for all $x \in S_n(G,k)$. Therefore, the composition $\pi \circ \phi$ is the identity on $T_n(g)$ and hence $T_n(g)$ is a direct summand of $T_n'(g)$ as a $k\overline{G}$-module, as needed. □

In degree 0, the endomorphism $\sigma_0(g)$ of $S_0(G,k) = kG$ maps any element $x \in kG$ onto $g^{-1}x$. Therefore,

$$T_0(g) = S_0(G,k)/\mathrm{im}(1 - \sigma_0(g)) = kG/(1 - g^{-1})kG = k\overline{G}$$

and hence we may consider the augmentation ε from $T_0(g) = k\overline{G}$ to k.

Lemma 4.46 *The chain complex*

$$k \xleftarrow{\ \varepsilon\ } T_0(g) \xleftarrow{\ \overline{\delta_1}\ } T_1(g) \xleftarrow{\ \overline{\delta_2}\ } \cdots \xleftarrow{\ \overline{\delta_n}\ } T_n(g) \xleftarrow{\ \overline{\delta_{n+1}}\ } \cdots$$

is a $k\overline{G}$-projective resolution of the trivial $k\overline{G}$-module k.

Proof. In view of Lemma 4.45, we know that $T_n(g)$ is a projective $k\overline{G}$-module for all $n \geq 0$. In order to prove exactness, we consider the short exact sequence of the augmented chain complexes

$$0 \longrightarrow \begin{pmatrix} (S(G,k), \delta') \\ \downarrow \\ 0 \end{pmatrix} \overset{1-\sigma(g)}{\longrightarrow} \begin{pmatrix} (S(G,k), \delta) \\ \varepsilon\downarrow \\ k \end{pmatrix} \longrightarrow \begin{pmatrix} (T(g), \overline{\delta}) \\ \varepsilon\downarrow \\ k \end{pmatrix} \longrightarrow 0 \ .$$

Since the first two complexes are acyclic (cf. Proposition D.6 of Appendix D), the associated long exact sequence in homology shows that the third complex is acyclic as well. $\qquad\square$

It is an immediate consequence of the definitions that the kG-linear endomorphism $\sigma_n(g)$ of $S_n(G,k)$ induces the k-linear endomorphism $\tau_n(g)$ of $\Gamma_n = \Gamma_n(g,k) = k \otimes_{kG} S_n(G,k)$; in other words, $\tau_n(g) = 1 \otimes \sigma_n(g)$ for all n. We know that $\tau_n(g)^{n+1}$ is the identity on Γ_n and hence $\tau_n(g)$ defines an action of the cyclic group of order $n+1$ on Γ_n. We note that

$$\begin{aligned} H_0(\mathbf{Z}/(n+1)\mathbf{Z}, \Gamma_n) &= \Gamma_n/\mathrm{im}(1 - \tau_n(g)) \\ &= k \otimes_{kG} (S_n(G,k)/\mathrm{im}(1 - \sigma_n(g))) \\ &= k \otimes_{kG} T_n(g) \ . \end{aligned} \qquad (4.12)$$

In general, the homology groups of $\mathbf{Z}/(n+1)\mathbf{Z}$ with coefficients in Γ_n may be computed from the n-th row of the double complex Γ_{**}.

Lemma 4.47 *There is a natural isomorphism $H_n(Tot\,\Gamma_{**}) \simeq H_n(\overline{G}, k)$ for all $n \geq 0$.*

Proof. Since k is a field of characteristic 0, Maschke's theorem implies that the homology groups of $\mathbf{Z}/(n+1)\mathbf{Z}$ with coefficients in Γ_n vanish in positive degrees (cf. Example D.8(ii) of Appendix D). It follows that the horizontal homology of the double complex Γ_{**} is concentrated in the 0-th column, where it is given by the groups $k \otimes_{kG} T_n(g)$, $n \geq 0$ (cf. (4.12)). Moreover, the differential induced on $k \otimes_{kG} T(g) = (k \otimes_{kG} T_n(g))_n$ from the differential $\beta(g) = 1 \otimes \delta$ of the 0-th column of Γ_{**} by passage to the quotient is equal to $1 \otimes \overline{\delta}$, in view of the identification of (4.12). Then, Corollary D.2(i) of Appendix D implies the existence of a natural isomorphism

$$H_n(\mathrm{Tot}\,\Gamma_{**}) \simeq H_n\big(k \otimes_{kG} T(g), 1 \otimes \overline{\delta}\big)$$

for all $n \geq 0$. Since $k \otimes_{kG} T(g) = k \otimes_{k\overline{G}} T(g)$, the proof follows by invoking Lemma 4.46. $\qquad\square$

In view of Lemma 4.47, in order to establish the existence of a natural isomorphism as in (4.8), in the case where $g \in G$ is a central element of infinite order, it only remains to identify the periodicity operator

$$S : H_n(\operatorname{Tot}\Gamma_{**}) \longrightarrow H_{n-2}(\operatorname{Tot}\Gamma_{**})$$

with the cap-product map

$$\alpha \cap {}_- : H_n(\overline{G}, k) \longrightarrow H_{n-2}(\overline{G}, k) \ ,$$

which is associated with the cohomology class $\alpha \in H^2(\overline{G}, \mathbf{Z})$ that classifies the central extension

$$1 \longrightarrow \mathbf{Z} \overset{g}{\longrightarrow} G \longrightarrow \overline{G} \longrightarrow 1 \ .$$

The proof of the latter assertion will be omitted, since it requires tools from homotopy theory that are beyond the scope of this book. The interested reader may consult on that matter the bibliographic sources listed at the end of the chapter.

4.2 The Nilpotency of Connes' Operator

We are now ready to use the results of the previous section, in order to study the idempotent conjectures for groups that satisfy a certain homological condition. This homological condition is equivalent to the nilpotency of Connes' operator on those components of the cyclic homology of the group algebra that correspond to conjugacy classes of elements of infinite order. It will turn out that the resulting class $\mathcal{C}$ of groups is closed under several group theoretic operations, thereby providing us with many examples of groups that satisfy the idempotent conjectures.

In §4.2.1, we explain the rationale behind the definition of class $\mathcal{C}$ and prove that groups that are residually contained in $\mathcal{C}$ satisfy Bass' conjecture. In the following subsection, we prove that $\mathcal{C}$ contains all abelian groups and is closed under subgroups, free products and finite direct products. We also examine the extent to which $\mathcal{C}$ is closed under extensions and establish the relevance of class $\mathcal{C}$ in the study of the idempotent conjecture.

4.2.1 Idempotent Conjectures and the Nilpotency of S

Let k be a subring of the field $\mathbf{C}$ of complex numbers, G a group and kG the corresponding group algebra. We are interested in idempotent $n \times n$ matrices E with entries in kG. In the case where the idempotent conjecture for a torsion-free group G is concerned, we have $k = \mathbf{C}$ and $n = 1$. In the case of Bass' conjecture, k is a subring of $\mathbf{C}$ with $k \cap \mathbf{Q} = \mathbf{Z}$, n any positive integer and G any group. For any element $g \in G$ with $g \neq 1$ we examine the vanishing of the complex number $r_g(E)$. The relevance of the equality $r_g(E) = 0$ for the idempotent conjecture stems from Proposition 3.15. In fact, we may restrict our attention to the case where the element g has infinite order. Indeed, if it is the idempotent conjecture that we are interested in, then G is a torsion-free group and hence any element $g \in G \setminus \{1\}$ has infinite order. On the other hand, we already know that $r_g(E) = 0$ if $g \neq 1$ is an element of finite order and k a subring of $\mathbf{C}$ with $k \cap \mathbf{Q} = \mathbf{Z}$ (cf. Corollary 3.35).

We are therefore lead to examine the vanishing of the additive map

$$r_g : K_0(kG) \longrightarrow k \,,$$

in the case where $g \in G$ is an element of infinite order. In view of the naturality of r_g with respect to coefficient ring homomorphisms (cf. Proposition 1.46), it suffices to prove the vanishing of the additive map

$$r_g : K_0(\mathbf{C}G) \longrightarrow \mathbf{C} \,.$$

We note that the latter map is the composition of the Hattori-Stallings rank $r_{HS} : K_0(\mathbf{C}G) \longrightarrow T(\mathbf{C}G)$, followed by the projection $\pi_{[g]}$ of the $\mathbf{C}$-vector space $T(\mathbf{C}G) = \mathbf{C}G/[\mathbf{C}G, \mathbf{C}G] = \bigoplus_{[x] \in \mathcal{C}(G)} \mathbf{C} \cdot [x]$ onto $\mathbf{C} \cdot [g] \simeq \mathbf{C}$. Then, Theorem 4.30(i) implies that r_g coincides with the composition

$$K_0(\mathbf{C}G) \xrightarrow{ch_0} HC_0(\mathbf{C}G) = T(\mathbf{C}G) \xrightarrow{\pi_{[g]}} \mathbf{C} \,.$$

For the remainder of this section, we use the following notational conventions, that apply to any element $g \in G$ of infinite order: We denote by C_g the centralizer of g in G and let N_g be the quotient of C_g by the infinite cyclic group generated by g. In addition, we denote by $\alpha_g \in H^2(N_g, \mathbf{Z})$ the cohomology class that classifies the central extension

$$1 \longrightarrow \mathbf{Z} \xrightarrow{g} C_g \longrightarrow N_g \longrightarrow 1$$

(cf. §D.2.2 of Appendix D). Using Theorems 4.30(ii) and 4.33, we obtain for all $n \geq 1$ the following commutative diagram

$$\begin{array}{ccccccc}
K_0(\mathbf{C}G) & \xrightarrow{ch_n} & HC_{2n}(\mathbf{C}G) & \xrightarrow{\pi_{[g]}} & X_n^{even}(g, \mathbf{C}) = H_{2n}(N_g, \mathbf{C}) \\
\| & & S^n \downarrow & & \downarrow & & \downarrow \alpha_g^n \cap \_ \\
K_0(\mathbf{C}G) & \xrightarrow{ch_0} & HC_0(\mathbf{C}G) & \xrightarrow{\pi_{[g]}} & \mathbf{C} & = & \mathbf{C}
\end{array}$$

(Note that the n-fold composition of cap-product maps with the cohomology class α_g is the cap-product map with the class α_g^n; cf. Corollary D.15(iii) of Appendix D.) Here, $\pi_{[g]}$ denotes also the projection of $HC_{2n}(\mathbf{C}G) = \bigoplus_{[x] \in \mathcal{C}(G)} X_n^{even}(x, \mathbf{C})$ onto the summand $X_n^{even}(g, \mathbf{C}) = H_{2n}(N_g, \mathbf{C})$. For later use, we record the following immediate consequence of the above discussion.

Observation 4.48 Let k be a subring of the field $\mathbf{C}$ of complex numbers, G a group and $g \in G$ an element of infinite order. Then, the map

$$r_g : K_0(kG) \longrightarrow k$$

is identically zero if there exists a positive integer n, such that the cap-product map

$$\alpha_g^n \cap \_ : H_{2n}(N_g, \mathbf{C}) \longrightarrow H_0(N_g, \mathbf{C}) \simeq \mathbf{C}$$

is the zero map.

Lemma 4.49 *Let N be a group and $\alpha \in H^i(N, \mathbf{Z})$ a cohomology class. Then, the following conditions are equivalent:*

(i) The image $\alpha_{\mathbf{Q}}$ of α in the cohomology group $H^i(N, \mathbf{Q})$ is zero.

(ii) The cap-product map $\alpha \cap \_ : H_i(N, \mathbf{Q}) \longrightarrow H_0(N, \mathbf{Q}) \simeq \mathbf{Q}$ vanishes.

(iii) The cap-product map $\alpha \cap \_ : H_i(N, \mathbf{C}) \longrightarrow H_0(N, \mathbf{C}) \simeq \mathbf{C}$ vanishes.

Proof. First of all, we note that the cap-product maps

$$\alpha \cap \_ : H_i(N, \mathbf{Q}) \longrightarrow H_0(N, \mathbf{Q}) \simeq \mathbf{Q}$$

and

$$\alpha_{\mathbf{Q}} \cap \_ : H_i(N, \mathbf{Q}) \longrightarrow H_0(N, \mathbf{Q}) \simeq \mathbf{Q}$$

coincide (cf. Proposition D.14(i) of Appendix D). Hence, the equivalence (i)↔(ii) follows since the dual $\mathbf{Q}$-vector space $\mathrm{Hom}_{\mathbf{Q}}(H_i(N, \mathbf{Q}), \mathbf{Q})$ is isomorphic with the cohomology group $H^i(N, \mathbf{Q})$, in such a way that the cap-product map

$$\alpha_{\mathbf{Q}} \cap \_ : H_i(N, \mathbf{Q}) \longrightarrow H_0(N, \mathbf{Q}) \simeq \mathbf{Q}$$

is identified with $\alpha_{\mathbf{Q}} \in H^i(N, \mathbf{Q})$ (cf. Corollary D.17 of Appendix D).

Since the trivial N-module $\mathbf{C}$ is a direct sum of copies of the trivial N-module $\mathbf{Q}$, the cap-product map

$$\alpha \cap \_ : H_i(N, \mathbf{C}) \longrightarrow H_0(N, \mathbf{C}) \simeq \mathbf{C}$$

is a direct sum of copies of the cap-product map

$$\alpha \cap \_ : H_i(N, \mathbf{Q}) \longrightarrow H_0(N, \mathbf{Q}) \simeq \mathbf{Q} \ .$$

The equivalence (ii)↔(iii) follows readily from this. □

For any group N the cup-product endows the $\mathbf{Q}$-vector space $H^\bullet(N, \mathbf{Q}) = \bigoplus_i H^i(N, \mathbf{Q})$ with the structure of an associative $\mathbf{Q}$-algebra (cf. Corollary D.13(i) of Appendix D). In view of the discussion above, we define a class $\mathcal{C}$ of groups, as follows:

Definition 4.50 *The class $\mathcal{C}$ consists of those groups G that satisfy the following condition: For any element $g \in G$ of infinite order the image $(\alpha_g)_{\mathbf{Q}}$ of the class α_g in the cohomology ring $H^\bullet(N_g, \mathbf{Q})$ is nilpotent.[3]*

---

[3] In this case, we say that α_g is rationally nilpotent.

Remarks 4.51 (i) One way of proving the rational nilpotency of a cohomology class is by expressing it as a pullback of another class, which is already known to be rationally nilpotent. More precisely, let us consider a morphism of central extensions

$$
\begin{array}{ccccccccc}
1 & \longrightarrow & \mathbf{Z} & \longrightarrow & C & \longrightarrow & N & \longrightarrow & 1 \\
& & \| & & \downarrow & & \downarrow f & & \\
1 & \longrightarrow & \mathbf{Z} & \longrightarrow & C' & \longrightarrow & N' & \longrightarrow & 1
\end{array}
$$

and let $\alpha \in H^2(N, \mathbf{Z})$ and $\alpha' \in H^2(N', \mathbf{Z})$ be the corresponding cohomology classes. Then, α is rationally nilpotent if this is the case for α'. Indeed, $\alpha = f^*\alpha'$ is the pullback of α' along f and hence $\alpha_{\mathbf{Q}} = (f^*\alpha')_{\mathbf{Q}} = f^*(\alpha'_{\mathbf{Q}})$. The nilpotency of $\alpha_{\mathbf{Q}}$ follows, since

$$
f^* : H^\bullet(N', \mathbf{Q}) \longrightarrow H^\bullet(N, \mathbf{Q})
$$

is a $\mathbf{Q}$-algebra homomorphism (cf. Corollary D.13(iii) of Appendix D).

(ii) A group G is contained in $\mathcal{C}$ if and only if for any subgroup $C \subseteq G$ and any central extension

$$
1 \longrightarrow \mathbf{Z} \longrightarrow C \longrightarrow N \longrightarrow 1
$$

the corresponding cohomology class $\alpha \in H^2(N, \mathbf{Z})$ is rationally nilpotent. It is clear that this condition is sufficient for G to be a group in $\mathcal{C}$. Conversely, let $G \in \mathcal{C}$ and consider a central extension as above. If $g \in C \subseteq G$ is the image of $1 \in \mathbf{Z}$, then C is contained in the centralizer C_g of g in G and hence there exists a morphism of extensions

$$
\begin{array}{ccccccccc}
1 & \longrightarrow & \mathbf{Z} & \overset{g}{\longrightarrow} & C & \longrightarrow & N & \longrightarrow & 1 \\
& & \| & & \downarrow & & \downarrow & & \\
1 & \longrightarrow & \mathbf{Z} & \overset{g}{\longrightarrow} & C_g & \longrightarrow & N_g & \longrightarrow & 1
\end{array}
$$

Since α_g is rationally nilpotent, it follows from (i) above that α is rationally nilpotent as well.

The relevance of class $\mathcal{C}$ in the study of the idempotent conjectures is illustrated by the next result. Recall that a group G is said to be residually contained in $\mathcal{C}$ if for any element $g \in G$ with $g \neq 1$ there exists a normal subgroup $K \trianglelefteq G$, such that $g \notin K$ and $G/K \in \mathcal{C}$.

Theorem 4.52 *Let G be a group.*
(i) If G is residually contained in $\mathcal{C}$, then G satisfies Bass' conjecture.
(ii) If $G \in \mathcal{C}$ is torsion-free, then G satisfies the idempotent conjecture.

Proof. (i) Assume on the contrary that G is residually contained in $\mathcal{C}$ and does not satisfy Bass' conjecture. Then, there exists a subring k of the field $\mathbf{C}$ of complex numbers with $k \cap \mathbf{Q} = \mathbf{Z}$, an idempotent matrix E with entries in kG and an element $g \in G \setminus \{1\}$, such that $r_g(E) \neq 0$. In view of Theorem 3.32(i), there exists an integer $u > 0$ such that g is conjugate to its n^u-th power for all $n \geq 1$. Let $K \trianglelefteq G$ be a normal subgroup, such that $g \notin K$ and $\overline{G} = G/K \in \mathcal{C}$. Then, the image $\overline{g} = gK$ of g in $\overline{G}$ is non-trivial and conjugate to its n^u-th power for all $n \geq 1$; in particular, $\overline{g}$ is an element of infinite order. We now consider the morphism of central extensions

$$
\begin{array}{ccccccccc}
1 & \longrightarrow & \mathbf{Z} & \overset{g}{\longrightarrow} & C_g & \longrightarrow & N_g & \longrightarrow & 1 \\
& & \| & & \downarrow & & \downarrow & & \\
1 & \longrightarrow & \mathbf{Z} & \overset{\overline{g}}{\longrightarrow} & C_{\overline{g}} & \longrightarrow & N_{\overline{g}} & \longrightarrow & 1
\end{array}
$$

where the vertical arrows are induced by the quotient map $G \longrightarrow \overline{G}$. Since $\overline{G} \in \mathcal{C}$, the cohomology class $\alpha_{\overline{g}}$ classifying the bottom row is rationally nilpotent. In view of Remark 4.51(i), the same is true for α_g; hence, there exists $n \gg 0$ such that

$$
(\alpha_g^n)_\mathbf{Q} = ((\alpha_g)_\mathbf{Q})^n = 0 \in H^{2n}(N_g, \mathbf{Q}) \, .
$$

We now invoke Lemma 4.49, in order to conclude that the cap-product map

$$
\alpha_g^n \cap \_ : H_{2n}(N_g, \mathbf{C}) \longrightarrow H_0(N_g, \mathbf{C}) \simeq \mathbf{C}
$$

is the zero map. Therefore, Observation 4.48 implies that the map

$$
r_g : K_0(kG) \longrightarrow k
$$

is the zero map as well. In particular, $r_g(E) = 0$ and this is the desired contradiction.

(ii) Let G be a torsion-free group contained in $\mathcal{C}$ and $e \in \mathbf{C}G$ an idempotent. We fix an element $g \in G \setminus \{1\}$ and note that g has infinite order. Since $G \in \mathcal{C}$, we may conclude (using Lemma 4.49 and Observation 4.48, as in the latter part of the proof of (i) above) that the map

$$
r_g : K_0(\mathbf{C}G) \longrightarrow \mathbf{C}
$$

is identically zero. In particular, $r_g(e) = 0$. Since this is the case for any element $g \in G \setminus \{1\}$, Proposition 3.15 implies that e is trivial. Hence, G satisfies the idempotent conjecture, as needed. $\qquad \square$

Remark 4.53 In the following subsection, we will complement Theorem 4.52(ii) and prove that the idempotent conjecture is also satisfied by groups that are residually contained in the class of torsion-free groups in $\mathcal{C}$ (cf. Proposition 4.56).

4.2.2 Closure Properties

Having introduced the class $\mathcal{C}$ of groups, we illustrated its importance in the study of idempotents in group algebras, by proving Theorem 4.52. In order to obtain specific examples of groups that satisfy the idempotent conjectures, we establish in the present subsection the closure of $\mathcal{C}$ under several group theoretic operations.

Proposition 4.54 *All torsion and all abelian groups are contained in* $\mathcal{C}$.

Proof. It is clear that torsion groups are contained in $\mathcal{C}$. On the other hand, let C be an abelian group and $g \in C$ an element of infinite order. Then, the extension

$$1 \longrightarrow \mathbf{Z} \xrightarrow{g} C \longrightarrow N \longrightarrow 1$$

is classified by a rationally trivial cohomology class $\alpha_g \in H^2(N, \mathbf{Z})$, i.e. we have $(\alpha_g)_{\mathbf{Q}} = 0 \in H^2(N, \mathbf{Q})$. Indeed, $(\alpha_g)_{\mathbf{Q}}$ classifies the central extension

$$1 \longrightarrow \mathbf{Q} \longrightarrow C' \longrightarrow N \longrightarrow 1 \, ,$$

where C' is the quotient of the direct product $\mathbf{Q} \times C$ by its normal subgroup $\{(-n, g^n) : n \in \mathbf{Z}\}$ (cf. §D.2.2 of Appendix D). Since the group C is abelian, the same is true for C'. In view of the divisibility of $\mathbf{Q}$, the above extension of abelian groups splits and hence the cohomology class $(\alpha_g)_{\mathbf{Q}}$ is trivial, as claimed. $\qquad\square$

Proposition 4.55 *The class* $\mathcal{C}$ *has the following properties:*
   (i) $\mathcal{C}$ *is closed under subgroups,*
   (ii) $\mathcal{C}$ *is closed under finite direct products and*
   (iii) $\mathcal{C}$ *is closed under free products.*

Proof. (i) This is an immediate consequence of Remark 4.51(ii).
   (ii) Let $(G_i)_i$ be a finite family of groups that are contained in $\mathcal{C}$ and $G = \prod_i G_i$ the corresponding direct product. If $g = (g_i)_i \in G$ is an element of infinite order, then $g_i \in G_i$ is an element of infinite order for some index i. We fix such an i and consider the morphism of central extensions

$$
\begin{array}{ccccccccc}
1 & \longrightarrow & \mathbf{Z} & \xrightarrow{g} & C_g & \longrightarrow & N_g & \longrightarrow & 1 \\
  &   & \| &   & \downarrow &   & \downarrow &   &   \\
1 & \longrightarrow & \mathbf{Z} & \xrightarrow{g_i} & C_{g_i} & \longrightarrow & N_{g_i} & \longrightarrow & 1
\end{array}
$$

Here, C_{g_i} denotes the centralizer of g_i in G_i, $N_{g_i} = C_{g_i}/\!<\!g_i\!>$ and the vertical arrows are induced by the i-th coordinate projection map $G \longrightarrow G_i$. Since $G_i \in \mathcal{C}$, the cohomology class $\alpha_{g_i} \in H^2(N_{g_i}, \mathbf{Z})$ classifying the bottom row is rationally nilpotent. In view of Remark 4.51(i), the same is true for the class α_g. This being the case for any element $g \in G$ of infinite order, we conclude that $G \in \mathcal{C}$.
   (iii) Let G_1, G_2 be two groups contained in $\mathcal{C}$ and $G = G_1 * G_2$ their free product. In order to show that $G \in \mathcal{C}$, we consider an element $g \in G$ of infinite order and distinguish two cases:

<u>Case 1:</u> The conjugacy class $[g]$ does not meet the union $G_1 \cup G_2$. Using the structure theorem for free products, one can show that, in this case, the centralizer C_g is infinite cyclic (cf. Exercise 4.3.3). It follows that the quotient $N_g = C_g/\!<\!g\!>$ is finite and hence the group $H^2(N_g, \mathbf{Q})$ is trivial (cf. Example D.8(ii) of Appendix D); in particular, $(\alpha_g)_{\mathbf{Q}} = 0$.

<u>Case 2:</u> The conjugacy class $[g]$ meets the union $G_1 \cup G_2$. Since the triple (C_g, N_g, α_g) depends, up to isomorphism, only upon the conjugacy class of g (cf. Remark 4.32), we can assume that $g \in G_1 \cup G_2$, say $g \in G_1$. Then, C_g coincides with the centralizer $C_{1,g}$ of g in G_1 and the quotient N_g coincides with the corresponding quotient $N_{1,g} = C_{1,g}/<g>$. In this way, α_g classifies the central extension

$$1 \longrightarrow \mathbf{Z} \overset{g}{\longrightarrow} C_{1,g} \longrightarrow N_{1,g} \longrightarrow 1 \ .$$

Since $G_1 \in \mathcal{C}$, it follows that α_g is rationally nilpotent.

Taking into account Cases 1 and 2, we conclude that the group G is contained in $\mathcal{C}$.

Now let $(G_i)_{i \in I}$ be any family of groups contained in $\mathcal{C}$ and $G = *_{i \in I} G_i$ the corresponding free product. Using induction on the cardinality of the index set I, one can show that $G \in \mathcal{C}$ if I is finite. In order to show that $G \in \mathcal{C}$ in the general case, we consider an element $g \in G$ of infinite order. Then, there is a finite subset $I' \subseteq I$, such that $g \in G' = *_{i \in I'} G_i$. Since C_g coincides with the centralizer C'_g of g in $G' \in \mathcal{C}$, we may conclude as before that the cohomology class α_g is rationally nilpotent. Therefore, it follows that $G \in \mathcal{C}$. $\square$

We are now ready to prove the generalization of Theorem 4.52(ii), that was promised in Remark 4.53. We denote by $\mathcal{C}_\infty$ the class consisting of those torsion-free groups that are contained in $\mathcal{C}$.

Proposition 4.56 *Let G be a group which is residually contained in $\mathcal{C}_\infty$. Then, G satisfies the idempotent conjecture.*

Proof. Let $e = \sum_g e_g g \in \mathbf{C}G$ be an idempotent, where $e_g \in \mathbf{C}$ for all $g \in G$. We consider the subset $\Lambda = \text{supp}\, e = \{g \in G : e_g \neq 0\} \subseteq G$. It is a finite set and hence there exists a *finite* family of normal subgroups $(K_i)_i$ of G, such that:

(i) $G/K_i \in \mathcal{C}_\infty$ for all i and

(ii) the set Λ maps injectively under the canonical group homomorphism $G \longrightarrow \prod_i G/K_i$.

In view of Proposition 4.55(ii), the product group $\prod_i G/K_i$ is contained in $\mathcal{C}_\infty$. Therefore, Theorem 4.52(ii) implies that the image of e in the complex group algebra of $\prod_i G/K_i$ is equal to either 0 or 1. It follows that $\text{card}(\Lambda) \leq 1$ and hence e must be itself equal to either 0 or 1. $\square$

We now examine the extent to which $\mathcal{C}$ is closed under group extensions. We note that a group G is said to have finite homological dimension over $\mathbf{Q}$ if there exists an integer n, such that $H_i(G, V) = 0$ for all $i > n$ and all $\mathbf{Q}G$-modules V. The smallest such n is the homological dimension $\text{hd}_{\mathbf{Q}} G$ of G over $\mathbf{Q}$ (cf. §D.2.1 of Appendix D).

Lemma 4.57 *Let C be a group and $A \trianglelefteq C$ a normal subgroup, such that $\text{hd}_{\mathbf{Q}}(C/A) < \infty$. If $g \in A$ is an element of infinite order, which is central in C, then the following conditions are equivalent:*

(i) The cohomology class $\alpha \in H^2(C/<g>, \mathbf{Z})$ classifying the extension

$$1 \longrightarrow \mathbf{Z} \overset{g}{\longrightarrow} C \longrightarrow C/<g> \longrightarrow 1$$

is rationally nilpotent.

(ii) The cohomology class $\alpha' \in H^2(A/<g>, \mathbf{Z})$ classifying the extension

$$1 \longrightarrow \mathbf{Z} \overset{g}{\longrightarrow} A \longrightarrow A/<g> \longrightarrow 1$$

is rationally nilpotent.

Proof. The implication (i)→(ii) follows from the principle of Remark 4.51(i). In order to show that (ii)→(i), we consider the extension of groups

$$1 \longrightarrow A/<g> \longrightarrow C/<g> \longrightarrow C/A \longrightarrow 1 .$$

Then, the associated Lyndon-Hochschild-Serre spectral sequence (cf. Theorem D.20 of Appendix D) provides us with a decreasing filtration $(F^p H^n)_p$ on the cohomology group $H^n(C/<g>, \mathbf{Q})$ for all $n \geq 0$, such that:

- $F^0 H^n = H^n(C/<g>, \mathbf{Q})$ and $F^{n+1} H^n = 0$,
- $F^p H^n/F^{p+1} H^n$ is a subquotient of $H^p(C/A, H^{n-p}(A/<g>, \mathbf{Q}))$,
- $F^1 H^n = \ker\left(H^n(C/<g>, \mathbf{Q}) \overset{res}{\longrightarrow} H^n(A/<g>, \mathbf{Q})\right)$ and
- if $\beta \in F^p H^n$ and $\beta' \in F^{p'} H^{n'}$, then $\beta \cup \beta' \in F^{p+p'} H^{n+n'}$.

If $d = \mathrm{hd}_{\mathbf{Q}}(C/A)$ then

$$H^p(C/A, H^q(A/<g>, \mathbf{Q})) = \mathrm{Hom}(H_p(C/A, H_q(A/<g>, \mathbf{Q})), \mathbf{Q}) = 0$$

for all $p > d$; here, the first equality follows applying Proposition D.16(ii) of Appendix D. It follows that $F^p H^n = F^{p+1} H^n$ for all $p > d$ and hence

$$F^{d+1} H^n = F^{d+2} H^n = F^{d+3} H^n = \cdots = 0$$

for all n. Since the cohomology class $\alpha' = \mathrm{res}(\alpha) \in H^2(A/<g>, \mathbf{Z})$ is assumed to be rationally nilpotent, there exists $n \gg 0$ such that

$$\mathrm{res}(\alpha_{\mathbf{Q}}^n) = \mathrm{res}(\alpha_{\mathbf{Q}})^n = \mathrm{res}(\alpha)_{\mathbf{Q}}^n = \alpha_{\mathbf{Q}}'^n = 0 \in H^{2n}(A/<g>, \mathbf{Q}) .$$

Then, $\alpha_{\mathbf{Q}}^n \in F^1 H^{2n}$ and hence $\alpha_{\mathbf{Q}}^{n(d+1)} = (\alpha_{\mathbf{Q}}^n)^{d+1} \in F^{d+1} H^{2n(d+1)} = 0.$ $\qquad \square$

Proposition 4.58 *Let G be a group and $K \trianglelefteq G$ a normal subgroup, such that both groups K and G/K are contained in C. If $\mathrm{hd}_{\mathbf{Q}}(G/K) < \infty$ then $G \in C$ as well.*

Proof. In order to show that $G \in C$, let us consider a subgroup $C \leq G$ and a central extension

$$1 \longrightarrow \mathbf{Z} \overset{g}{\longrightarrow} C \longrightarrow N \longrightarrow 1 .$$

We have to show that the corresponding cohomology class $\alpha \in H^2(N, \mathbf{Z})$ is rationally nilpotent (cf. Remark 4.51(ii)). We let $\bar{g} = gK$ be the image of g in $\bar{G} = G/K$ and distinguish three cases:

<u>Case 1:</u> Assume that $\bar{g} = \bar{1} \in \bar{G}$, i.e. that $g \in K$. Then, $A = K \cap C$ is a normal subgroup of C containing g and we may consider the morphism of central extensions

$$
\begin{array}{ccccccccc}
1 & \longrightarrow & \mathbf{Z} & \overset{g}{\longrightarrow} & A & \longrightarrow & A/<g> & \longrightarrow & 1 \\
& & \| & & \downarrow & & \downarrow & & \\
1 & \longrightarrow & \mathbf{Z} & \overset{g}{\longrightarrow} & C & \longrightarrow & C/<g> & \longrightarrow & 1
\end{array}
$$

Let $\alpha' \in H^2(A/<g>, \mathbf{Z})$ be the element classifying the top row. Then, α' is rationally nilpotent since $A \le K \in \mathcal{C}$. We note that the group $C/A = C/(K \cap C)$ has finite homological dimension over $\mathbf{Q}$, being isomorphic with a subgroup of $\bar{G}$ (cf. Proposition D.9(i) of Appendix D). Therefore, Lemma 4.57 implies that the cohomology class α is rationally nilpotent as well.

<u>Case 2:</u> Assume that $\bar{g} \in \bar{G}$ is an element of finite order n. Since $g^n \in K$, it follows from Case 1 above that the class $\beta \in H^2(C/<g^n>, \mathbf{Z})$ that classifies the central extension

$$
1 \longrightarrow \mathbf{Z} \overset{g^n}{\longrightarrow} C \longrightarrow C/<g^n> \longrightarrow 1
$$

is rationally nilpotent. We now consider the morphism of central extensions

$$
\begin{array}{ccccccccc}
1 & \longrightarrow & \mathbf{Z} & \overset{g^n}{\longrightarrow} & C & \longrightarrow & C/<g^n> & \longrightarrow & 1 \\
& & n \downarrow & & \| & & \downarrow \pi & & \\
1 & \longrightarrow & \mathbf{Z} & \overset{g}{\longrightarrow} & C & \longrightarrow & N & \longrightarrow & 1
\end{array}
$$

and note that $\pi^*\alpha = n\beta$ (cf. Proposition D.10 of Appendix D); it follows that the class $\pi^*\alpha$ is rationally nilpotent as well. We note that the group homomorphism

$$
\pi : C/<g^n> \longrightarrow N
$$

is surjective with kernel a cyclic group of order n; hence, the induced map

$$
\pi^* : H^i(N, \mathbf{Q}) \longrightarrow H^i(C/<g^n>, \mathbf{Q})
$$

is an isomorphism for all i (cf. Proposition D.21 of Appendix D). Since $\pi^*(\alpha_{\mathbf{Q}}) = (\pi^*\alpha)_{\mathbf{Q}}$, we conclude that the cohomology class α is rationally nilpotent.

<u>Case 3:</u> Assume that $\bar{g} \in \bar{G}$ is an element of infinite order. In this case, we consider the morphism of central extensions

$$
\begin{array}{ccccccc}
1 & \longrightarrow & \mathbf{Z} & \overset{g}{\longrightarrow} & C & \longrightarrow & N & \longrightarrow & 1 \\
& & \| & & \downarrow & & \downarrow & & \\
1 & \longrightarrow & \mathbf{Z} & \overset{\bar{g}}{\longrightarrow} & C_{\bar{g}} & \longrightarrow & N_{\bar{g}} & \longrightarrow & 1
\end{array}
$$

where the vertical arrows are induced by the projection $G \longrightarrow \overline{G}$. Since $\overline{G} \in \mathcal{C}$, the cohomology class $\alpha_{\overline{g}}$ classifying the bottom row is rationally nilpotent. Taking into account Remark 4.51(i), we conclude that α is rationally nilpotent as well.   □

Having proved Proposition 4.58, we now consider the class consisting of those groups $G \in \mathcal{C}$ with $\mathrm{hd}_\mathbf{Q} G < \infty$.

Definition 4.59 *The class $\mathcal{E}$ consists of those groups G that satisfy the following two conditions:*
(i) $\mathrm{hd}_\mathbf{Q} G < \infty$ and
(ii) $\mathrm{hd}_\mathbf{Q} N_g < \infty$ for any element $g \in G$ of infinite order.

Proposition 4.60 *The class $\mathcal{E}$ consists of those groups $G \in \mathcal{C}$ for which $\mathrm{hd}_\mathbf{Q} G < \infty$.*

Proof. Let G be a group contained in $\mathcal{E}$. Then, by assumption, the homological dimension of G over $\mathbf{Q}$ is finite. In order to prove that $G \in \mathcal{C}$, let us fix an element $g \in G$ of infinite order. Since $\mathrm{hd}_\mathbf{Q} N_g < \infty$, the homology group $H_{2n}(N_g, \mathbf{Q})$ is trivial for $n \gg 0$ and hence $H^{2n}(N_g, \mathbf{Q}) = \mathrm{Hom}(H_{2n}(N_g, \mathbf{Q}), \mathbf{Q}) = 0$ for $n \gg 0$ (cf. Corollary D.17 of Appendix D). In particular, the cohomology class α_g is rationally nilpotent. Since this is the case for any element $g \in G$ of infinite order, we conclude that $G \in \mathcal{C}$.

Conversely, let G be a group contained in $\mathcal{C}$ with $\mathrm{hd}_\mathbf{Q} G < \infty$. In order to prove that $G \in \mathcal{E}$, we consider an element $g \in G$ of infinite order. Then, there exists an integer $n \gg 0$, such that $\mathrm{hd}_\mathbf{Q} G \leq 2n$ and $(\alpha_g)_\mathbf{Q}^n = 0 \in H^{2n}(N_g, \mathbf{Q})$. We shall prove that $\mathrm{hd}_\mathbf{Q} N_g \leq 2n - 1$, i.e. that $H_i(N_g, V) = 0$ for all $\mathbf{Q}N_g$-modules V and all $i \geq 2n$. Let us fix a $\mathbf{Q}N_g$-module V; we denote by V' the $\mathbf{Q}C_g$-module obtained from V by restriction of scalars along the quotient map $\mathbf{Q}C_g \longrightarrow \mathbf{Q}N_g$. Being a subgroup of G, the group C_g has finite homological dimension over $\mathbf{Q}$; in fact, $\mathrm{hd}_\mathbf{Q} C_g \leq \mathrm{hd}_\mathbf{Q} G \leq 2n$ (cf. Proposition D.9(i) of Appendix D). It follows that the homology group $H_i(C_g, V')$ is trivial for all $i \geq 2n + 1$. On the other hand, Proposition D.22 of Appendix D shows that there are exact sequences

$$H_i(C_g, V') \longrightarrow H_i(N_g, V) \xrightarrow{\alpha_g \cap -} H_{i-2}(N_g, V) \longrightarrow H_{i-1}(C_g, V')$$

for all $i \geq 0$. Therefore, the cap-product map

$$\alpha_g \cap - : H_i(N_g, V) \longrightarrow H_{i-2}(N_g, V)$$

is an isomorphism for all $i \geq 2n + 2$. It follows that the composition

$$H_{i+2n}(N_g, V) \xrightarrow{\alpha_g \cap -} H_{i+2n-2}(N_g, V) \xrightarrow{\alpha_g \cap -} \cdots \xrightarrow{\alpha_g \cap -} H_i(N_g, V) ,$$

is an isomorphism for all $i \geq 2n$. This composition coincides with the map

$$\alpha_g^n \cap - : H_{i+2n}(N_g, V) \longrightarrow H_i(N_g, V)$$

(cf. Proposition D.14(iii) of Appendix D) and hence with the map

$$(\alpha_g)_{\mathbf{Q}}^n \cap - : H_{i+2n}(N_g, V) \longrightarrow H_i(N_g, V)$$

(cf. Proposition D.14(i) of Appendix D). Then, our assumption about the vanishing of the class $(\alpha_g)_{\mathbf{Q}}^n$ implies that $H_i(N_g, V) = 0$ for all $i \geq 2n$, as needed. $\qquad\square$

Corollary 4.61 *(i) An abelian group G is contained in $\mathcal{E}$ if and only if $hd_{\mathbf{Q}} G < \infty$.*

(ii) The class $\mathcal{E}$ is closed under subgroups, extensions and free products of families of groups of uniformly bounded homological dimension over $\mathbf{Q}$.

Proof. (i) Since abelian groups are contained in $\mathcal{C}$ (cf. Proposition 4.54), this is an immediate consequence of Proposition 4.60.

(ii) We note that the group operations under consideration preserve the finiteness of the homological dimension (cf. Proposition D.9 and Corollary D.19 of Appendix D). Hence, the result follows from Proposition 4.60, in view of Propositions 4.55 and 4.58. $\qquad\square$

We conclude with a few explicit examples.

Examples 4.62 (i) If G is an abelian group of finite rank, then $hd_{\mathbf{Q}} G$ is finite (cf. Exercise D.3.1 of Appendix D). Therefore, Corollary 4.61 implies that $\mathcal{E}$ contains all solvable groups with finite Hirsch number. In this way, Theorem 4.52(i) provides us with an alternative proof of Corollary 3.44.

(ii) The inclusion $\mathcal{E} \subseteq \mathcal{C}$ is strict, since there are abelian groups of infinite homological dimension over $\mathbf{Q}$; cf. Exercise D.3.1 of Appendix D. In fact, the group $\mathcal{H}$ that was constructed at the end of §3.2.3 is an example of a *finitely generated* solvable group, which is contained in $\mathcal{C}$ (cf. Exercise 4.3.5), while having infinite homological dimension over $\mathbf{Q}$ (as it contains an infinite direct sum of copies of $\mathbf{Q}$ as a subgroup).

(iii) If G is a finitely generated metabelian group, then $G \in \mathcal{C}$ (in view of Propositions 4.54 and 4.58). In particular, Theorem 4.52(i) implies that G satisfies Bass' conjecture. Invoking Exercise 1.3.9(ii), we conclude that any metabelian group satisfies Bass' conjecture.

4.3 Exercises

1. We are interested in chain bicomplexes $C = (C_{ij})_{i,j \geq 0}$, which are 2-periodic in the horizontal direction; by this, we mean that the chain complexes $D_i = (C_{ij})_j$ and $D_{i+2} = (C_{i+2\,j})_j$ are identical for all $i \geq 0$, whereas the horizontal differentials $d^h : C_{i+1\,j} \longrightarrow C_{ij}$ and $d^h : C_{i+3\,j} \longrightarrow C_{i+2\,j}$ coincide for all $i, j \geq 0$.

(i) Show that for any chain bicomplex C, which is 2-periodic in the horizontal direction, there is a chain bicomplex map which vanishes on the first two columns and maps identically the i-th column onto the $(i-2)$-th one for all $i \geq 2$. We denote by S the induced endomorphism of degree -2 of the homology of the associated chain complex $\operatorname{Tot} C$.

(ii) Let C, C' be two chain bicomplexes, which are 2-periodic in the horizontal direction, and $\varphi : C \longrightarrow C'$ a chain bicomplex map. Show that the induced maps in homology are such that $S' \circ \varphi_n = \varphi_{n-2} \circ S$ for all n. Here, we denote by S, S' the homology endomorphisms that correspond to C and C' respectively.

2. Let k be a commutative ring.

(i) Consider a group H, an H-set X, an H-invariant subset $Y \subseteq X$ and the permutation kH-modules $M = k[X] = \bigoplus_{x \in X} k \cdot x$ and $N = k[Y] = \bigoplus_{y \in Y} k \cdot y$. Show that N is a direct summand of M.

(ii) Let G be a group, $H \subseteq G$ a subgroup and n a non-negative integer. Show that the inclusion $S_n(H, k) \subseteq S_n(G, k)$ is a split monomorphism of kH-modules, where $S_n(H, k) = k[H^{n+1}]$ and $S_n(G, k) = k[G^{n+1}]$.

3. Let G_1, G_2 be two groups, $G = G_1 * G_2$ their free product and $g \in G$ an element which is not conjugate to any element of G_1 or G_2. Show that the centralizer C_g of g in G is an infinite cyclic group.

4. Let G be a group. For any $g \in G$ we denote by C_g the centralizer of g in G and consider the quotient $N_g = C_g / {<}g{>}$. Show that the following two conditions are equivalent:

(i) $G \in \mathcal{E}$ (cf. Definition 4.59),

(ii) $\operatorname{hd}_{\mathbf{Q}} N_g < \infty$ for any element $g \in G$.

5. Let $\mathcal{H}$ be the group that was constructed at the end of §3.2.3. Recall that $\mathcal{H}$ is an extension of a finitely generated torsion-free metabelian group Λ by an abelian group V. The goal of this Exercise is to prove that $\mathcal{H} \in \mathcal{C}$. To that end, let us fix an element $g \in \mathcal{H}$ of infinite order.

(i) If $g \in V$ show that the cohomology class α_g is rationally trivial. (*Hint:* Use property (δ) at the very end of Chap. 3.)

(ii) Show that for any element $\lambda \in \Lambda$ of infinite order the class α_λ is rationally nilpotent.

(iii) Assume that $g \notin V$. Show that the image of g in the quotient group $\Lambda = \mathcal{H}/V$ is an element of infinite order and conclude that the cohomology class α_g is rationally nilpotent.

Notes and Comments on Chap. 4. The cyclic (co-)homology of complex algebras and the characters from the corresponding K_0-group were introduced by A. Connes in [15]. The theory was subsequently developed for algebras over an arbitrary commutative ground ring by J.L. Loday and D. Quillen [45]. The proof of the Morita invariance of Hochschild and cyclic homology given here follows R. MacCarthy [47]. The computation of the cyclic homology of group algebras is due to D. Burghelea [10] (see also [1,40]). Burghelea's proof uses tools from homotopy theory; the algebraic proof given in §4.1.3 is due to Z. Marciniak [51]. For more details on the subject, the reader is referred to Loday's book [44]. The applicability of cyclic homology in the study of the idempotent conjectures was first noticed by B. Eckmann [19] and Z. Marciniak [50]. The class $\mathcal{E}$ was introduced by Eckmann in [19] and studied, independently, by R. Ji [35] and G. Chadha and I.B.S. Passi [11]. A relative version of Eckmann's method was studied by J. Schafer in [63]. The class $\mathcal{C}$ was introduced and studied in [22], as a generalization of class $\mathcal{E}$, whereas [27] pursues this approach one step further, by considering a homological condition that takes into account the arithmetic properties of the ground ring.

5

Completions of $\mathbf{C}G$

5.1 The Integrality of the Trace Conjecture

Let G be a torsion-free group. The idempotent conjecture for the complex group algebra $\mathbf{C}G$ can be strengthened to the conjecture about the triviality of idempotents in the reduced group C^*-algebra C_r^*G. This latter conjecture can be further strengthened to a conjecture about the integrality of the values of the additive map

$$\tau_* : K_0(C_r^*G) \longrightarrow \mathbf{C} ,$$

which is induced by the canonical trace τ on C_r^*G. Some evidence for the validity of the integrality of the trace conjecture is provided by Zaleskii's theorem (cf. Theorem 3.19), which asserts that for any group G (possibly with torsion) the values of τ_* on K-theory classes that come from the group algebra $\mathbf{C}G$ are rational. Our goal in this section is to prove the integrality of the trace conjecture in the cases where G is a torsion-free abelian or a free group. We note that the idempotent conjecture for the complex group algebra in these two cases was taken care of in §1.2.3.

After formulating the integrality of the trace conjecture in §5.1.1, we consider the case of an abelian group G in §5.1.2. In that case, the C^*-algebra C_r^*G is commutative and can be identified with the algebra of continuous complex-valued functions on the dual group $\widehat{G}$. In this way, both the idempotent and the integrality of the trace conjectures are seen to be equivalent to the connectedness of the dual group. It will turn out that $\widehat{G}$ is connected if and only if the abelian group G is torsion-free. This approach places both conjectures (in the abelian group case) into a more geometric perspective and should be compared to the proof of Bass' conjecture for abelian groups that was given in §2.1. In §5.1.3 we consider the case where G is a free group. In that case, there is a tree X on which the group G acts freely. The representations of C_r^*G that are associated with the actions of G to the sets of vertices and edges of X respectively, define a certain unital and dense subalgebra $\mathcal{A} \subseteq C_r^*G$. Then, the integrality of the trace on $K_0(C_r^*G)$ follows from the integrality of its values on the K-theory classes coming from the subalgebra $\mathcal{A}$, since the inclusion of $\mathcal{A}$ into C_r^*G will turn out to induce an isomorphism between the respective K_0-groups.

5.1.1 Formulation of the Conjecture

Let G be a group and C_r^*G the corresponding reduced group C^*-algebra; recall that C_r^*G is the norm-closure of the complex group algebra $\mathbf{C}G$ under the left regular representation L of the latter on the Hilbert space $\ell^2 G$ (cf. §1.1.2.III). We note that C_r^*G is a subalgebra of the von Neumann algebra $\mathcal{N}G$ and hence we may consider the canonical trace

$$\tau : C_r^*G \longrightarrow \mathbf{C} ,$$

which is defined by letting $\tau(a) = <a(\delta_1), \delta_1>$ for all $a \in C_r^*G$ (cf. Proposition 3.11). It is clear that τ is continuous, whereas its restriction to the subalgebra $\mathbf{C}G \simeq L(\mathbf{C}G) \subseteq C_r^*G$ is the trace functional r_1. In view of Proposition 1.40(ii), the composition

$$K_0(\mathbf{C}G) \xrightarrow{K_0(L)} K_0(C_r^*G) \xrightarrow{\tau_*} \mathbf{C} ,$$

where the first arrow is the additive map between the K_0-groups induced by the algebra homomorphism L and the second one the additive map induced by the trace τ (cf. §1.1.4.I), coincides with the additive map r_{1*} induced by the trace r_1.

Remarks 5.1 (i) Let G be a group and $H \leq G$ a finite subgroup of order n. Then, $e = \frac{1}{n}\sum\{g : g \in H\} \in \mathbf{C}G$ is an idempotent and $r_1(e) = \frac{1}{n}$. It follows that the image of r_{1*} (and, a fortiori, that of τ_*) contains the subgroup of $\mathbf{C}$ generated by the inverses of the orders of the finite subgroups of G. In particular, we have

$$\mathbf{Z} \subseteq \operatorname{im}\left[K_0(\mathbf{C}G) \xrightarrow{r_{1*}} \mathbf{C}\right] \subseteq \operatorname{im}\left[K_0(C_r^*G) \xrightarrow{\tau_*} \mathbf{C}\right] \subseteq \mathbf{C} .$$

(ii) If G is a group and E an idempotent matrix with entries in $\mathbf{C}G$, then Zaleskii's theorem (Theorem 3.19) asserts that $r_1(E)$ is a rational number. Therefore,

$$\operatorname{im}\left[K_0(\mathbf{C}G) \xrightarrow{r_{1*}} \mathbf{C}\right] \subseteq \mathbf{Q} .$$

(iii) Since the reduced C^*-algebra C_r^*G of a group G is a subalgebra of the von Neumann algebra $\mathcal{N}G$, we may invoke Kaplansky's positivity theorem (Theorem 3.12) in order to conclude that the image of the additive map $\tau_* : K_0(C_r^*G) \longrightarrow \mathbf{C}$ is a subgroup of $(\mathbf{R}, +)$.

The following conjecture provides a prediction for the image of the additive map τ_*, at least in the case of a torsion-free group. (The situation is more complicated for groups with torsion; see the Notes at the end of the chapter.)

The integrality of the trace conjecture: If G is a torsion-free group then $\mathrm{im}\left[K_0(C_r^*G) \xrightarrow{\tau_*} \mathbf{C}\right] = \mathbf{Z} \subseteq \mathbf{C}$.

In view of Kaplansky's positivity theorem, the above conjecture is stronger than the idempotent conjecture for the reduced C^*-algebra of a torsion-free group, as we now explain.

Proposition 5.2 *If G is a torsion-free group satisfying the integrality of the trace conjecture, then the C^*-algebra C_r^*G has no idempotents $\neq 0, 1$.*

Proof. If $e \in C_r^*G$ is an idempotent then $\tau(e)$ is a real number contained in the interval $[0, 1]$, in view of Theorem 3.12. Since $\tau(e) \in \mathrm{im}\,\tau_* = \mathbf{Z}$, it follows that $\tau(e) = 0$ or 1; then, $e = 0$ or 1, in view of the final assertion of loc.cit. □

5.1.2 The Case of an Abelian Group

Our first goal is to prove that torsion-free abelian groups satisfy the integrality of the trace conjecture. We consider an abelian group G that will remain fixed throughout this subsection. The strategy of the proof consists in reformulating the conjecture in terms of the dual group of G.

I. THE C^*-ALGEBRA C_r^*G AND THE DUAL GROUP. Let $\widehat{G}$ be the set of group homomorphisms from G to the circle group $S^1 = \{z \in \mathbf{C} : |z| = 1\}$, i.e. define

$$\widehat{G} = \mathrm{Hom}(G, S^1) \ .$$

Then, $\widehat{G}$ is a group with multiplication defined pointwise; in fact, $\widehat{G}$ is a subgroup of the direct product group $(S^1)^G = \prod_{g \in G} S^1$. The usual topology on S^1 induces the structure of a compact topological group on $(S^1)^G$ (Tychonoff's theorem). We note that for any two elements $g, g' \in G$ the subgroup $\Lambda_{g,g'} = \{f \in (S^1)^G : f(gg') = f(g)f(g')\} \subseteq (S^1)^G$ is closed (and hence compact). It follows that $\widehat{G} = \bigcap_{g,g' \in G} \Lambda_{g,g'}$ is a compact topological group; as such, $\widehat{G}$ is referred to as the dual group or character group of G. The elements $\chi \in \widehat{G}$ are the characters of G.

Examples 5.3 (i) The dual group $\widehat{\mathbf{Z}}$ of the infinite cyclic group $\mathbf{Z}$ can be identified with S^1, by means of the map $\chi \mapsto \chi(1)$, $\chi \in \widehat{\mathbf{Z}}$.

(ii) Let n be a positive integer. Then, the dual group $(\mathbf{Z}/n\mathbf{Z})\widehat{}$ of the finite cyclic group $\mathbf{Z}/n\mathbf{Z}$ can be identified with the subgroup of S^1 consisting of the n-th roots of unity (which is itself isomorphic with $\mathbf{Z}/n\mathbf{Z}$), by means of the map $\chi \mapsto \chi(\overline{1})$, $\chi \in (\mathbf{Z}/n\mathbf{Z})\widehat{}$.

(iii) Let $(G_i)_i$ be a family of abelian groups and $G = \bigoplus_i G_i$ the corresponding direct sum. Then, the dual group $\widehat{G}$ can be identified with the direct product $\prod_i \widehat{G_i}$ of the family $\left(\widehat{G_i}\right)_i$. Under this identification, a character χ of G corresponds to the family $(\chi_i)_i$, where χ_i is the restriction of χ

to the subgroup $G_i \subseteq G$ for all i. Using this principle, together with (i) and (ii) above, one can determine the dual group of any finitely generated abelian group.

(iv) Let $(G_i)_i$ be an inductive system of abelian groups, $G = \lim\limits_{\longrightarrow i} G_i$ the corresponding direct limit and $(\lambda_i : G_i \longrightarrow G)_i$ the canonical maps. Then, the dual group $\widehat{G}$ can be identified with the inverse limit $\lim\limits_{\longleftarrow i} \widehat{G_i}$ of the projective system $\left(\widehat{G_i}\right)_i$. Under this identification, a character χ of G corresponds to the compatible family $(\chi_i)_i$, where $\chi_i = \chi \circ \lambda_i$ for all i. Using this principle, together with (iii) above, one can determine the dual group of any abelian group, by expressing it as the directed union of its finitely generated subgroups (cf. Exercise 5.3.1).

Let $C\left(\widehat{G}\right)$ be the algebra of continuous complex-valued functions on the dual group $\widehat{G}$. We endow $\widehat{G}$ with its normalized Haar measure μ. The measure μ is the unique regular Borel probability measure on $\widehat{G}$, which is translation invariant, in the sense that

$$\int_{\widehat{G}} f(\chi)\, d\mu(\chi) = \int_{\widehat{G}} f(\chi_0 \chi)\, d\mu(\chi)$$

for all $f \in C\left(\widehat{G}\right)$ and $\chi_0 \in \widehat{G}$. The existence of such a measure μ on $\widehat{G}$ is a key result, which, besides being important in its own right, will turn out to be very useful for our purposes. The construction of μ is a standard topic, which can be found in books on Functional Analysis such as Rudin's (cf. [60, Theorem 5.14]).

For any element $g \in G$ we consider the evaluation map $\mathrm{ev}_g : \widehat{G} \longrightarrow S^1$, which is given by $\chi \mapsto \chi(g)$, $\chi \in \widehat{G}$. It is clear that ev_g is a continuous group homomorphism. We denote by $\mathrm{Hom}\left(\widehat{G}, S^1\right)$ the set of continuous group homomorphisms from $\widehat{G}$ to S^1. This set is a group with multiplication defined pointwise, whereas the map

$$\mathrm{ev} : G \longrightarrow \mathrm{Hom}\left(\widehat{G}, S^1\right), \tag{5.1}$$

which is given by $g \mapsto \mathrm{ev}_g$, $g \in G$, is a group homomorphism.

Lemma 5.4 *(i) For any continuous group homomorphism $\omega : \widehat{G} \longrightarrow S^1$ the value of the integral $\int_{\widehat{G}} \omega(\chi)\, d\mu(\chi)$ is 1 (resp. 0) if $\omega = 1$ (resp. if $\omega \neq 1$).*

(ii) The set $\mathrm{Hom}\left(\widehat{G}, S^1\right)$ of all continuous group homomorphisms from $\widehat{G}$ to S^1 is an orthonormal subset of the Hilbert space $L^2\left(\widehat{G}\right)$ of square-integrable functions on $\widehat{G}$ (with respect to the Haar measure μ).

Proof. (i) If $\omega = 1$ then $\omega(\chi) = 1$ for all $\chi \in \widehat{G}$ and the assertion follows since μ is a probability measure. If $\omega \neq 1$ then $\omega(\chi_0) \neq 1$ for some $\chi_0 \in \widehat{G}$ and hence

$$\int_{\widehat{G}} \omega(\chi) \, d\mu(\chi) = \int_{\widehat{G}} \omega(\chi_0 \chi) \, d\mu(\chi)$$
$$= \int_{\widehat{G}} \omega(\chi_0) \omega(\chi) \, d\mu(\chi)$$
$$= \omega(\chi_0) \int_{\widehat{G}} \omega(\chi) \, d\mu(\chi) \, .$$

It follows that $\int_{\widehat{G}} \omega(\chi) \, d\mu(\chi) = 0$, as needed.

(ii) If $\omega_1, \omega_2 : \widehat{G} \longrightarrow S^1$ are continuous group homomorphisms, we compute

$$<\omega_1, \omega_2> = \int_{\widehat{G}} \omega_1(\chi) \overline{\omega_2(\chi)} \, d\mu(\chi)$$
$$= \int_{\widehat{G}} \omega_1(\chi) \omega_2(\chi)^{-1} \, d\mu(\chi)$$
$$= \int_{\widehat{G}} (\omega_1 \omega_2^{-1})(\chi) \, d\mu(\chi) \, .$$

Therefore, the result follows from (i) above.   □

We now consider the composition

$$G \xrightarrow{ev} \mathrm{Hom}\left(\widehat{G}, S^1\right) \hookrightarrow C\left(\widehat{G}\right),$$

which extends uniquely to an algebra homomorphism

$$\mathcal{F}_0 : \mathbf{C}G \longrightarrow C\left(\widehat{G}\right) . \tag{5.2}$$

Since $\chi(g^{-1}) = \chi(g)^{-1} = \overline{\chi(g)}$ for all $\chi \in \widehat{G}$ and all $g \in G$, it follows that $\mathcal{F}_0$ is a $*$-algebra homomorphism.

Lemma 5.5 *Let $\mathcal{F}_0$ be the $*$-algebra homomorphism defined above.*

(i) The image $\mathcal{F}_0(\mathbf{C}G)$ is uniformly dense in $C\left(\widehat{G}\right)$.

(ii) The functions $ev_g = \mathcal{F}_0(g)$, $g \in G$, form an orthonormal basis of the Hilbert space $L^2\left(\widehat{G}\right)$ of square-integrable functions on $\widehat{G}$.

(iii) If $L : \mathbf{C}G \longrightarrow B(\ell^2 G)$ is the $*$-algebra homomorphism associated with the action of $\mathbf{C}G$ on the Hilbert space $\ell^2 G$ by left translations, then $\| L_a \| = \| \mathcal{F}_0(a) \|_\infty$ for any $a \in \mathbf{C}G$.

Proof. (i) Since $\mathcal{F}_0(\mathbf{C}G)$ is a $*$-subalgebra of $C\left(\widehat{G}\right)$, which contains the constant functions and separates the points of $\widehat{G}$, the result follows from the Stone-Weierstrass theorem.

(ii) We note that $C\left(\widehat{G}\right)$ is a dense subspace of $L^2\left(\widehat{G}\right)$. Since the supremum norm of a continuous function dominates its L^2-norm, it follows from (i) above that $\mathcal{F}_0(\mathbf{C}G)$ is dense in $L^2\left(\widehat{G}\right)$ as well. Lemma 5.4(ii) implies that $\{ev_g : g \in G\} \subseteq \mathrm{Hom}\left(\widehat{G}, S^1\right) \subseteq L^2\left(\widehat{G}\right)$ is an orthonormal set. Its linear span being dense in $L^2\left(\widehat{G}\right)$, this set is an orthonormal basis.

(iii) In view of (ii) above, we may consider the isometry

$$U : \ell^2 G \longrightarrow L^2\left(\widehat{G}\right),$$

which is given by $\delta_g \mapsto ev_g$, $g \in G$; here, $(\delta_g)_g$ is the canonical orthonormal basis of $\ell^2 G$. For any elements $g, x \in G$ we compute

$$U L_g U^{-1}(ev_x) = U L_g(\delta_x) = U(\delta_{gx}) = ev_{gx} = ev_g ev_x = M_{ev_g}(ev_x),$$

where M_{ev_g} denotes the multiplication operator on $L^2\left(\widehat{G}\right)$ associated with the continuous function $ev_g \in C\left(\widehat{G}\right)$. It follows that the operators $U L_g U^{-1}$ and M_{ev_g} agree on the orthonormal basis $\{ev_x : x \in G\}$ of $L^2\left(\widehat{G}\right)$; hence, these operators are equal for all $g \in G$. By linearity, it follows that $U L_a U^{-1} = M_{\mathcal{F}_0(a)}$ for any element $a \in \mathbf{C}G$. Since U is an isometry, we have $\| L_a \| = \| U L_a U^{-1} \| = \| M_{\mathcal{F}_0(a)} \|$ for any $a \in \mathbf{C}G$. On the other hand, for any $f \in C\left(\widehat{G}\right)$ the associated multiplication operator M_f on $L^2\left(\widehat{G}\right)$ has norm equal to the supremum norm $\| f \|_\infty$ of f (cf. Exercise 5.3.2). It follows that $\| L_a \| = \| \mathcal{F}_0(a) \|_\infty$ for all $a \in \mathbf{C}G$, as needed. $\square$

As a first consequence, we obtain the following duality result.

Theorem 5.6 *(Pontryagin) The group homomorphism ev of (5.1) is an isomorphism.*

Proof. The surjectivity of ev is an immediate consequence of Lemmas 5.4(ii) and 5.5(ii). Therefore, it only remains to prove that the map ev is injective. To that end, we note that the group S^1 contains a copy of all cyclic groups. In particular, for any element $g \in G \setminus \{1\}$ there is a group homomorphism ψ from the cyclic subgroup $< g > \subseteq G$ to S^1 with $\psi(g) \neq 1$. Since S^1 is a divisible abelian group, we can extend ψ to a character χ of G. Then, $ev_g(\chi) = \chi(g) = \psi(g) \neq 1$ and hence $ev_g \neq 1$. $\square$

We can now provide a concrete description of the reduced C^*-algebra of G, in terms of the dual group $\widehat{G}$.

Theorem 5.7 *The $*$-algebra homomorphism $\mathcal{F}_0$ of (5.2) induces a C^*-algebra isomorphism $\mathcal{F} : C_r^* G \longrightarrow C\left(\widehat{G}\right)$.*

Proof. This is an immediate consequence of Lemma 5.5(i),(iii), which implies that the composition

$$L(\mathbf{C}G) \xrightarrow{L^{-1}} \mathbf{C}G \xrightarrow{\mathcal{F}_0} C\left(\widehat{G}\right)$$

is an isometric $*$-algebra homomorphism with dense image. $\square$

II. THE CONNECTEDNESS OF THE DUAL GROUP. The C^*-algebra isomorphism of Theorem 5.7 reduces the study of idempotent elements of C_r^*G to the study of clopen subsets of $\widehat{G}$. Indeed, it is clear that the idempotent elements of the algebra $C\left(\widehat{G}\right)$ are precisely the characteristic functions of the clopen subsets $Y \subseteq \widehat{G}$. In particular, C_r^*G has no idempotents other than 0 and 1 if and only if $\widehat{G}$ has no clopen subsets other than $\emptyset$ and $\widehat{G}$. We are therefore lead to examine the connectedness of $\widehat{G}$.

We recall that a topological space X is called totally disconnected if its connected components are singletons or, equivalently, if any subspace $Y \subseteq X$ with more than one element is not connected. The space X is called 0-dimensional if there is a basis of its topology consisting of clopen sets. It is easily seen that a 0-dimensional Hausdorff space is totally disconnected. We prove a partial converse of that assertion in Corollary 5.9 below.

Lemma 5.8 *Let X be a topological space which is compact and Hausdorff.*

(i) Assume that $Y \subseteq X$ is a closed subspace and let $x \in X$ be a point which can be separated from any point $y \in Y$ by a clopen set. Then, there is a clopen set Z with $Y \subseteq Z$, such that $x \notin Z$.

(ii) For any $x \in X$ consider the subspace $W = W(x) \subseteq X$ that consists of all points $w \in X$ that can't be separated from x by a clopen set. Then, W is connected.

Proof. (i) For any $y \in Y$ there is a clopen set Z_y with $y \in Z_y$ and $x \notin Z_y$. The open cover $(Z_y)_{y \in Y}$ of the closed (and hence compact) subspace Y has a finite subcover; hence, there are finitely many elements $y_1, \ldots, y_n \in Y$, such that Y is a subset of the union $Z = \bigcup_{i=1}^n Z_{y_i}$. Of course, Z is a clopen set and $x \notin Z$.

(ii) It is easily seen that W is a closed subspace containing x. Assuming that W is not connected, we can find two non-empty disjoint subsets W_1 and W_2 which are closed (in W and hence in X), such that $W = W_1 \cup W_2$. Without any loss of generality, we may assume that $x \in W_1$. Being compact and Hausdorff, the topological space X is normal; therefore, there is an open set U, such that $W_1 \subseteq U$ and $W_2 \cap \overline{U} = \emptyset$. It follows that the closed set $\overline{U} \setminus U$ intersects W trivially and hence any point $y \in \overline{U} \setminus U$ can be separated from x by a clopen set. In view of (i) above, there is a clopen set Z, such that $\overline{U} \setminus U \subseteq Z$ and $x \notin Z$. Then, the set $U \setminus Z = \overline{U} \setminus Z$ is clopen, contains x and is disjoint from W_2. This is a contradiction, since the points of the non-empty set $W_2 \subseteq W$ cannot be separated from x by a clopen set. $\square$

Corollary 5.9 *A compact Hausdorff and totally disconnected space X is 0-dimensional.*

Proof. We fix an element $x \in X$ and let V be an open neighborhood of it. Since the connected component of x is $\{x\}$, it follows from Lemma 5.8(ii), applied to the case of the compact space $\overline{V}$, that any point $y \in \overline{V}$ with $y \neq x$ can be separated from x by a set $Z_y \subseteq \overline{V}$, which is clopen in $\overline{V}$. We now invoke Lemma 5.8(i), applied to the case of the closed subset $\overline{V} \setminus V$ of the compact space $\overline{V}$, in order to conclude that there is a subset $Z \subseteq \overline{V}$, which is clopen in $\overline{V}$, such that $\overline{V} \setminus V \subseteq Z$ and $x \notin Z$. Then, $U = \overline{V} \setminus Z = V \setminus Z$ is a neighborhood of x with $U \subseteq V$, which is clopen in X. Hence, clopen sets form a basis for the topology of X, as needed.   □

Before specializing to the case of $\widehat{G}$, we state and prove a few general results about topological groups.

Lemma 5.10 *Let Γ be a topological group.*

(i) Assume that K, U are subsets of Γ with K compact, U open and $K \subseteq U$. Then, there exists an open neighborhood V of the identity, such that $KV \subseteq U$.

(ii) Assume that Δ is a subgroup of Γ containing a non-empty open subset V. Then, Δ is open.

(iii) If $U \subseteq \Gamma$ is an open and compact neighborhood of the identity, then U contains an open subgroup Δ.

(iv) If $\Delta \trianglelefteq \Gamma$ is a closed normal subgroup, then the quotient group $\overline{\Gamma} = \Gamma / \Delta$, endowed with the quotient topology, is Hausdorff.

Proof. (i) For any $\gamma \in K$ let W_γ be an open neighborhood of the identity with $\gamma W_\gamma \subseteq U$. We choose an open neighborhood V_γ of the identity with $V_\gamma^2 \subseteq W_\gamma$ and consider the open cover $(\gamma V_\gamma)_\gamma$ of K. By compactness, there are finitely many elements $\gamma_1, \dots, \gamma_n \in K$, such that $K \subseteq \bigcup_{i=1}^n \gamma_i V_{\gamma_i}$. Then, $V = \bigcap_{i=1}^n V_{\gamma_i}$ is an open neighborhood of the identity with

$$ KV \subseteq \bigcup_{i=1}^n \gamma_i V_{\gamma_i} V \subseteq \bigcup_{i=1}^n \gamma_i V_{\gamma_i}^2 \subseteq \bigcup_{i=1}^n \gamma_i W_{\gamma_i} \subseteq U \,, $$

as needed.

(ii) Let us fix an element $\gamma_0 \in V$. Then, $V_0 = \gamma_0^{-1} V$ is an open neighborhood of the identity and for any $\gamma \in \Delta$ we have $\gamma V_0 = \gamma \gamma_0^{-1} V \subseteq \gamma \gamma_0^{-1} \Delta = \Delta$; hence, Δ is open.

(iii) In view of (i) above, we can choose an open neighborhood V of the identity with $V = V^{-1}$, $V \subseteq U$ and $UV \subseteq U$. Then, $V^2 \subseteq UV \subseteq U$ and an inductive argument shows that $V^n = V^{n-1} V \subseteq UV \subseteq U$ for all $n \geq 1$. Therefore, $\Delta = \bigcup_{n=1}^\infty V^n$ is a subgroup contained in U. Finally, (ii) shows that Δ is open.

(iv) Let $\pi : \Gamma \longrightarrow \overline{\Gamma}$ be the quotient map, $e \in \Gamma$ the identity element and $\gamma \in \Gamma \setminus \Delta$. We have to prove that the elements $\pi(e)$ and $\pi(\gamma)$ of $\overline{\Gamma}$ can be separated by disjoint open sets. To that end, we consider an open neighborhood V of e in Γ, such that $V^{-1} V \subseteq \Gamma \setminus \gamma \Delta$. Then, the open subsets

$V\Delta$ and $V\gamma\Delta$ are disjoint. Indeed, if $\gamma_1, \gamma_2 \in V$ and $\delta_1, \delta_2 \in \Delta$ are such that $\gamma_1\delta_1 = \gamma_2\gamma\delta_2$, then $\gamma_2^{-1}\gamma_1 = \gamma\delta_2\delta_1^{-1}$ is an element in the intersection $V^{-1}V \cap \gamma\Delta$, which is assumed to be empty. Since $V\Delta \cap V\gamma\Delta = \emptyset$, the subsets $\pi(V)$ and $\pi(V\gamma)$ of $\overline{\Gamma}$ are disjoint. The proof is finished, since $\pi(V)$ (resp. $\pi(V\gamma)$) is an open neighborhood of $\pi(e)$ (resp. of $\pi(\gamma)$) in $\overline{\Gamma}$. $\square$

Using the above results, we can determine the structure of the connected component of the identity in a topological group.

Proposition 5.11 *Let Γ be a topological group and consider the connected component Γ_0 of the identity element $e \in \Gamma$. Then:*

  (i) Γ_0 is a closed normal subgroup of Γ.

  (ii) The quotient group $\overline{\Gamma} = \Gamma/\Gamma_0$ is Hausdorff and totally disconnected.

  (iii) If Γ is compact then Γ_0 is the intersection of the open subgroups Δ of Γ.

Proof. (i) Since inversion is a homeomorphism of Γ, the set Γ_0^{-1} is connected. Moreover, we have $e \in \Gamma_0^{-1}$ and hence $\Gamma_0^{-1} \subseteq \Gamma_0$. Therefore, for any element $\gamma \in \Gamma_0$ we also have $\gamma^{-1} \in \Gamma_0$ and hence the connected set $\gamma\Gamma_0$ contains e. But then $\gamma\Gamma_0 \subseteq \Gamma_0$ and hence Γ_0 is a subgroup. In order to show that Γ_0 is a normal subgroup, we consider an element $x \in \Gamma$ and note that conjugation by x is a homeomorphism. It follows that $x\Gamma_0 x^{-1}$ is a connected subset containing e; hence, $x\Gamma_0 x^{-1} \subseteq \Gamma_0$. Finally, being a connected component, Γ_0 is closed.

(ii) Taking into account Lemma 5.10(iv), it follows from (i) above that the quotient group $\overline{\Gamma}$ is Hausdorff. We consider the projection map $\pi : \Gamma \longrightarrow \overline{\Gamma}$ and let X be a subspace of Γ, such that $\pi(X)$ contains strictly the singleton $\{\pi(e)\}$. Then, $X\Gamma_0$ is a subspace of Γ that contains strictly Γ_0 and hence $X\Gamma_0$ is not connected. Therefore, there exist open subsets $V_1, V_2 \subseteq \Gamma$, such that $X\Gamma_0 \subseteq V_1 \cup V_2$, with $X\Gamma_0 \cap V_1 \neq \emptyset$, $X\Gamma_0 \cap V_2 \neq \emptyset$ and $X\Gamma_0 \cap V_1 \cap V_2 = \emptyset$. Then, $\pi(X)$ meets non-trivially the open subsets $\pi(V_i)$, $i = 1, 2$, and is contained in their union. In order to show that $\pi(X)$ is not connected, it remains to show that the intersection $\pi(X) \cap \pi(V_1) \cap \pi(V_2)$ is empty. To that end, we note that for any $\gamma \in X$ the connected set $\gamma\Gamma_0$ is contained in the union $V_1 \cup V_2$ and hence we have $\gamma\Gamma_0 \subseteq V_i$ for some i. Therefore, both sets $X\Gamma_0 \cap V_1$ and $X\Gamma_0 \cap V_2$ are unions of Γ_0-cosets. Since the intersection $X\Gamma_0 \cap V_1 \cap V_2$ is empty, it follows easily that the same is true for the intersection $\pi(X) \cap \pi(V_1) \cap \pi(V_2)$.

(iii) If Δ is an open subgroup of Γ then Δ and $\bigcup\{\gamma\Delta : \gamma \notin \Delta\}$ are disjoint open subsets covering the connected set Γ_0. Since e is contained in both Δ and Γ_0, we must have $\Gamma_0 \subseteq \Delta$. We now consider an element $\gamma \in \Gamma$ with $\gamma \notin \Gamma_0$. Then, $\pi(\gamma) \neq \pi(e) \in \overline{\Gamma} = \Gamma/\Gamma_0$, where $\pi : \Gamma \longrightarrow \overline{\Gamma}$ is the projection map. In view of the compactness assumption on Γ and (ii) above, the space $\overline{\Gamma}$ is Hausdorff, compact and totally disconnected; hence, it is 0-dimensional (cf. Corollary 5.9). Therefore, there exists a clopen neighborhood U of $\pi(e)$ with $\pi(\gamma) \notin U$. Since U is compact (being closed), Lemma 5.10(iii) implies the existence of an open subgroup of $\overline{\Gamma}$ that doesn't contain $\pi(\gamma)$. The inverse image of that subgroup under the homomorphism π is an open subgroup Δ of Γ with $\gamma \notin \Delta$. □

We now return to our previous discussion about the abelian group G and its dual group $\widehat{G}$. The following result relates the triviality of torsion elements in G to the connectedness of $\widehat{G}$ and hence to the triviality of idempotents in the C^*-algebra $C_r^* G$.

Theorem 5.12 *The following conditions are equivalent:*

(i) *G is torsion-free,*

(ii) *$\widehat{G}$ is connected and*

(iii) *$C_r^* G$ has no non-trivial idempotents.*

Proof. (i)→(ii): Assume that the dual group $\widehat{G}$ is not connected. Then, Proposition 5.11(iii) implies the existence of a proper open subgroup $\Delta \leq \widehat{G}$. Since Δ is open, the quotient group $\widehat{G}/\Delta$ is discrete. On the other hand, being an epimorphic image of the compact group $\widehat{G}$, the quotient $\widehat{G}/\Delta$ is compact. It follows that the non-trivial abelian group $\widehat{G}/\Delta$ is finite; hence, it admits an epimorphism onto the group $\mathbf{Z}/n\mathbf{Z}$ for some $n > 1$. Embedding $\mathbf{Z}/n\mathbf{Z}$ into S^1 as the group of n-th roots of unity, we obtain a continuous homomorphism $\omega : \widehat{G} \longrightarrow S^1$, as the composition

$$\widehat{G} \longrightarrow \widehat{G}/\Delta \longrightarrow \mathbf{Z}/n\mathbf{Z} \hookrightarrow S^1 \; .$$

It is clear that $\omega \in \mathrm{Hom}\left(\widehat{G}, S^1\right)$ is an element of order n. In view of the Pontryagin duality theorem (Theorem 5.6), there exists an element $g \in G$ of order n, such that $\omega = \mathrm{ev}_g$. This is a contradiction, since G is assumed to be torsion-free.

(ii)→(i): We now assume that the dual group $\widehat{G}$ is connected and consider an element $g \in G$ of finite order n. Let Δ be the kernel of the continuous homomorphism $\mathrm{ev}_g : \widehat{G} \longrightarrow S^1$. The group Δ consists of those characters $\chi \in \widehat{G}$, for which the subgroup $\{\chi(g^i) : i = 0, 1, \ldots, n-1\} \subseteq S^1$ is trivial. It is easily seen that a subgroup $A \subseteq S^1$ is trivial if and only if $|z - 1| < \sqrt{3}$ for any $z \in A$ (cf. Exercise 5.3.3). Therefore, we conclude that

$$\Delta = \{\chi \in \widehat{G} : |\chi(g^i) - 1| < \sqrt{3} \text{ for all } i = 0, 1, \ldots, n-1\} \; .$$

It follows that Δ is an open subgroup of $\widehat{G}$. Since $\widehat{G}$ is connected, we must have $\Delta = \widehat{G}$ and hence $\mathrm{ev}_g = 1$. Then $g = 1$, in view of Theorem 5.6.

As we have already noted before, the equivalence (ii)↔(iii) follows from the existence of the isomorphism $C_r^* G \simeq C\left(\widehat{G}\right)$ of Theorem 5.7. □

Remark 5.13 In view of the Pontryagin duality theorem, the topological group $\widehat{G}$ determines (and is, of course, determined by) the abelian group G. In this way, every algebraic property of G corresponds to a certain property of the topological group $\widehat{G}$. It is in that spirit that one should view the equivalence of assertions (i) and (ii) in Theorem 5.12.[1]

III. THE INTEGRALITY OF THE TRACE. Having verified the C^*-algebraic version of the idempotent conjecture for torsion-free abelian groups, we now prove that these groups satisfy the integrality of the trace conjecture. To that end, we note that the Haar measure μ on the dual group $\widehat{G}$ induces a continuous linear functional

$$\overline{\tau} : C\left(\widehat{G}\right) \longrightarrow \mathbf{C} ,$$

which is given by $f \mapsto \int_{\widehat{G}} f(\chi)\, d\mu(\chi)$, $f \in C\left(\widehat{G}\right)$. Since the algebra $C\left(\widehat{G}\right)$ is commutative, $\overline{\tau}$ is a trace. The following result shows that the isomorphism of Theorem 5.7 identifies $\overline{\tau}$ with the canonical trace τ.

Proposition 5.14 *The linear functionals $\overline{\tau} \circ \mathcal{F}$ and τ are equal.*

Proof. We recall that $\mathcal{F}(L_g) = \mathrm{ev}_g$ for all $g \in G$. The C^*-algebra $C_r^* G$ being the closed linear span of the L_g's, it suffices to verify that $(\overline{\tau} \circ \mathcal{F})(L_g) = \tau(L_g)$, i.e. that $\overline{\tau}(\mathrm{ev}_g) = \tau(L_g)$ for all $g \in G$. Since $\mathrm{ev}_g = 1$ if and only if $g = 1$ (cf. Theorem 5.6), Lemma 5.4(i) implies that $\overline{\tau}(\mathrm{ev}_g) = \int_{\widehat{G}} \mathrm{ev}_g(\chi)\, d\mu(\chi)$ is equal to 1 if $g = 1$ and vanishes if $g \neq 1$, as needed. □

Corollary 5.15 *The additive map $\tau_* : K_0(C_r^* G) \longrightarrow \mathbf{C}$, which is induced by the canonical trace τ, coincides with the composition*

$$K_0(C_r^* G) \xrightarrow{K_0(\mathcal{F})} K_0\left(C\left(\widehat{G}\right)\right) \xrightarrow{\overline{\tau}_*} \mathbf{C} ,$$

where $K_0(\mathcal{F})$ is the isomorphism between the K_0-groups induced by the C^*-algebra isomorphism $\mathcal{F}$ and $\overline{\tau}_*$ the additive map induced by the trace $\overline{\tau}$.

Proof. This is an immediate consequence of Proposition 5.14, in view of Proposition 1.40(ii). □

Having established the equality $\tau_* = \overline{\tau}_* \circ K_0(\mathcal{F})$, we can prove the main result of this subsection.

Theorem 5.16 *Torsion-free abelian groups satisfy the integrality of the trace conjecture.*

Proof. Let G be a torsion-free abelian group. In order to show that the image of the additive map $\tau_* : K_0(C_r^* G) \longrightarrow \mathbf{C}$ is the subgroup $\mathbf{Z} \subseteq \mathbf{C}$, it suffices (in view of Corollary 5.15) to show that this is the case for the image of the additive map $\overline{\tau}_* : K_0\left(C\left(\widehat{G}\right)\right) \longrightarrow \mathbf{C}$. To that end, we fix an idempotent

---

[1] As another illustration of this principle, one can prove that an abelian group is a torsion group if and only if its dual group is 0-dimensional (cf. [33, Theorem 24.26]).

matrix $E = (e_{ij})_{i,j}$ with entries in $C\left(\widehat{G}\right)$. For any $\chi \in \widehat{G}$ the matrix $E(\chi) = (e_{ij}(\chi))_{i,j}$ is an idempotent matrix with entries in **C** and hence its trace $\mathrm{tr}(E(\chi))$ is a non-negative integer. Therefore, the continuous function $\mathrm{tr}(E) = \sum_i e_{ii} \in C\left(\widehat{G}\right)$ maps $\widehat{G}$ into the discrete space **Z**. In view of Theorem 5.12, the space $\widehat{G}$ is connected; hence, the map $\mathrm{tr}(E)$ must be constant. It follows that there exists an integer n, such that $\mathrm{tr}(E(\chi)) = n$ for all $\chi \in \widehat{G}$. We now compute

$$\overline{\tau}_*[E] = \overline{\tau}(\mathrm{tr}(E)) = \int_{\widehat{G}} \mathrm{tr}(E(\chi))\, d\mu(\chi) = \int_{\widehat{G}} n\, d\mu(\chi) = n \int_{\widehat{G}} d\mu(\chi) = n \in \mathbf{Z}$$

and this finishes the proof. ☐

5.1.3 The Case of a Free Group

Our main goal in the present subsection is to prove that free groups satisfy the integrality of the trace conjecture as well. To that end, we consider a free action of such a group on a tree and study the induced representations of the reduced group C^*-algebra.

I. GRAPHS, TREES AND GROUP ACTIONS. An oriented graph X consists of a set V (whose elements are called vertices), a set E^{or} (whose elements are called oriented edges) and maps

$$o : E^{or} \longrightarrow V, \ t : E^{or} \longrightarrow V \ \text{and} \ r : E^{or} \longrightarrow E^{or},$$

which are such that $r \circ r = \mathrm{id}$ and $t \circ r = o$. For any oriented edge $e \in E^{or}$ the vertex $o(e)$ (resp. $t(e)$) is called the origin (resp. terminus) of e. The involution r is said to reverse orientation and the oriented edges e and $r(e)$ are said to have opposite orientation (or, simply, to be opposite to each other). The set E of un-oriented edges of X is the quotient of E^{or} modulo the equivalence relation generated by the relations $e \sim r(e)$, $e \in E^{or}$.

A path p on the graph X is a sequence of edges $p = (e_1, \ldots, e_n)$, such that $t(e_i) = o(e_{i+1})$ for all $i = 1, \ldots, n-1$. We say that a path p as above has origin $o(e_1)$, terminus $t(e_n)$ and passes through the vertices $t(e_i) = o(e_{i+1})$, $i = 1, \ldots, n-1$. The path $p = (e_1, \ldots, e_n)$ is reduced if $e_{i+1} \neq r(e_i)$ for all $i = 1, \ldots, n-1$. A loop around a vertex $v \in V$ is a path with origin and terminus at v. The graph X is called connected if for any two vertices $v, v' \in V$ there is a path (or, equivalently, a reduced path) p with origin v and terminus v'. A graph X is called a tree if it is connected and has no reduced loops. Equivalently, a graph X is a tree if and only if for any two vertices $v, v' \in V$ with $v \neq v'$ there is a unique reduced path p with origin v and terminus v'; this path, denoted by $[v, v']$, is called the geodesic joining v and v'.

The number of vertices of a tree exceeds the number of un-oriented edges of it by 1. In order to make that assertion precise, we let X be a tree and fix a vertex $v_0 \in V$. For any vertex $v \in V$ with $v \neq v_0$ we consider the geodesic $[v_0, v] = (e_1, \ldots, e_n)$ and define the map

$$\lambda : V \setminus \{v_0\} \longrightarrow E \,,$$

by letting $\lambda(v)$ be the equivalence class of the oriented edge e_n.

Lemma 5.17 *Let X be a tree and fix a vertex $v_0 \in V$.*

(i) The map λ defined above is bijective.

(ii) Let $v_0' \in V$ be another vertex and consider the corresponding map $\lambda' : V \setminus \{v_0'\} \longrightarrow E$. Then, $\lambda(v) = \lambda'(v)$ for all but finitely many vertices $v \in V \setminus \{v_0, v_0'\}$.

Proof. (i) We define a map

$$\kappa : E \longrightarrow V \setminus \{v_0\}$$

as follows: We consider an oriented edge $e \in E^{or}$ and let $v = o(e)$ and $v' = t(e)$. Since e is not a loop, at least one of the vertices v, v' is different than v_0. If one of these vertices is v_0, then we let the other vertex be the image of the un-oriented edge $\{e, r(e)\}$ under κ. We now assume that none of the vertices v, v' is v_0 and consider the geodesics $[v_0, v]$ and $[v_0, v']$. It is easily seen that the absence of reduced loops in X implies that precisely one of the following two conditions is satisfied:

- $[v_0, v]$ consists of the geodesic $[v_0, v']$ followed by $r(e)$ or
- $[v_0, v']$ consists of the geodesic $[v_0, v]$ followed by e.

In the former (resp. the latter) case, we let v (resp. v') be the image of the un-oriented edge $\{e, r(e)\}$ under κ. It is an immediate consequence of the definitions that the compositions $\kappa \circ \lambda$ and $\lambda \circ \kappa$ are the identity maps of the sets $V \setminus \{v_0\}$ and E respectively. In particular, λ is bijective.

(ii) Assume that the geodesic $[v_0, v_0']$ joining v_0 and v_0' passes through the vertices $v_1, \ldots, v_{n-1}$. We fix a vertex $v \in V \setminus \{v_0, v_1, \ldots, v_{n-1}, v_0'\}$ and note that the absence of reduced loops in X implies that precisely one of the following three conditions is satisfied:

- the geodesic $[v_0, v]$ consists of the geodesic $[v_0, v_0']$ followed by $[v_0', v]$ or
- the geodesic $[v_0', v]$ consists of the geodesic $[v_0', v_0]$ followed by $[v_0, v]$ or
- there is an index $i \in \{1, \ldots, n-1\}$, such that the geodesic $[v_0, v]$ consists of the geodesic $[v_0, v_i]$ followed by $[v_i, v]$, whereas the geodesic $[v_0', v]$ consists of the geodesic $[v_0', v_i]$ followed by $[v_i, v]$.

In any case, the geodesics $[v_0, v]$ and $[v_0', v]$ have the same final edge and hence $\lambda(v) = \lambda'(v)$. □

Let X_1, X_2 be two graphs. Then, a morphism

$$\alpha : X_1 \longrightarrow X_2$$

is a pair of maps

$$\alpha_V : V_1 \longrightarrow V_2 \quad \text{and} \quad \alpha_{E^{or}} : E_1^{or} \longrightarrow E_2^{or}$$

between the sets of vertices and oriented edges of the two graphs respectively, such that $o \circ \alpha_{E^{or}} = \alpha_V \circ o$, $t \circ \alpha_{E^{or}} = \alpha_V \circ t$ and $r \circ \alpha_{E^{or}} = \alpha_{E^{or}} \circ r$. In other words, if $e_1 \in E_1^{or}$ is an oriented edge of X_1 with origin v_1, terminus v_1' and opposite edge ε_1, then the oriented edge $\alpha_{E^{or}}(e_1) \in E_2^{or}$ has origin $\alpha_V(v_1)$, terminus $\alpha_V(v_1')$ and opposite edge $\alpha_{E^{or}}(\varepsilon_1)$. It is clear that $\alpha_{E^{or}}$ induces, by passage to the quotient, a map

$$\alpha_E : E_1 \longrightarrow E_2$$

between the un-oriented edges of the two graphs. The composition of two morphisms of graphs is also a morphism of graphs. A morphism of graphs α as above is an isomorphism of graphs if both maps α_V and $\alpha_{E^{or}}$ are bijective. An automorphism of a graph X is an isomorphism of X onto itself; it is clear that composition endows the set of these automorphisms with the structure of a group.

Let $\alpha : X_1 \longrightarrow X_2$ be an isomorphism of trees, fix a vertex $v_1 \in V_1$ of X_1 and consider the associated bijection $\lambda_1 : V_1 \setminus \{v_1\} \longrightarrow E_1$. We also consider the vertex $v_2 = \alpha_V(v_1) \in V_2$ of X_2 and the corresponding bijection $\lambda_2 : V_2 \setminus \{v_2\} \longrightarrow E_2$. Then, it is easily seen that

$$\alpha_E \circ \lambda_1 = \lambda_2 \circ \alpha_V' , \tag{5.3}$$

where α_V' denotes the restriction of α_V to the subset $V_1 \setminus \{v_1\} \subseteq V_1$.

Let G be a group and X a graph. We say that G acts on X if we are given a homomorphism of G into the group of automorphisms of X. Equivalently, the group G acts on X if it acts on the sets V and E^{or} of vertices and oriented edges of X respectively, in such a way that $o(g \cdot e) = g \cdot o(e)$, $t(g \cdot e) = g \cdot t(e)$ and $r(g \cdot e) = g \cdot r(e)$ for any group element $g \in G$ and any oriented edge $e \in E^{or}$. In that case, the group G acts on the set E of un-oriented edges of X, in such a way that the quotient map $E^{or} \longrightarrow E$ is G-equivariant. The action of G on X is said to be free if G acts freely on the sets V and E of vertices and un-oriented edges of X respectively.

Example 5.18 Let G be a group and consider a subset $S \subseteq G$. Then, the Cayley graph $X = X(G, S)$ of G with respect to S is defined as follows: The set V of vertices of X coincides with G, whereas the set E^{or} of oriented edges is the subset of the Cartesian product $G \times G$, consisting of those pairs (g, g') for which $g^{-1}g' \in S$ or $g'^{-1}g \in S$; in other words, $(g, g') \in E^{or}$ if and only if $g' = gs$ for some element $s \in G$ with $s \in S \cup S^{-1}$. Any oriented edge

$(g, g') \in E^{or}$ has origin $o(g, g') = g$ and terminus $t(g, g') = g'$, whereas its opposite edge is defined by letting $r(g, g') = (g', g)$. The following assertions are easily verified:

(i) If $1 \in G$ is the identity element, then the set of vertices that can be joined by a path with the vertex 1 consists of the vertices $g \in <S>$, where $<S>$ is the subgroup of G generated by S. In particular, the Cayley graph is connected if and only if G can be generated by S.

(ii) The set of loops around a fixed vertex of the Cayley graph is in bijective correspondence with the set of sequences $(s_1, \varepsilon_1, s_2, \varepsilon_2, \ldots, s_n, \varepsilon_n)$, where $s_i \in S$, $\varepsilon_i = \pm 1 \in \mathbf{Z}$ for all i and $\prod_{i=1}^{n} s_i^{\varepsilon_i} = 1 \in G$. In this way, the set of reduced loops corresponds to the set of sequences as above, which are such that $(s_{i+1}, \varepsilon_{i+1}) \neq (s_i, -\varepsilon_i)$ for all i. In particular, the Cayley graph has no reduced loops if and only if the set S generates freely the subgroup $<S> \subseteq G$.

(iii) It follows from (i) and (ii) above that the Cayley graph is a tree if and only if the group G is free on S.

For any pair (G, S) as above, the group G acts on the Cayley graph $X = X(G, S)$ by left translations. More precisely, for any group element $g \in G$ and any vertex $v = g'$ (resp. any oriented edge $e = (g', g'')$) we let $g \cdot v = gg'$ (resp. $g \cdot e = (gg', gg'')$). If the group G has no element of order 2, this action is easily seen to be free. In particular, we conclude that a free group acts freely on a tree.[2]

Proposition 5.19 *Let X be a tree, $v_0 \in V$ a vertex of X and*

$$\lambda : V \setminus \{v_0\} \longrightarrow E$$

the associated bijection. We consider a group G acting on X and fix an element $g \in G$. Then, $g \cdot \lambda(v) = \lambda(g \cdot v)$ for all but finitely many $v \in V \setminus \{v_0, g^{-1} \cdot v_0\}$.

Proof. We consider the vertex $g^{-1} \cdot v_0$ and let

$$\lambda' : V \setminus \{g^{-1} \cdot v_0\} \longrightarrow E$$

be the associated bijection. Since the element $g \in G$ induces an automorphism of the tree X, which maps the vertex $g^{-1} \cdot v_0$ onto v_0, it follows from (5.3) that $g \cdot \lambda'(v) = \lambda(g \cdot v)$ for all $v \in V \setminus \{g^{-1} \cdot v_0\}$. This finishes the proof, since $\lambda(v) = \lambda'(v)$ for all but finitely many vertices $v \in V \setminus \{v_0, g^{-1} \cdot v_0\}$ (cf. Lemma 5.17(ii)). $\qquad\square$

II. FREE ACTIONS AND THE ASSOCIATED REPRESENTATIONS OF $C_r^* G$. Let G be a group acting on a graph X. Then, we may consider the unitary representations U_V and U_E of G on the Hilbert spaces $\ell^2 V$ and $\ell^2 E$ respectively, which are defined by letting $U_V(g)(\delta_v) = \delta_{g \cdot v}$ and $U_E(g)(\delta_e) = \delta_{g \cdot e}$ for all

---

[2] In fact, it can be shown that any group G that acts freely on a tree is free (cf. [65]).

$g \in G$, $v \in V$ and $e \in E$; here, we denote by $(\delta_v)_{v \in V}$ and $(\delta_e)_{e \in E}$ the canonical orthonormal bases of the Hilbert spaces $\ell^2 V$ and $\ell^2 E$ respectively. Let
$V = \bigcup_i V_i$ and $E = \bigcup_j E_j$ be the orbit decompositions of the G-sets V and
E and consider the induced orthogonal decompositions of Hilbert spaces

$$\ell^2 V = \bigoplus_i \ell^2 V_i \quad \text{and} \quad \ell^2 E = \bigoplus_j \ell^2 E_j \;.$$

Then, the operators $U_V(g)$ and $U_E(g)$ restrict to unitary operators

$$U_{V_i}(g) : \ell^2 V_i \longrightarrow \ell^2 V_i \quad \text{and} \quad U_{E_j}(g) : \ell^2 E_j \longrightarrow \ell^2 E_j$$

for all i, j, in such a way that $U_V(g) = \bigoplus_i U_{V_i}(g)$ and $U_E(g) = \bigoplus_j U_{E_j}(g)$
for all $g \in G$.

In the special case where the action of G on X is free, the orbits V_i and
E_j are isomorphic as G-sets with G for all i, j. Then, the Hilbert spaces $\ell^2 V_i$
and $\ell^2 E_j$ are isomorphic with $\ell^2 G$, in such a way that the operators $U_{V_i}(g)$
and $U_{E_j}(g)$ are identified with $L_g \in \mathcal{B}(\ell^2 G)$ for all indices i, j and all $g \in G$.
It follows that the unitary representations U_V and U_E induce, in this case,
unique $*$-algebra representations

$$\pi_V : C_r^* G \longrightarrow \mathcal{B}(\ell^2 V) \quad \text{and} \quad \pi_E : C_r^* G \longrightarrow \mathcal{B}(\ell^2 E) \;,$$

such that $\pi_V(L_g) = U_V(g)$ and $\pi_E(L_g) = U_E(g)$ for all $g \in G$. Since the
representations π_V and π_E are direct sums of the regular representation of
$C_r^* G$ on $\ell^2 G$, we have

$$\| \pi_V(a) \| = \| a \| \quad \text{and} \quad \| \pi_E(a) \| = \| a \| \tag{5.4}$$

for all $a \in C_r^* G$. Furthermore, we may invoke Remark 3.10, in order to conclude that

$$< \pi_V(a)(\delta_v), \delta_v > = \tau(a) \quad \text{and} \quad < \pi_E(a)(\delta_e), \delta_e > = \tau(a) \tag{5.5}$$

for all $a \in C_r^* G$, $v \in V$ and $e \in E$, where τ is the canonical trace on $C_r^* G$.

We now consider a tree X and fix a vertex $v_0 \in V$. If $\lambda : V \setminus \{v_0\} \longrightarrow E$
is the associated bijection, we consider the continuous linear operator

$$P = P_{v_0} : \ell^2 V \longrightarrow \ell^2 E \;,$$

which is defined by letting $P(\delta_v) = \delta_{\lambda(v)}$ if $v \neq v_0$ and $P(\delta_{v_0}) = 0$.

Lemma 5.20 *Let G be a group acting on a tree X and consider the associated
unitary representations U_V and U_E of G on $\ell^2 V$ and $\ell^2 E$ respectively. We fix
a vertex $v_0 \in V$ and let $P : \ell^2 V \longrightarrow \ell^2 E$ be the linear operator defined above.
Then:*

*(i) $PP^* = 1$ and $P^* P = 1 - p_0$, where $p_0 \in \mathcal{B}(\ell^2 V)$ is the orthogonal
projection onto the 1-dimensional subspace $\mathbf{C} \cdot \delta_{v_0} \subseteq \ell^2 V$.*

(ii) The operator $U_V(g) - P^*U_E(g)P \in \mathcal{B}(\ell^2 V)$ is of finite rank for all $g \in G$.

(iii) Assume that G acts freely on X and let π_V and π_E be the induced representations of C_r^*G on $\ell^2 V$ and $\ell^2 E$ respectively. Then, the operator $\pi_V(a) - P^*\pi_E(a)P \in \mathcal{B}(\ell^2 V)$ is compact for all $a \in C_r^*G$.

Proof. (i) This is straightforward, since the operator $P^* : \ell^2 E \longrightarrow \ell^2 V$ maps δ_e onto $\delta_{\lambda^{-1}(e)}$ for any un-oriented edge $e \in E$.

(ii) We note that $U_V(g)(\delta_v) = \delta_{g \cdot v}$ for all $v \in V$ and $[P^*U_E(g)P](\delta_v) = \delta_{\lambda^{-1}(g \cdot \lambda(v))}$ for all $v \in V \setminus \{v_0\}$. Since

$$\lambda^{-1}(g \cdot \lambda(v)) = g \cdot v \Longleftrightarrow g \cdot \lambda(v) = \lambda(g \cdot v)$$

for all $v \in V \setminus \{v_0, g^{-1} \cdot v_0\}$, Proposition 5.19 shows that the operator $U_V(g) - P^*U_E(g)P$ vanishes on δ_v for all but finitely many vertices $v \in V$. In particular, $U_V(g) - P^*U_E(g)P$ is of finite rank.

(iii) It is an immediate consequence of (ii) above that the operator $\pi_V(a) - P^*\pi_E(a)P \in \mathcal{B}(\ell^2 V)$ has finite rank (and is, therefore, compact) for all $a \in L(\mathbf{C}G)$. Since the ideal of compact operators is closed in $\mathcal{B}(\ell^2 V)$, the continuity of the map $a \mapsto \pi_V(a) - P^*\pi_E(a)P$, $a \in C_r^*G$, finishes the proof. $\square$

III. FREE ACTIONS AND THE SUBALGEBRA $\mathcal{A} \subseteq C_r^*G$. Let G be a group acting freely on a tree X and consider the associated representations π_V and π_E of C_r^*G on $\ell^2 V$ and $\ell^2 E$ respectively. We fix a vertex $v_0 \in V$ and let $P : \ell^2 V \longrightarrow \ell^2 E$ be the linear operator constructed above. We denote the operator $P^*\pi_E(a)P$ by $\widetilde{\pi_E}(a)$ for all $a \in C_r^*G$; then,

$$\widetilde{\pi_E} : C_r^*G \longrightarrow \mathcal{B}(\ell^2 V)$$

is easily seen to be a *-homomorphism of non-unital algebras. Moreover, since the operators P and P^* have norm 1, we have

$$\| \widetilde{\pi_E}(a) \| \leq \| a \| \tag{5.6}$$

for all $a \in C_r^*G$. We recall that the ideal $\mathcal{L}^1(\ell^2 V)$ of trace-class operators on $\ell^2 V$ consists of those bounded operators $f \in \mathcal{B}(\ell^2 V)$ for which the family $(<| f | (\delta_v), \delta_v >)_{v \in V}$ is summable (cf. Theorem 1.12) and define

$$\mathcal{A} = \mathcal{A}_{v_0} = \{a \in C_r^*G : \pi_V(a) - \widetilde{\pi_E}(a) \in \mathcal{L}^1(\ell^2 V)\} .$$

In other words, the linear subspace $\mathcal{A} \subseteq C_r^*G$ fits into a pullback diagram

$$\begin{array}{ccc} \mathcal{A} & \hookrightarrow & C_r^*G \\ {\scriptstyle \pi_V - \widetilde{\pi_E}} \downarrow & & \downarrow {\scriptstyle \pi_V - \widetilde{\pi_E}} \\ \mathcal{L}^1(\ell^2 V) & \hookrightarrow & \mathcal{B}(\ell^2 V) \end{array}$$

We note that there is a linear functional

$$\text{Tr} : \mathcal{L}^1(\ell^2 V) \longrightarrow \mathbf{C} \,,$$

which maps any trace-class operator $f \in \mathcal{L}^1(\ell^2 V)$ onto $\sum_{v \in V} < f(\delta_v), \delta_v >$ (loc.cit.), and consider the induced linear functional

$$\tau' : \mathcal{A} \longrightarrow \mathbf{C} \,,$$

which is defined by letting $\tau'(a) = \text{Tr}\,[\pi_V(a) - \widetilde{\pi_E}(a)]$ for all $a \in \mathcal{A}$.

Proposition 5.21 *Let G be a group acting freely on a tree X, fix a vertex $v_0 \in V$ and consider the pair $(\mathcal{A}, \tau')$ defined above. Then:*

(i) $\mathcal{A}$ is a subalgebra of $C_r^* G$ containing $L(\mathbf{C}G)$; in particular, $\mathcal{A}$ is unital and dense in $C_r^* G$.

(ii) The linear functional τ' coincides with the restriction to $\mathcal{A}$ of the canonical trace τ on $C_r^* G$. In particular, τ' is a trace on $\mathcal{A}$.

Proof. (i) For any $a, a' \in \mathcal{A}$ we have

$$\begin{aligned}
\pi_V(aa') - \widetilde{\pi_E}(aa') &= \pi_V(a)\pi_V(a') - \widetilde{\pi_E}(a)\widetilde{\pi_E}(a') \\
&= \pi_V(a)(\pi_V(a') - \widetilde{\pi_E}(a')) + (\pi_V(a) - \widetilde{\pi_E}(a))\widetilde{\pi_E}(a') \,.
\end{aligned}$$

Since $\mathcal{L}^1(\ell^2 V)$ is an ideal in $\mathcal{B}(\ell^2 V)$, it follows that $aa' \in \mathcal{A}$ and hence $\mathcal{A}$ is a subalgebra of $C_r^* G$. Finite rank operators being contained in $\mathcal{L}^1(\ell^2 V)$, Lemma 5.20(ii) implies that $\pi_V(L_g) - \widetilde{\pi_E}(L_g) \in \mathcal{L}^1(\ell^2 V)$ for all $g \in G$. It follows that $L_g \in \mathcal{A}$ for all $g \in G$ and hence $L(\mathbf{C}G) \subseteq \mathcal{A}$.

(ii) We consider an element $a \in \mathcal{A}$ and compute

$$\begin{aligned}
\tau'(a) &= \text{Tr}\,[\pi_V(a) - \widetilde{\pi_E}(a)] \\
&= \sum_{v \in V} < [\pi_V(a) - \widetilde{\pi_E}(a)](\delta_v), \delta_v > \\
&= \sum_{v \in V} < \pi_V(a)(\delta_v) - \widetilde{\pi_E}(a)(\delta_v), \delta_v > \\
&= \sum_{v \in V} (< \pi_V(a)(\delta_v), \delta_v > - < \widetilde{\pi_E}(a)(\delta_v), \delta_v >) \\
&= \sum_{v \in V} (< \pi_V(a)(\delta_v), \delta_v > - < P^* \pi_E(a) P(\delta_v), \delta_v >) \\
&= \sum_{v \in V} (< \pi_V(a)(\delta_v), \delta_v > - < \pi_E(a) P(\delta_v), P(\delta_v) >)
\end{aligned}$$

In view of (5.5), we have $< \pi_V(a)(\delta_v), \delta_v > = \tau(a)$ for all $v \in V$, whereas the inner product $< \pi_E(a) P(\delta_v), P(\delta_v) >$ is equal to $\tau(a)$ if $v \in V \setminus \{v_0\}$ and vanishes if $v = v_0$. Therefore, it follows from the computation above that $\tau'(a) = \tau(a)$. □

Corollary 5.22 *Let G be a group acting freely on a tree X, fix a vertex $v_0 \in V$ and consider the pair $(\mathcal{A}, \tau')$ defined above. Then, the additive map $\tau'_* : K_0(\mathcal{A}) \longrightarrow \mathbf{C}$, which is induced by the trace τ', coincides with the composition*

$$K_0(\mathcal{A}) \xrightarrow{K_0(\iota)} K_0(C_r^* G) \xrightarrow{\tau_*} \mathbf{C} \,,$$

where $K_0(\iota)$ is the additive map induced by the inclusion $\iota : \mathcal{A} \hookrightarrow C_r^* G$.

Proof. This is an immediate consequence of Proposition 5.21(ii), in view of Proposition 1.40(ii). $\square$

In view of the Corollary above, the following result reduces the study of the additive map τ_* to the study of τ'_*.

Proposition 5.23 *Let G be a group acting freely on a tree X, fix a vertex $v_0 \in V$ and consider the subalgebra $\mathcal{A} \subseteq C_r^*G$ defined above.*

(i) If $U \in \mathbf{M}_n(\mathcal{A})$ is a matrix with entries in $\mathcal{A}$, which is invertible in $\mathbf{M}_n(C_r^*G)$, then U is invertible in $\mathbf{M}_n(\mathcal{A})$.

(ii) The inclusion $\iota : \mathcal{A} \hookrightarrow C_r^*G$ induces an isomorphism of groups $K_0(\iota) : K_0(\mathcal{A}) \xrightarrow{\sim} K_0(C_r^*G)$.

Proof. (i) Let $\mathcal{B} = \mathbf{M}_n(\mathcal{B}(\ell^2 V)) = \mathcal{B}(\ell^2 V \oplus \cdots \oplus \ell^2 V)$ be the algebra of bounded linear operators on the direct sum $\ell^2 V \oplus \cdots \oplus \ell^2 V$ of n copies of $\ell^2 V$ and consider the ideal $\mathcal{L} = \mathbf{M}_n(\mathcal{L}^1(\ell^2 V)) \subseteq \mathcal{B}$. The $*$-representation π_V of C_r^*G on $\ell^2 V$ induces a $*$-representation

$$\Pi_V : \mathbf{M}_n(C_r^*G) \longrightarrow \mathcal{B}$$

of $\mathbf{M}_n(C_r^*G)$ on the n-fold direct sum $\ell^2 V \oplus \cdots \oplus \ell^2 V$. In the same way, the $*$-homomorphism of non-unital algebras $\widetilde{\pi_E}$ induces a $*$-homomorphism of non-unital algebras

$$\widetilde{\Pi_E} : \mathbf{M}_n(C_r^*G) \longrightarrow \mathcal{B} \ .$$

Since $\mathbf{M}_n(\mathcal{A})$ consists of those matrices $A \in \mathbf{M}_n(C_r^*G)$ for which $\Pi_V(A) - \widetilde{\Pi_E}(A) \in \mathcal{L}$, in order to prove that the inverse $U^{-1} \in \mathbf{M}_n(C_r^*G)$ of U is contained in $\mathbf{M}_n(\mathcal{A})$, we have to show that $\Pi_V(U^{-1}) - \widetilde{\Pi_E}(U^{-1}) \in \mathcal{L}$. It is easily seen that $\Pi_V(U^{-1}) - \widetilde{\Pi_E}(U^{-1})$ is equal to

$$-\Pi_V(U^{-1})\left[\Pi_V(U) - \widetilde{\Pi_E}(U)\right]\widetilde{\Pi_E}(U^{-1}) + \Pi_V(U^{-1})\left[\Pi_V(I_n) - \widetilde{\Pi_E}(I_n)\right] ,$$

where I_n is the identity $n \times n$ matrix. This finishes the proof, since $\mathcal{L} \subseteq \mathcal{B}$ is an ideal, whereas $I_n \in \mathbf{M}_n(\mathcal{A})$ (cf. Proposition 5.21(i)).

(ii) Let $\| \cdot \|_1$ be the Schatten 1-norm on the ideal $\mathcal{L}^1(\ell^2 V)$ of trace-class operators (cf. Theorem 1.12(ii)) and define a new norm on $\mathcal{A}$, by letting

$$||| a ||| = \| a \| + \| \pi_V(a) - \widetilde{\pi_E}(a) \|_1$$

for any $a \in \mathcal{A}$, where $\| \cdot \|$ is the norm of the C^*-algebra C_r^*G. We note that for any $a, a' \in \mathcal{A}$ we have

$$\begin{aligned}
\|\pi_V(aa') - \widetilde{\pi_E}(aa')\|_1 &= \|\pi_V(a)\pi_V(a') - \widetilde{\pi_E}(a)\widetilde{\pi_E}(a')\|_1 \\
&\leq \|\pi_V(a)\pi_V(a') - \pi_V(a)\widetilde{\pi_E}(a')\|_1 \\
&\quad + \|\pi_V(a)\widetilde{\pi_E}(a') - \widetilde{\pi_E}(a)\widetilde{\pi_E}(a')\|_1 \\
&= \|\pi_V(a)(\pi_V(a') - \widetilde{\pi_E}(a'))\|_1 \\
&\quad + \|(\pi_V(a) - \widetilde{\pi_E}(a))\widetilde{\pi_E}(a')\|_1 \\
&\leq \|\pi_V(a)\| \cdot \|\pi_V(a') - \widetilde{\pi_E}(a')\|_1 \\
&\quad + \|\widetilde{\pi_E}(a')\| \cdot \|\pi_V(a) - \widetilde{\pi_E}(a)\|_1 \\
&\leq \|a\| \cdot \|\pi_V(a') - \widetilde{\pi_E}(a')\|_1 \\
&\quad + \|a'\| \cdot \|\pi_V(a) - \widetilde{\pi_E}(a)\|_1 \ .
\end{aligned}$$

The second inequality above is a consequence of Theorem 1.12(iii), whereas the last one follows from (5.4) and (5.6). Since $\|aa'\| \leq \|a\| \cdot \|a'\|$, it follows that $\||aa'\|| \leq \||a\|| \cdot \||a'\||$ and hence $(\mathcal{A}, \||\cdot\||)$ is a normed algebra. Moreover, the ideal of trace-class operators being complete under the Schatten 1-norm, it is easily seen that $(\mathcal{A}, \||\cdot\||)$ is a Banach algebra. The inclusion

$$\iota : (\mathcal{A}, \||\cdot\||) \longrightarrow (C_r^* G, \|\cdot\|)$$

is continuous, since the $\||\cdot\||$-norm dominates the $\|\cdot\|$-norm, and has dense image (cf. Proposition 5.21(i)). In view of (i) above, we may finish the proof invoking the Karoubi density theorem (Theorem 1.33).   □

Proposition 5.24 *Let G be a group acting freely on a tree X, fix a vertex $v_0 \in V$ and consider the pair $(\mathcal{A}, \tau')$ defined above and the induced additive map $\tau'_* : K_0(\mathcal{A}) \longrightarrow \mathbf{C}$. Then, $\operatorname{im}\tau'_* = \mathbf{Z} \subseteq \mathbf{C}$.*

Proof. The class of the unit element $1 \in \mathcal{A}$ in the group $K_0(\mathcal{A})$ gets mapped under τ'_* onto $\tau'(1) = \operatorname{Tr}(1 - P^*P) = \operatorname{Tr}(p_0) = 1 \in \mathbf{C}$; here, $p_0 \in \mathcal{B}(\ell^2 V)$ is the orthogonal projection onto the 1-dimensional subspace $\mathbf{C} \cdot \delta_{v_0}$ (cf. Lemma 5.20(i)). It follows that $\mathbf{Z} \subseteq \operatorname{im}\tau'_*$. In order to prove the reverse inclusion, we note that the image of the additive map

$$\operatorname{Tr}_* : K_0(\mathcal{L}^1(\ell^2 V)) \longrightarrow \mathbf{C} \ ,$$

which is induced by the trace Tr on $\mathcal{L}^1(\ell^2 V)$, is the group $\mathbf{Z}$ of integers (cf. Example 1.43). Since $\operatorname{Tr}(ab) = \operatorname{Tr}(ba)$ for all $a \in \mathcal{L}^1(\ell^2 V)$ and $b \in \mathcal{B}(\ell^2 V)$ (cf. Theorem 1.12(iv)), the result follows invoking Proposition 1.44(ii), applied to the case of the morphisms of non-unital algebras

$$\pi_V, \widetilde{\pi_E} : \mathcal{A} \longrightarrow \mathcal{B}(\ell^2 V)$$

and the ideal $\mathcal{L}^1(\ell^2 V)$, endowed with the trace Tr.   □

IV. THE INTEGRALITY OF THE TRACE. We can now prove the main result of this subsection.

Theorem 5.25 *Free groups satisfy the integrality of the trace conjecture.*

Proof. We consider a free group G and fix a tree X on which the group acts freely; for example, X may be the Cayley graph $X(G, S)$ of G associated with a set S of free generators (cf. Example 5.18). We choose a vertex $v_0 \in V$ and consider the corresponding pair $(\mathcal{A}, \tau')$. In view of Proposition 5.23(ii), the inclusion $\iota : \mathcal{A} \hookrightarrow C_r^* G$ induces an isomorphism between the respective K_0-groups. Therefore, in order to show that the image of the additive map $\tau_* : K_0(C_r^* G) \longrightarrow \mathbf{C}$ is the subgroup $\mathbf{Z} \subseteq \mathbf{C}$, it suffices to show that this is the case for the image of the composition

$$ K_0(\mathcal{A}) \xrightarrow{K_0(\iota)} K_0(C_r^* G) \xrightarrow{\tau_*} \mathbf{C} \,. $$

In view of Corollary 5.22, the latter composition coincides with the additive map $\tau_*' : K_0(\mathcal{A}) \longrightarrow \mathbf{C}$, which is induced by the trace τ'. Therefore, the result follows from Proposition 5.24.   □

Corollary 5.26 *If G is a free group, then the C^*-algebra $C_r^* G$ has no non-trivial idempotents.*

Proof. This is an immediate consequence of Theorem 5.25, taking into account Proposition 5.2.   □

5.2 Induced Modules over $\mathcal{N}G$

Let k be a subring of the field $\mathbf{C}$ of complex numbers with $k \cap \mathbf{Q} = \mathbf{Z}$ and K its field of fractions. We consider a finite group G and let P be a finitely generated projective kG-module. Then, Swan's theorem (Theorem 2.38) asserts that the induced KG-module $KG \otimes_{kG} P$ is free. Since KG is a subring of the complex group algebra $\mathbf{C}G$, it follows that the $\mathbf{C}G$-module $\mathbf{C}G \otimes_{kG} P$ is free as well. In this section, we prove a generalization of that result to the case of a possibly infinite group G, which is due to Eckmann. More precisely, let us consider a subring $k \subseteq \mathbf{C}$ with $k \cap \mathbf{Q} = \mathbf{Z}$, a group G and a finitely generated projective kG-module P. The von Neumann algebra $\mathcal{N}G$ of G is an algebra containing an isomorphic copy of the complex group algebra $\mathbf{C}G$ and hence of the group ring kG as well. Then, Eckmann's theorem asserts that the $\mathcal{N}G$-module $\mathcal{N}G \otimes_{kG} P$ is free. This generalizes Swan's theorem, since $\mathcal{N}G \simeq \mathbf{C}G$ if the group G is finite.

The proof of Eckmann's theorem is based on certain properties of the center-valued trace t on $\mathcal{N}G$. We construct the trace t and prove the properties that we need in §5.2.1. In the following subsection, we extend t to matrix algebras with entries in $\mathcal{N}G$ and obtain explicit formulae for the image under t of idempotent matrices with entries in $\mathbf{C}G$, in terms of their Hattori-Stallings rank. Finally, we prove Eckmann's result using Linnell's theorem (Theorem 3.32).

5.2.1 The Center-Valued Trace on $\mathcal{N}G$

Let G be a countable group that will remain fixed throughout this subsection and consider the action of the group algebra $\mathbf{C}G$ on the Hilbert space $\ell^2 G$ by left translations. The associated algebra homomorphism

$$L : \mathbf{C}G \longrightarrow \mathcal{B}(\ell^2 G)$$

identifies $\mathbf{C}G$ with the self-adjoint algebra $L(\mathbf{C}G)$ of operators on $\ell^2 G$, in such a way that $L_g^* = L_{g^{-1}}$ for any element $g \in G$. We recall (cf. §1.1.2.II) that the weak operator topology (WOT) on the algebra $\mathcal{B}(\ell^2 G)$ is the locally convex topology defined by the family of semi-norms $(P_{\xi,\eta})_{\xi,\eta\in\ell^2 G}$, where

$$P_{\xi,\eta}(a) = |<a(\xi),\eta>|$$

for all $\xi, \eta \in \ell^2 G$ and $a \in \mathcal{B}(\ell^2 G)$. In other words, a net $(a_\lambda)_\lambda$ of bounded operators on $\ell^2 G$ is WOT-convergent to $0 \in \mathcal{B}(\ell^2 G)$ if and only if the net of complex numbers $(<a_\lambda(\xi),\eta>)_\lambda$ converges to $0 \in \mathbf{C}$ for all vectors $\xi, \eta \in \ell^2 G$. Then, the group von Neumann algebra $\mathcal{N}G$ is defined as the WOT-closure of $L(\mathbf{C}G)$ in $\mathcal{B}(\ell^2 G)$. In view of von Neumann's bicommutant theorem (Theorem 1.18), $\mathcal{N}G$ coincides with the bicommutant $(L(\mathbf{C}G))''$ of the self-adjoint algebra $L(\mathbf{C}G)$ in $\mathcal{B}(\ell^2 G)$. We note (cf. Proposition 3.11) that there is a WOT-continuous, positive, faithful and normalized trace functional

$$\tau : \mathcal{N}G \longrightarrow \mathbf{C} ,$$

which is defined by letting $\tau(a) = <a(\delta_1),\delta_1>$ for all $a \in \mathcal{N}G$. Moreover, for any positive integer n there is an associated WOT-continuous, positive and faithful trace functional τ_n on the matrix algebra $\mathbf{M}_n(\mathcal{N}G)$, which maps a matrix $(a_{ij})_{i,j} \in \mathbf{M}_n(\mathcal{N}G)$ onto $\sum_i \tau(a_{ii})$.

Let $\mathcal{Z}G$ be the center of the von Neumann algebra $\mathcal{N}G$; it is clear that $\mathcal{Z}G = \mathcal{N}G \cap (\mathcal{N}G)'$, being WOT-closed, is itself a von Neumann algebra of operators on $\ell^2 G$. Our goal in this subsection is to construct a trace

$$t = t_G : \mathcal{N}G \longrightarrow \mathcal{Z}G ,$$

which is WOT-continuous on bounded sets, maps $\mathcal{Z}G$ identically onto itself and is closely related to the trace functional τ. The importance of t in the study of idempotent matrices with entries in $\mathcal{N}G$ will be illustrated by proving (in the following subsection) that the induced additive map

$$t_* : K_0(\mathcal{N}G) \longrightarrow \mathcal{Z}G$$

is injective.

I. THE TRACE ON $\mathbf{C}G$. We begin by defining t on the group algebra $\mathbf{C}G$. To that end, we consider the subset $G_f \subseteq G$, consisting of all elements $g \in G$ that have finitely many conjugates. Since the cardinality of the conjugacy class of any element $g \in G$ is equal to the index of the centralizer C_g of g in G, we conclude that $G_f = \{g \in G : [G : C_g] < \infty\}$. We denote by $\mathcal{C}_f(G)$ the subset of $\mathcal{C}(G)$ that consists of those conjugacy classes $[g]$, for which $g \in G_f$.

Lemma 5.27 *Let G_f and $\mathcal{C}_f(G)$ be the sets defined above. Then:*

(i) G_f is a characteristic (and hence normal) subgroup of G.

(ii) For any commutative ring k the center $Z(kG)$ of the group algebra kG is a free k-module with basis consisting of the elements $\zeta_{[g]} = \sum\{x : x \in [g]\}$, $[g] \in \mathcal{C}_f(G)$.

Proof. (i) It is clear that G_f is non-empty, since $1 \in G_f$. We note that for any two elements $g_1, g_2 \in G$ the intersection $C_{g_1} \cap C_{g_2}$ is contained in the centralizer of the product $g_1 g_2$. In particular, if $g_1, g_2 \in G_f$ then

$$[G : C_{g_1 g_2}] \leq [G : C_{g_1} \cap C_{g_2}] \leq [G : C_{g_1}][G : C_{g_2}] < \infty$$

and hence $g_1 g_2 \in G_f$. For any element $g \in G$ we have $C_g = C_{g^{-1}}$; therefore, $g^{-1} \in G_f$ if $g \in G_f$. We have proved that G_f is a subgroup of G. In order to prove that G_f is characteristic in G, let us consider an automorphism $\sigma : G \longrightarrow G$. Then, σ restricts to a bijection between the conjugacy classes $[g]$ and $[\sigma(g)]$ for any element $g \in G$. In particular, $g \in G_f$ if and only if $\sigma(g) \in G_f$.

(ii) It is clear that the subset $\{\zeta_{[g]} : [g] \in \mathcal{C}_f(G)\} \subseteq kG$ is linearly independent over k. Moreover, $x\zeta_{[g]}x^{-1} = \zeta_{[g]}$ for all $x \in G$ and hence $\zeta_{[g]} \in Z(kG)$ for all $[g] \in \mathcal{C}_f(G)$. In order to show that the $\zeta_{[g]}$'s form a basis of $Z(kG)$, let us consider a central element $a = \sum_{g \in G} a_g g \in kG$, where $a_g \in k$ for all $g \in G$. Then, $a = xax^{-1}$ for all $x \in G$ and hence $a_g = a_{x^{-1}gx}$ for all $g, x \in G$. Therefore, the function $g \mapsto a_g$, $g \in G$, is constant on conjugacy classes. Since its support is finite, that function must vanish on the infinite conjugacy classes. It follows that a is a linear combination of the $\zeta_{[g]}$'s, as needed. □

We now define the linear map

$$t_0 : \mathbf{C}G \longrightarrow Z(\mathbf{C}G) ,$$

by letting $t_0(g) = 0$ if $g \notin G_f$ and $t_0(g) = \frac{1}{[G:C_g]}\zeta_{[g]}$ if $g \in G_f$.[3]

Proposition 5.28 *Let $t_0 : \mathbf{C}G \longrightarrow Z(\mathbf{C}G)$ be the $\mathbf{C}$-linear map defined above. Then:*

(i) t_0 is a trace with values in $Z(\mathbf{C}G)$,

(ii) $t_0(a) = a$ for all $a \in Z(\mathbf{C}G)$,

(iii) $t_0(aa') = at_0(a')$ for all $a \in Z(\mathbf{C}G)$ and $a' \in \mathbf{C}G$ (i.e. t_0 is $Z(\mathbf{C}G)$-linear) and

---

[3] This definition is imposed by the requirement that t_0 extends to a trace on the von Neumann algebra $\mathcal{N}G$ with values in $\mathcal{Z}G$, which is WOT-continuous on bounded sets and maps $\mathcal{Z}G$ identically onto itself; cf. Exercise 5.3.5.

(iv) the trace functional r_1 on $\mathbf{C}G$ factors as the composition

$$\mathbf{C}G \xrightarrow{t_0} Z(\mathbf{C}G) \xrightarrow{r_1'} \mathbf{C},$$

where r_1' is the restriction of r_1 to the center $Z(\mathbf{C}G)$.

Proof. (i) Since t_0 is $\mathbf{C}$-linear, it suffices to show that $t_0(g) = t_0(g')$ whenever $[g] = [g'] \in \mathcal{C}(G)$. But this is an immediate consequence of the definition of t_0.

(ii) We consider an element $g \in G_f$ with $[G : C_g] = n$ and let $[g] = \{g_1, \ldots, g_n\}$. Then, $t_0(g_i) = t_0(g)$ for all $i = 1, \ldots, n$ and hence

$$t_0(\zeta_{[g]}) = t_0\left(\sum\nolimits_{i=1}^{n} g_i\right) = \sum\nolimits_{i=1}^{n} t_0(g_i) = nt_0(g) = \zeta_{[g]} .$$

Since t_0 is $\mathbf{C}$-linear, the proof is finished by invoking Lemma 5.27(ii).

(iii) We consider an element $g \in G_f$ with $[G : C_g] = n$ and let $[g] = \{g_1, \ldots, g_n\}$; then, $g_i \in G_f$ for all $i = 1, \ldots, n$. If $g' \in G$ is an element with $g' \notin G_f$, then (G_f being a subgroup of G, in view of Lemma 5.27(i)) $g_i g' \notin G_f$ for all $i = 1, \ldots, n$. In particular,

$$t_0(\zeta_{[g]}g') = t_0\left(\sum\nolimits_{i=1}^{n} g_i g'\right) = \sum\nolimits_{i=1}^{n} t_0(g_i g') = 0 = \zeta_{[g]} t_0(g') .$$

We now assume that $g' \in G_f$ and consider the conjugacy class $[g'] = \{g_1', \ldots, g_m'\}$, where $m = [G : C_{g'}]$. Then, for any $j = 1, \ldots, m$ there exists an element $x_j \in G$, such that $g_j' = x_j g' x_j^{-1}$. Since $\zeta_{[g]}$ is central in $\mathbf{C}G$, we have $\zeta_{[g]}g_j' = x_j \zeta_{[g]} g' x_j^{-1}$ and hence (t_0 being a trace, in view of (i) above) $t_0(\zeta_{[g]}g_j') = t_0(\zeta_{[g]}g')$ for all $j = i, \ldots, m$. It follows that

$$\zeta_{[g]}\zeta_{[g']} = t_0(\zeta_{[g]}\zeta_{[g']}) = t_0\left(\sum\nolimits_{j=1}^{m} \zeta_{[g]}g_j'\right) = \sum\nolimits_{j=1}^{m} t_0(\zeta_{[g]}g_j') = mt_0(\zeta_{[g]}g') ,$$

where the first equality is a consequence of (ii) above, since the element $\zeta_{[g]}\zeta_{[g']}$ is central in $\mathbf{C}G$. We conclude that

$$t_0(\zeta_{[g]}g') = \frac{1}{m}\zeta_{[g]}\zeta_{[g']} = \zeta_{[g]}t_0(g')$$

in this case as well. Therefore, we have proved that $t_0(\zeta_{[g]}g') = \zeta_{[g]}t_0(g')$ for all $g' \in G$. Since this is the case for any $g \in G_f$, the linearity of t_0, combined with Lemma 5.27(ii), finishes the proof.

(iv) It suffices to verify that the linear functionals $r_1' \circ t_0$ and r_1 agree on g for all $g \in G$. But this follows immediately from the definitions. $\square$

II. THE WOT-CONTINUITY OF THE TRACE ON $L(\mathbf{C}G)$. In order to extend the trace t_0 defined above to the von Neumann algebra $\mathcal{N}G$, we consider the linear maps

$$\Delta : \mathbf{C}G \longrightarrow \mathbf{C}G_f \quad \text{and} \quad c : \mathbf{C}G_f \longrightarrow Z(\mathbf{C}G),$$

which are defined by letting Δ map any group element $g \in G$ onto g (resp. onto 0) if $g \in G_f$ (resp. if $g \notin G_f$) and c map any $g \in G_f$ onto $\frac{1}{[G:C_g]}\zeta_{[g]}$. Then, t_0 can be expressed as the composition

$$\mathbf{C}G \xrightarrow{\Delta} \mathbf{C}G_f \xrightarrow{c} Z(\mathbf{C}G). \tag{5.7}$$

Viewing the algebras above as algebras of operators acting on $\ell^2 G$ by left translations, we study the continuity properties of Δ and c and show that both of them extend to the respective WOT-closures.

The map Δ of (5.7). We begin by considering a (possibly infinite) family $(\mathcal{H}_s)_{s \in S}$ of Hilbert spaces and define $\mathcal{H}$ to be the corresponding Hilbert space direct sum. Then, $\mathcal{H} = \bigoplus_{s \in S} \mathcal{H}_s$ consists of those elements $\xi = (\xi_s)_s \in \prod_{s \in S} \mathcal{H}_s$, for which the series $\sum_{s \in S} \|\xi_s\|_s^2$ is convergent. (Here, we denote for any $s \in S$ by $\|\cdot\|_s$ the norm of the Hilbert space $\mathcal{H}_s$.) The inner product on $\mathcal{H}$ is defined by letting $<\xi,\eta> = \sum_{s \in S} <\xi_s, \eta_s>_s$ for any two vectors $\xi = (\xi_s)_s$ and $\eta = (\eta_s)_s$ of $\mathcal{H}$, where $<\_,\_>_s$ denotes the inner product of $\mathcal{H}_s$ for all $s \in S$. The Hilbert spaces $\mathcal{H}_s$, $s \in S$, admit isometric embeddings as closed orthogonal subspaces of $\mathcal{H}$ by means of the operators $\iota_s : \mathcal{H}_s \longrightarrow \mathcal{H}$, which map an element $\xi_s \in \mathcal{H}_s$ onto the element $\iota_s(\xi_s) = (\eta_{s'})_{s'} \in \mathcal{H}$ with $\eta_s = \xi_s$ and $\eta_{s'} = 0$ for $s' \neq s$. Then, the Hilbert space $\mathcal{H}$ is the closed linear span of the subspaces $\iota_s(\mathcal{H}_s)$, $s \in S$. For any index $s \in S$ we also consider the projection $P_s : \mathcal{H} \longrightarrow \mathcal{H}_s$, which maps an element $\xi = (\xi_{s'})_{s'} \in \mathcal{H}$ onto $\xi_s \in \mathcal{H}_s$. It is clear that P_s is a continuous linear map with $\|P_s\| \leq 1$ for all $s \in S$. Moreover, for any vectors $\xi \in \mathcal{H}$ and $\eta_s \in \mathcal{H}_s$ we have $<P_s(\xi), \eta_s>_s = <\xi, \iota_s(\eta_s)>$; therefore, $P_s = \iota_s^*$ is the adjoint of ι_s for all $s \in S$.

Let us consider a bounded operator $a \in \mathcal{B}(\mathcal{H})$ and a vector $\xi = (\xi_s)_s \in \mathcal{H}$. Then, the family $(P_s a \iota_s(\xi_s))_s \in \prod_{s \in S} \mathcal{H}_s$ is also a vector in $\mathcal{H}$, since

$$
\begin{aligned}
\sum_{s \in S} \| P_s a \iota_s(\xi_s)\|_s^2 &\leq \sum_{s \in S} \| a \iota_s(\xi_s)\|^2 \\
&\leq \|a\|^2 \sum_{s \in S} \|\iota_s(\xi_s)\|^2 \\
&= \|a\|^2 \sum_{s \in S} \|\xi_s\|_s^2 \\
&= \|a\|^2 \|\xi\|^2.
\end{aligned}
\tag{5.8}
$$

This is the case for any $\xi \in \mathcal{H}$ and hence we may consider the map

$$\Delta(a) : \mathcal{H} \longrightarrow \mathcal{H},$$

which maps an element $\xi = (\xi_s)_s \in \mathcal{H}$ onto $\Delta(a)(\xi) = (P_s a \iota_s(\xi_s))_s \in \mathcal{H}$. It is clear that the map $\Delta(a)$ is linear. Moreover, it follows from (5.8) that $\Delta(a)$ is a bounded operator; in fact, we have $\| \Delta(a) \| \leq \| a \|$. Therefore, we may consider the map

$$\Delta : \mathcal{B}(\mathcal{H}) \longrightarrow \mathcal{B}(\mathcal{H}),$$

which is given by $a \mapsto \Delta(a)$, $a \in \mathcal{B}(\mathcal{H})$. The map Δ is linear and continuous with respect to the operator norm topology on $\mathcal{B}(\mathcal{H})$; in fact, $\|\Delta\| \leq 1$.[4] It is easily seen that

$$\Delta(a)\iota_s = \iota_s P_s a \iota_s \qquad (5.9)$$

for all $a \in \mathcal{B}(\mathcal{H})$ and all indices $s \in S$. Since Δ is a contraction, it induces by restriction to the r-ball a map

$$\Delta_r : (\mathcal{B}(\mathcal{H}))_r \longrightarrow (\mathcal{B}(\mathcal{H}))_r$$

for any radius r. Of course, Δ_r is continuous with respect to the operator norm topology on $(\mathcal{B}(\mathcal{H}))_r$.

Lemma 5.29 *The map Δ_r defined above is WOT-continuous for any r.*

Proof. Let $(a_\lambda)_\lambda$ be a bounded net of operators in $\mathcal{B}(\mathcal{H})$, which is WOT-convergent to 0. In order to show that the net $(\Delta(a_\lambda))_\lambda$ of operators in $\mathcal{B}(\mathcal{H})$ is WOT-convergent to 0 as well, it suffices, in view of Proposition 1.14, to show that $\lim_\lambda <\Delta(a_\lambda)(\xi), \eta> = 0$, whenever there are two indices $s, s' \in S$ and vectors $\xi_s \in \mathcal{H}_s$ and $\eta_{s'} \in \mathcal{H}_{s'}$, such that $\xi = \iota_s(\xi_s)$ and $\eta = \iota_{s'}(\eta_{s'})$. Since

$$\Delta(a_\lambda)(\xi) = \Delta(a_\lambda)\iota_s(\xi_s) = \iota_s P_s a_\lambda \iota_s(\xi_s)$$

(cf. (5.9)), the inner product $<\Delta(a_\lambda)\xi, \eta> = <\Delta(a_\lambda)\xi, \iota_{s'}(\eta_{s'})>$ vanishes if $s \neq s'$. On the other hand, if $s = s'$ we have

$$\begin{aligned} <\Delta(a_\lambda)(\xi), \eta> &= <\iota_s P_s a_\lambda \iota_s(\xi_s), \iota_s(\eta_s)> \\ &= <P_s a_\lambda \iota_s(\xi_s), \eta_s>_s \\ &= <a_\lambda \iota_s(\xi_s), \iota_s(\eta_s)>, \end{aligned}$$

where the last equality follows since $P_s = \iota_s^*$. Since WOT-$\lim_\lambda a_\lambda = 0$, we conclude that $\lim_\lambda <\Delta(a_\lambda)(\xi), \eta> = 0$ in this case as well. $\qquad \square$

In order to apply the conclusion of Lemma 5.29, we consider the group G and a subgroup $H \leq G$. If S is a set of representatives of the left cosets of H in G, then the decomposition of G into the disjoint union of the cosets Hs, $s \in S$, induces a Hilbert space decomposition $\ell^2 G = \bigoplus_{s \in S} \ell^2(Hs)$. We consider the operator

$$\Delta : \mathcal{B}(\ell^2 G) \longrightarrow \mathcal{B}(\ell^2 G),$$

which is associated with that decomposition as above. In particular, let us fix an element $g \in G$ and try to identify the operator $\Delta(L_g) \in \mathcal{B}(\ell^2 G)$. For any $x \in G$ there is a unique $s = s(x) \in S$, such that $x \in Hs$. Then,

---

[4] The decomposition $\mathcal{H} = \bigoplus_{s \in S} \mathcal{H}_s$ identifies the algebra $\mathcal{B}(\mathcal{H})$ with a certain algebra of $S \times S$ matrices whose (s, s')-entry consists of bounded operators from $\mathcal{H}_{s'}$ to $\mathcal{H}_s$ for all $s, s' \in S$. Under this identification, the linear map Δ maps any $a = (a_{ss'})_{s,s' \in S}$ onto the diagonal matrix $\mathrm{diag}\{a_{ss} : s \in S\}$.

$$\Delta(L_g)(\delta_x) = \Delta(L_g)\iota_s(\delta_x) = \iota_s P_s L_g \iota_s(\delta_x) = \iota_s P_s L_g(\delta_x) = \iota_s P_s(\delta_{gx}) \,,$$

where the second equality follows from (5.9). We note that $gx \in Hs$ if and only if $g \in H$ and hence $\Delta(L_g)(\delta_x)$ is equal to $\iota_s(\delta_{gx}) = \delta_{gx}$ if $g \in H$ and vanishes if $g \notin H$. Since this is the case for all $x \in G$, we conclude that

$$\Delta(L_g) = \begin{cases} L_g \text{ if } g \in H \\ 0 \text{ if } g \notin H \end{cases}$$

In particular, $\Delta(L_g)$ is an element of the subalgebra $L(\mathbf{C}H) \subseteq \mathcal{B}(\ell^2 G)$. (We note that here $L(\mathbf{C}H)$ is viewed as an algebra of operators acting on $\ell^2 G$.) Hence, Δ restricts to a linear map

$$\Delta : L(\mathbf{C}G) \longrightarrow L(\mathbf{C}H) \subseteq \mathcal{B}(\ell^2 G) \,.$$

Corollary 5.30 *Let H be a subgroup of G and consider the linear operator $\Delta : L(\mathbf{C}G) \longrightarrow L(\mathbf{C}H) \subseteq \mathcal{B}(\ell^2 G)$, which is defined above. Then:*
(i) The operator Δ is a contraction.
(ii) The map

$$\Delta_r : (L(\mathbf{C}G))_r \longrightarrow (L(\mathbf{C}H))_r \subseteq (\mathcal{B}(\ell^2 G))_r \,,$$

induced from Δ by restriction to the respective r-balls, is WOT-continuous for any r. $\square$

The map c of (5.7). Having established Corollary 5.30 (that will be applied in the special case where $H = G_f$), we turn our attention to the map c of (5.7). We begin by considering a group N together with an automorphism $\phi : N \longrightarrow N$. Then, ϕ extends by linearity to an automorphism of the complex group algebra $\mathbf{C}N$, which will be still denoted (by an obvious abuse of notation) by ϕ. We also consider the associated automorphism L_ϕ of the algebra of operators $L(\mathbf{C}N) \subseteq \mathcal{B}(\ell^2 N)$, which is defined by letting $L_\phi(L_a) = L_{\phi(a)}$ for all $a \in \mathbf{C}N$. On the other hand, there is a unitary operator $\Phi \in \mathcal{B}(\ell^2 N)$, such that $\Phi(\delta_x) = \delta_{\phi(x)}$ for all $x \in N$; here, we denote by $(\delta_x)_{x \in N}$ the canonical orthonormal basis of $\ell^2 N$.

Lemma 5.31 *Let N be a group and ϕ an automorphism of N.*
(i) The associated isometry Φ of the Hilbert space $\ell^2 N$ is such that $L_{\phi(a)} \circ \Phi = \Phi \circ L_a \in \mathcal{B}(\ell^2 N)$ for all $a \in \mathbf{C}N$.
(ii) The automorphism L_ϕ of the algebra $L(\mathbf{C}N)$ is norm-preserving and WOT-continuous.

Proof. (i) By linearity, it suffices to verify that $L_{\phi(x)} \circ \Phi = \Phi \circ L_x$ for all $x \in N$. For any element $y \in N$ we have

$$(L_{\phi(x)} \circ \Phi)(\delta_y) = L_{\phi(x)}(\delta_{\phi(y)}) = \delta_{\phi(x)\phi(y)} = \delta_{\phi(xy)} = \Phi(\delta_{xy}) = (\Phi \circ L_x)(\delta_y) \,.$$

Since the bounded operators $L_{\phi(x)} \circ \Phi$ and $\Phi \circ L_x$ agree on the orthonormal basis $\{\delta_y : y \in N\}$ of the Hilbert space $\ell^2 N$, they are equal.

(ii) For any $a \in \mathbf{C}N$ we have $L_{\phi(a)} = \Phi \circ L_a \circ \Phi^{-1}$, in view of (i) above. Since Φ is unitary, it follows that $\| L_{\phi(a)} \| = \| L_a \|$ for all $a \in \mathbf{C}N$ and hence L_ϕ is norm-preserving. On the other hand, the map $b \mapsto \Phi \circ b \circ \Phi^{-1}$, $b \in \mathcal{B}(\ell^2 N)$, is WOT-continuous (cf. Remark 1.13(ii)). Being a restriction of it, L_ϕ is WOT-continuous as well. □

We now assume that N is a group on which the group G acts by automorphisms. Then, for any $g \in G$ we are given an automorphism $\phi_g : N \longrightarrow N$, in such a way that $\phi_g \circ \phi_{g'} = \phi_{gg'}$ for all $g, g' \in G$. There is an induced action of G by automorphisms $(\phi_g)_g$ on the complex group algebra $\mathbf{C}N$ and a corresponding action of G by automorphisms $(L_{\phi_g})_g$ on the algebra of operators $L(\mathbf{C}N) \subseteq \mathcal{B}(\ell^2 N)$. More precisely, for any $g \in G$ the automorphism $\phi_g : \mathbf{C}N \longrightarrow \mathbf{C}N$ is the linear extension of $\phi_g \in \mathrm{Aut}(N)$, whereas $L_{\phi_g} : L(\mathbf{C}N) \longrightarrow L(\mathbf{C}N)$ maps L_a onto $L_{\phi_g(a)}$ for all $a \in \mathbf{C}N$.

If the G-action on N is such that all orbits are finite (equivalently, if for any element $x \in N$ the stabilizer subgroup Stab_x has finite index in G), then we define the linear map

$$c : L(\mathbf{C}N) \longrightarrow L(\mathbf{C}N) ,$$

as follows: For any element $x \in N$ with G-orbit $\{x_1, \ldots, x_m\} \subseteq N$, where $m = m(x) = [G : \mathrm{Stab}_x]$, we let $c(L_x) = \frac{1}{m} \sum_{i=1}^m L_{x_i} \in L(\mathbf{C}N)$.

Lemma 5.32 *Assume that G acts on a group N by automorphisms, in such a way that all orbits are finite, and consider the linear endomorphism c of the algebra $L(\mathbf{C}N)$ defined above.*

(i) Let x be an element of N and $H \leq G$ a subgroup of finite index with $H \subseteq \mathrm{Stab}_x$. If $[G : H] = k$ and $\{g_1, \ldots, g_k\}$ is a set of representatives of the right H-cosets $\{gH : g \in G\}$, then $c(L_x) = \frac{1}{k} \sum_{i=1}^k L_{\phi_{g_i}(x)}$.

(ii) The operator c is a contraction.

(iii) The map

$$c_r : (L(\mathbf{C}N))_r \longrightarrow (L(\mathbf{C}N))_r ,$$

induced from c by restriction to the r-balls, is WOT-continuous for any r.

Proof. (i) Since H is contained in the stabilizer Stab_x, we have $\phi_g(x) = \phi_{g'}(x) \in N$ if $gH = g'H$. Therefore, the right hand side of the equality to be proved doesn't depend upon the choice of the set of representatives of the cosets $\{gH : g \in G\}$. Let $\{s_1, \ldots, s_m\}$ be a set of representatives of the cosets $\{g \, \mathrm{Stab}_x : g \in G\}$, where $m = m(x) = [G : \mathrm{Stab}_x]$. Then, the G-orbit of x is the set $\{\phi_{s_1}(x), \ldots, \phi_{s_m}(x)\}$ and hence

$$c(L_x) = \frac{1}{m} \sum_{i=1}^m L_{\phi_{s_i}(x)} .$$

If $\{u_1, \ldots, u_l\}$ is a set of representatives of the cosets $\{gH : g \in \mathrm{Stab}_x\}$, where $l = [\mathrm{Stab}_x : H]$, then the set $\{s_i u_j : 1 \leq i \leq m , \ 1 \leq j \leq l\}$

is a set of representatives of the the cosets $\{gH : g \in G\}$. In particular, $k = [G : H] = [G : \mathrm{Stab}_x] \cdot [\mathrm{Stab}_x : H] = ml$. Since the u_j's stabilize x, we have $\phi_{s_i u_j}(x) = \phi_{s_i}(x)$ for all i, j and hence

$$c(L_x) = \frac{1}{m}\sum_{i=1}^{m} L_{\phi_{s_i}(x)} = \frac{l}{k}\sum_{i=1}^{m} L_{\phi_{s_i}(x)} = \frac{1}{k}\sum_{i=1}^{m}\sum_{j=1}^{l} L_{\phi_{s_i u_j}(x)},$$

as needed.

(ii) Let $a = \sum_{i=1}^{r} a_i x_i \in \mathbf{C}N$, where $a_i \in \mathbf{C}$ and $x_i \in N$ for all $i = 1, \dots, r$. We consider the subgroup $H = \bigcap_{i=1}^{r} \mathrm{Stab}_{x_i}$, which has finite index in G, and fix a set of representatives $\{g_1, \dots, g_k\}$ of the cosets $\{gH : g \in G\}$. We note that $L_a = \sum_{i=1}^{r} a_i L_{x_i}$, whereas $L_{\phi_{g_j}(a)} = \sum_{i=1}^{r} a_i L_{\phi_{g_j}(x_i)}$ for all $j = 1, \dots, k$. Hence, it follows from (i) above that

$$c(L_a) = \sum_{i=1}^{r} a_i c(L_{x_i}) = \sum_{i=1}^{r} a_i \frac{1}{k}\sum_{j=1}^{k} L_{\phi_{g_j}(x_i)} = \frac{1}{k}\sum_{j=1}^{k} L_{\phi_{g_j}(a)}.$$

Since $\| L_{\phi_{g_j}(a)} \| = \| L_a \|$ for all $j = 1, \dots, k$ (cf. Lemma 5.31(ii)), we may conclude that $\| c(L_a) \| \leq \| L_a \|$ and hence c is a contraction.

(iii) Let $(a_\lambda)_\lambda$ be a net of elements in the group algebra $\mathbf{C}N$, such that the net of operators $(L_{a_\lambda})_\lambda$ is bounded and WOT-convergent to $0 \in \mathcal{B}(\ell^2 N)$. For any index λ we write $a_\lambda = \sum_{x \in N} a_{\lambda,x} x$, where the $a_{\lambda,x}$'s are complex numbers, and note that

$$< L_{a_\lambda}(\delta_1), \delta_x > = <\sum_{x' \in N} a_{\lambda,x'} \delta_{x'}, \delta_x > = a_{\lambda,x}$$

for all $x \in N$; in particular, it follows that $\lim_\lambda a_{\lambda,x} = 0$ for all $x \in N$. In order to show that the bounded net $(c(L_{a_\lambda}))_\lambda$ of operators in $L(\mathbf{C}N) \subseteq \mathcal{B}(\ell^2 N)$ is WOT-convergent to 0 as well, it suffices to show that

$$\lim_\lambda <c(L_{a_\lambda})(\delta_y), \delta_z > = 0$$

for all $y, z \in N$ (cf. Proposition 1.14). For any pair of elements $x, x' \in N$ we write $x \sim x'$ if and only if x and x' are in the same orbit under the G-action, whereas $m(x)$ denotes the cardinality of the G-orbit of x. Then,

$$\begin{aligned} c(L_{a_\lambda}) &= \sum_{x \in N} a_{\lambda,x} c(L_x) \\ &= \sum_{x \in N} a_{\lambda,x} \frac{1}{m(x)}\sum \{L_{x'} : x' \sim x\} \\ &= \sum_{x' \in N}\sum \{a_{\lambda,x} \tfrac{1}{m(x)} : x \sim x'\} L_{x'} \end{aligned}$$

and hence

$$<c(L_{a_\lambda})(\delta_y), \delta_z > = \sum \{a_{\lambda,x} \tfrac{1}{m(x)} : x \sim zy^{-1}\}.$$

Since $\lim_\lambda a_{\lambda,x} = 0$ for each one of the finitely many x's in the G-orbit of zy^{-1}, we conclude that $\lim_\lambda <c(L_{a_\lambda})(\delta_y), \delta_z > = 0$. $\qquad\square$

Let $\mathcal{H}$ be a Hilbert space, S a non-empty index set and $\mathcal{H}^{(S)}$ the Hilbert space direct sum of the constant family of Hilbert spaces $(\mathcal{H}_s)_{s\in S}$ with $\mathcal{H}_s = \mathcal{H}$ for all $s \in S$. For any bounded operator $a \in \mathcal{B}(\mathcal{H})$ there is an associated linear operator $a^{(S)} : \mathcal{H}^{(S)} \longrightarrow \mathcal{H}^{(S)}$, which maps an element $(\xi_s)_s \in \mathcal{H}^{(S)}$ onto $(a(\xi_s))_s$. The map $a^{(S)}$ is well-defined, since for any $(\xi_s)_s \in \mathcal{H}^{(S)}$ we have

$$\sum\nolimits_{s\in S} \| a(\xi_s) \|^2 \le \sum\nolimits_{s\in S} \| a \|^2 \| \xi_s \|^2 = \| a \|^2 \sum\nolimits_{s\in S} \| \xi_s \|^2 < \infty .$$

It follows that the operator $a^{(S)}$ is bounded and $\| a^{(S)} \| \le \| a \|$. In fact, we may fix an index $s \in S$ and consider the restriction of $a^{(S)}$ on the subspace $\iota_s(\mathcal{H})$, in order to conclude that $\| a^{(S)} \| = \| a \|$. Hence, the linear map

$$\nu : \mathcal{B}(\mathcal{H}) \longrightarrow \mathcal{B}\left(\mathcal{H}^{(S)}\right) ,$$

which is given by $a \mapsto a^{(S)}$, $a \in \mathcal{B}(\mathcal{H})$, is an isometry and we may consider its restriction to the r-balls

$$\nu_r : (\mathcal{B}(\mathcal{H}))_r \longrightarrow \left(\mathcal{B}\left(\mathcal{H}^{(S)}\right)\right)_r .$$

Then, a net $(a_\lambda)_\lambda$ in $(\mathcal{B}(\mathcal{H}))_r$ is WOT-convergent to 0 if and only if this is the case for the associated net $\left(a_\lambda^{(S)}\right)_\lambda$ of operators on $\mathcal{H}^{(S)}$. Indeed, if WOT-$\lim_\lambda a_\lambda^{(S)} = 0$, then we may consider the restriction of the $a_\lambda^{(S)}$'s on the subspace $\iota_s(\mathcal{H}) \subseteq \mathcal{H}^{(S)}$, for some index $s \in S$, in order to conclude that WOT-$\lim_\lambda a_\lambda = 0$. Conversely, assume that the bounded net $(a_\lambda)_\lambda$ of operators in $\mathcal{B}(\mathcal{H})$ is WOT-convergent to 0. Then, for any pair of indices $s, s' \in S$ and any vectors $\xi, \xi' \in \mathcal{H}$, we have

$$< a_\lambda^{(S)} \iota_s(\xi), \iota_{s'}(\xi') > = < \iota_s a_\lambda(\xi), \iota_{s'}(\xi') > = \begin{cases} < a_\lambda(\xi), \xi' > & \text{if } s = s' \\ 0 & \text{if } s \ne s' \end{cases}$$

where the first equality follows since $a^{(S)} \iota_s = \iota_s a$ for any operator $a \in \mathcal{B}(\mathcal{H})$. In any case, we conclude that $\lim_\lambda < a_\lambda^{(S)} \iota_s(\xi), \iota_{s'}(\xi') > = 0$ and hence the bounded net $\left(a_\lambda^{(S)}\right)_\lambda$ is WOT-convergent to 0 (cf. Proposition 1.14).

Corollary 5.33 *Assume that G acts on a group N by automorphisms, in such a way that all orbits are finite. We consider a group N' containing N as a subgroup and let c be the linear endomorphism of the algebra $L(\mathbf{C}N) \subseteq L(\mathbf{C}N') \subseteq \mathcal{B}(\ell^2 N')$, which is defined as in the paragraph before Lemma 5.32. Then:*

(i) The operator c is a contraction.

(ii) The map

$$c_r : (L(\mathbf{C}N))_r \longrightarrow (L(\mathbf{C}N))_r ,$$

induced from c by restriction to the r-balls, is continuous with respect to the weak operator topology on $(L(\mathbf{C}N))_r \subseteq (\mathcal{B}(\ell^2 N'))_r$ for any r.

Proof. For any element $a \in \mathbf{C}N$ we denote by L_a (resp. L'_a) the left translation induced by a on the Hilbert space $\ell^2 N$ (resp. $\ell^2 N'$). If $S \subseteq N'$ is a set of representatives of the cosets $\{Nx : x \in N'\}$, then the Hilbert space $\ell^2 N' = \bigoplus_{s \in S} \ell^2(Ns)$ is identified with $(\ell^2 N)^{(S)}$, in such a way that L'_a is identified with $L_a^{(S)}$ for all $a \in \mathbf{C}N$. Therefore, assertions (i) and (ii) are immediate consequences of Lemma 5.32(ii),(iii), in view of the discussion above.
□

The continuity of t on $L(\mathbf{C}G)$. Since $L : \mathbf{C}G \longrightarrow L(\mathbf{C}G)$ is an algebra isomorphism, it follows that the center $Z(L(\mathbf{C}G))$ of $L(\mathbf{C}G)$ coincides with $L(Z(\mathbf{C}G))$, where $Z(\mathbf{C}G)$ is the center of $\mathbf{C}G$. In particular, the linear map $t_0 : \mathbf{C}G \longrightarrow Z(\mathbf{C}G)$ of Proposition 5.28 induces a linear map

$$ t : L(\mathbf{C}G) \longrightarrow Z(L(\mathbf{C}G)) \,, $$

by letting $t(L_a) = L_{t_0(a)}$ for any $a \in \mathbf{C}G$. Using the results obtained above, we can now establish certain key continuity properties of t.

Proposition 5.34 *Let* $t : L(\mathbf{C}G) \longrightarrow Z(L(\mathbf{C}G))$ *be the linear map defined above. Then:*

(i) t is a contraction and its restriction

$$ t_r : (L(\mathbf{C}G))_r \longrightarrow (Z(L(\mathbf{C}G)))_r $$

to the respective r-balls is WOT-continuous for any r,

(ii) t is a trace with values in $Z(L(\mathbf{C}G))$,

(iii) $t(L_a) = L_a$ for all $L_a \in Z(L(\mathbf{C}G))$,

(iv) $t(L_a L_{a'}) = L_a t(L_{a'})$ for all $L_a \in Z(L(\mathbf{C}G))$ and $L_{a'} \in L(\mathbf{C}G)$ (i.e. t is $Z(L(\mathbf{C}G))$-linear) and

(v) the canonical trace functional τ on $L(\mathbf{C}G)$ factors as the composition

$$ L(\mathbf{C}G) \overset{t}{\longrightarrow} Z(L(\mathbf{C}G)) \overset{\tau'}{\longrightarrow} \mathbf{C} \,, $$

where τ' is the restriction of τ to the center $Z(L(\mathbf{C}G))$.

Proof. (i) Let $G_f \trianglelefteq G$ be the normal subgroup consisting of those elements $g \in G$ that have finitely many conjugates and consider the linear map

$$ \Delta : L(\mathbf{C}G) \longrightarrow L(\mathbf{C}G_f) \,, $$

which is defined on the generators L_g, $g \in G$, by letting $\Delta(L_g) = L_g$ if $g \in G_f$ and $\Delta(L_g) = 0$ if $g \notin G_f$. The orbit of an element $g \in G_f$ under the conjugation action of G is the conjugacy class $[g] \in \mathcal{C}(G)$, a finite set with $[G : C_g]$ elements. We consider the linear map

$$ c : L(\mathbf{C}G_f) \longrightarrow L(\mathbf{C}G_f) \,, $$

which maps L_g onto $\frac{1}{[G:C_g]} \sum \{L_x : x \in [g]\}$ for all $g \in G_f$. It is clear that the composition

$$L(\mathbf{C}G) \xrightarrow{\Delta} L(\mathbf{C}G_f) \xrightarrow{c} L(\mathbf{C}G_f)$$

coincides with the composition

$$L(\mathbf{C}G) \xrightarrow{t} Z(L(\mathbf{C}G)) \hookrightarrow L(\mathbf{C}G_f) \, .$$

Therefore, (i) is a consequence of Corollaries 5.30 and 5.33. The proof of assertions (ii), (iii), (iv) and (v) follows readily from Proposition 5.28.   □

III. THE CENTER OF $\mathcal{N}G$. Our next goal is to identify the WOT-closure of the center $Z(L(\mathbf{C}G))$ of $L(\mathbf{C}G)$ with the center $\mathcal{Z}G = \mathcal{N}G \cap (\mathcal{N}G)'$ of the von Neumann algebra $\mathcal{N}G$. We note that

$$Z(L(\mathbf{C}G)) = L(\mathbf{C}G) \cap (L(\mathbf{C}G))' \subseteq L(\mathbf{C}G)'' \cap (L(\mathbf{C}G))''' = \mathcal{N}G \cap (\mathcal{N}G)' \, .$$

Hence, $\mathcal{Z}G$ being WOT-closed, we have $\overline{Z(L(\mathbf{C}G))}^{WOT} \subseteq \mathcal{Z}G$. In order to prove the reverse inclusion, we need a couple of auxiliary results.

Lemma 5.35 *Let $a \in \mathcal{Z}G$ be an operator in the center of the von Neumann algebra $\mathcal{N}G$. Then:*

(i) For all $g, h \in G$ we have $<a(\delta_1), \delta_{g^{-1}hg}> = <a(\delta_1), \delta_h>$.

(ii) The inner product $<a(\delta_1), \delta_g>$ depends only upon the conjugacy class $[g] \in \mathcal{C}(G)$ and vanishes if $g \notin G_f$.

(iii) For any $g \in G$ we have

$$a(\delta_g) = \sum_{[x] \in \mathcal{C}_f(G)} <a(\delta_1), \delta_x> L_{\zeta_{[x]}}(\delta_g) \in \ell^2 G \, ,$$

where $\mathcal{C}_f(G) = \{[x] \in \mathcal{C}(G) : x \in G_f\}$ and $\zeta_{[x]} = \sum \{x' : x' \in [x]\}$ for all $[x] \in \mathcal{C}_f(G)$.

Proof. (i) We fix the elements $g, h \in G$ and compute

$$\begin{aligned}
<a(\delta_1), \delta_{g^{-1}hg}> &= <a(\delta_1), L_{g^{-1}}(\delta_{hg})> \\
&= <L_{g^{-1}}^* a(\delta_1), \delta_{hg}> \\
&= <L_g a(\delta_1), \delta_{hg}> \\
&= <a L_g(\delta_1), \delta_{hg}> \\
&= <a(\delta_g), \delta_{hg}> \\
&= <a(\delta_1), \delta_h>
\end{aligned}$$

In the above chain of equalities, the third one follows since $L_{g^{-1}}^* = L_g$, the fourth one since a commutes with L_g, whereas the last one was established in Lemma 3.9(i).

(ii) It follows from (i) that the function $g \mapsto <a(\delta_1), \delta_g>$, $g \in G$, is constant on conjugacy classes. Being square-summable, that function must vanish on those elements $g \in G$ with infinitely many conjugates.

(iii) It follows from (i) and (ii) above that

$$
\begin{aligned}
a(\delta_1) &= \sum_{[x]\in\mathcal{C}_f(G)} <a(\delta_1),\delta_x> \sum\{\delta_{x'} : x' \in [x])\} \\
&= \sum_{[x]\in\mathcal{C}_f(G)} <a(\delta_1),\delta_x> L_{\zeta_{[x]}}(\delta_1) \, .
\end{aligned}
\tag{5.10}
$$

On the other hand, for any $g \in G$ the operator L_g commutes with a (since $a \in \mathcal{Z}G$) and $L_{\zeta_{[x]}}$ for any $x \in G_f$ (since the $L_{\zeta_{[x]}}$'s are central in $L(\mathbf{C}G)$; cf. Lemma 5.27(ii)). Therefore, we have

$$
\begin{aligned}
a(\delta_g) &= aL_g(\delta_1) \\
&= L_g a(\delta_1) \\
&= \sum_{[x]\in\mathcal{C}_f(G)} <a(\delta_1),\delta_x> L_g L_{\zeta_{[x]}}(\delta_1) \\
&= \sum_{[x]\in\mathcal{C}_f(G)} <a(\delta_1),\delta_x> L_{\zeta_{[x]}} L_g(\delta_1) \\
&= \sum_{[x]\in\mathcal{C}_f(G)} <a(\delta_1),\delta_x> L_{\zeta_{[x]}}(\delta_g) \, .
\end{aligned}
$$

In the above chain of equalities, the third one follows from (5.10), in view of the continuity of L_g. $\qquad\square$

Corollary 5.36 *Let $a \in \mathcal{Z}G$ be an operator in the center of the von Neumann algebra $\mathcal{N}G$ and $b \in (Z(L(\mathbf{C}G)))'$ an operator in the commutant of $Z(L(\mathbf{C}G))$ in $\mathcal{B}(\ell^2 G)$. Then, for any two elements $g, h \in G$ the family of complex numbers $(<a(\delta_1),\delta_x> \cdot <b(\delta_g),\delta_{x^{-1}h}>)_{x\in G}$ is summable and*

$$
\sum_{x\in G} <a(\delta_1),\delta_x> \cdot <b(\delta_g),\delta_{x^{-1}h}> \, = \, <ba(\delta_g),\delta_h> \, .
$$

Proof. In view of the continuity of b, Lemma 5.35(iii) implies that

$$
\begin{aligned}
ba(\delta_g) &= \sum_{[x]\in\mathcal{C}_f(G)} <a(\delta_1),\delta_x> bL_{\zeta_{[x]}}(\delta_g) \\
&= \sum_{[x]\in\mathcal{C}_f(G)} <a(\delta_1),\delta_x> L_{\zeta_{[x]}}b(\delta_g) \\
&= \sum_{x\in G} <a(\delta_1),\delta_x> L_x b(\delta_g) \, .
\end{aligned}
$$

In the above chain of equalities, the second one follows since b commutes with $L_{\zeta_{[x]}} \in Z(L(\mathbf{C}G))$ for all $[x] \in \mathcal{C}_f(G)$ (cf. Lemma 5.27(ii)), whereas the last one is a consequence of Lemma 5.35(ii). Therefore, we have

$$
\begin{aligned}
<ba(\delta_g),\delta_h> &= \sum_{x\in G} <a(\delta_1),\delta_x> \cdot <L_x b(\delta_g),\delta_h> \\
&= \sum_{x\in G} <a(\delta_1),\delta_x> \cdot <b(\delta_g),L_x^*(\delta_h)> \\
&= \sum_{x\in G} <a(\delta_1),\delta_x> \cdot <b(\delta_g),L_{x^{-1}}(\delta_h)> \\
&= \sum_{x\in G} <a(\delta_1),\delta_x> \cdot <b(\delta_g),\delta_{x^{-1}h}> \, ,
\end{aligned}
$$

where the first equality follows from the continuity of the inner product map $<\_,\delta_h>$ and the third one from the equalities $L_x^* = L_{x^{-1}}$, $x \in G$. $\qquad\square$

We are now ready to prove the following result, describing the center of the von Neumann algebra $\mathcal{N}G$.

Proposition 5.37 *The center $\mathcal{Z}G$ of the von Neumann algebra $\mathcal{N}G$ is the WOT-closure of the center $Z(L(\mathbf{C}G))$ of the algebra $L(\mathbf{C}G)$ in the algebra $\mathcal{B}(\ell^2 G)$ of bounded operators on $\ell^2 G$.*

Proof. As we have already noted, the von Neumann algebra $\mathcal{Z}G$ contains the WOT-closure of $Z(L(\mathbf{C}G))$. On the other hand, the WOT-closure of the self-adjoint algebra $Z(L(\mathbf{C}G))$ coincides with its bicommutant in $\mathcal{B}(\ell^2 G)$ (cf. Theorem 1.18). Hence, it only remains to show that $\mathcal{Z}G \subseteq (Z(L(\mathbf{C}G)))''$, i.e. that any $a \in \mathcal{Z}G$ commutes with any $b \in (Z(L(\mathbf{C}G)))'$. Let us fix such a pair of operators a, b. Since $a \in \mathcal{Z}G \subseteq \mathcal{N}G$, we have

$$< a(\xi), \delta_h > = \sum_{x \in G} < a(\delta_1), \delta_x > \cdot < \xi, \delta_{x^{-1}h} >$$

for all $\xi \in \ell^2 G$ and $h \in G$ (cf. Lemma 3.9(ii)). In particular, we have

$$< ab(\delta_g), \delta_h > = \sum_{x \in G} < a(\delta_1), \delta_x > \cdot < b(\delta_g), \delta_{x^{-1}h} >$$

for all $g, h \in G$. Therefore, Corollary 5.36 implies that

$$< ab(\delta_g), \delta_h > = < ba(\delta_g), \delta_h >$$

for all $g, h \in G$ and hence $ab = ba$, as needed. $\square$

IV. THE CONSTRUCTION OF t ON $\mathcal{N}G$. Using the results obtained above, we can construct the center-valued trace t on the von Neumann algebra $\mathcal{N}G$ of the countable group G. We note that the countability of G implies that the Hilbert space $\ell^2 G$ is separable. For any radius r we consider the r-ball $(\mathcal{B}(\ell^2 G))_r$ of the algebra of bounded operators on $\ell^2 G$. Then, the space $((\mathcal{B}(\ell^2 G))_r, \mathrm{WOT})$ is compact and metrizable; in fact, we can choose for any r a metric d_r on $((\mathcal{B}(\ell^2 G))_r, \mathrm{WOT})$, in such a way that

$$d_r(a, a') = d_{2r}(a' - a, 0) \tag{5.11}$$

for all $a, a' \in (\mathcal{B}(\ell^2 G))_r$ (cf. Theorem 1.15 and its proof). In view of Kaplansky's density theorem (Theorem 1.20), the r-ball $(\mathcal{N}G)_r$ is the WOT-closure of the r-ball $(L(\mathbf{C}G))_r$. It follows that $((\mathcal{N}G)_r, \mathrm{WOT})$ is also a compact metric space; in particular, it is a complete metric space. In fact, $((\mathcal{N}G)_r, \mathrm{WOT})$ can be identified with the completion of its dense subspace $((L(\mathbf{C}G))_r, \mathrm{WOT})$. As an immediate consequence of the discussion above, we note that any operator in $\mathcal{N}G$ is the WOT-limit of a bounded sequence of operators in $L(\mathbf{C}G)$. Using a similar argument, combined with Proposition 5.37, we may identify the complete metric space $((\mathcal{Z}G)_r, \mathrm{WOT})$ with the completion of its dense subspace $((Z(L(\mathbf{C}G)))_r, \mathrm{WOT})$. It follows that any operator in $\mathcal{Z}G$ is the WOT-limit of a bounded sequence of operators in $Z(L(\mathbf{C}G))$.

We now consider the linear map $t : L(\mathbf{C}G) \longrightarrow Z(L(\mathbf{C}G))$ of Proposition 5.34. We know that t is a contraction, whereas its restriction t_r to the respective r-balls is WOT-continuous for all r. Having fixed the radius r, we note that the continuity of t_{2r} at 0 implies that for any $\varepsilon > 0$ there is $\delta = \delta(r, \varepsilon) > 0$, such that

$$d_{2r}(a, 0) < \delta \Longrightarrow d_{2r}(t(a), 0) < \varepsilon$$

for all $a \in (L(\mathbf{C}G))_{2r}$. Taking into account the linearity of t and (5.11), it follows that

$$d_r(a, a') < \delta \Longrightarrow d_r(t(a), t(a')) < \varepsilon$$

for all $a, a' \in (L(\mathbf{C}G))_r$. Therefore, the map

$$t_r : ((L(\mathbf{C}G))_r, \text{WOT}) \longrightarrow ((Z(L(\mathbf{C}G)))_r, \text{WOT})$$

is uniformly continuous and hence admits a unique extension to a continuous map between the completions

$$t_r : ((\mathcal{N}G)_r, \text{WOT}) \longrightarrow ((\mathcal{Z}G)_r, \text{WOT}) . \tag{5.12}$$

Taking into account the uniqueness of these extensions, it follows that there is a well-defined map

$$t : \mathcal{N}G \longrightarrow \mathcal{Z}G ,$$

which is contractive, extends $t : L(\mathbf{C}G) \longrightarrow Z(L(\mathbf{C}G))$ and its restriction to the respective r-balls is the WOT-continuous map t_r of (5.12) for all r.

Theorem 5.38 *Let $\mathcal{Z}G$ be the center of the von Neumann algebra $\mathcal{N}G$ and $t : \mathcal{N}G \longrightarrow \mathcal{Z}G$ the map defined above. Then:*

(i) t extends the trace $t_0 : \mathbf{C}G \longrightarrow Z(\mathbf{C}G)$, in the sense that the following diagram is commutative

$$\begin{array}{ccc} \mathbf{C}G & \xrightarrow{t_0} & Z(\mathbf{C}G) \\ {\scriptstyle L}\downarrow & & \downarrow{\scriptstyle L} \\ \mathcal{N}G & \xrightarrow{t} & \mathcal{Z}G \end{array}$$

(ii) t is contractive and its restriction to bounded sets is WOT-continuous,
(iii) t is $\mathbf{C}$-linear,
(iv) t is a trace with values in $\mathcal{Z}G$,
(v) $t(a) = a$ for all $a \in \mathcal{Z}G$,
(vi) $t(aa') = at(a')$ for all $a \in \mathcal{Z}G$ and $a' \in \mathcal{N}G$ (i.e. t is $\mathcal{Z}G$-linear),
(vii) the canonical trace functional τ on $\mathcal{N}G$ factors as the composition

$$\mathcal{N}G \xrightarrow{t} \mathcal{Z}G \xrightarrow{\tau'} \mathbf{C} ,$$

where τ' is the restriction of τ on $\mathcal{Z}G$.

The trace t will be referred to as the center-valued trace on $\mathcal{N}G$.

Proof. Assertions (i) and (ii) follow from the construction of t.

(iii) As we have already noted, for any $a, a' \in \mathcal{N}G$ there are bounded sequences $(a_n)_n$ and $(a'_n)_n$ in $L(\mathbf{C}G)$, such that WOT-$\lim_n a_n = a$ and WOT-$\lim_n a'_n = a'$. Then, for any $\lambda, \lambda' \in \mathbf{C}$ the sequence $(\lambda a_n + \lambda' a'_n)_n$ is bounded and WOT-convergent to $\lambda a + \lambda' a'$. In view of the linearity of t on $L(\mathbf{C}G)$, we have $t(\lambda a_n + \lambda' a'_n) = \lambda t(a_n) + \lambda' t(a'_n)$ for all n. Since t is WOT-continuous on bounded sets, it follows that $t(\lambda a + \lambda' a') = \lambda t(a) + \lambda' t(a')$.

(iv) We recall that multiplication in $\mathcal{B}(\ell^2 G)$ is separately continuous in the weak operator topology (cf. Remark 1.13(ii)). For any element $a \in L(\mathbf{C}G)$ the map $a' \mapsto t(aa') - t(a'a)$, $a' \in \mathcal{N}G$, is WOT-continuous on bounded sets and vanishes on $L(\mathbf{C}G)$, in view of Proposition 5.34(ii). Therefore, approximating any operator of $\mathcal{N}G$ by a bounded sequence in $L(\mathbf{C}G)$, we conclude that $t(aa') = t(a'a)$ for all $a' \in \mathcal{N}G$. We now fix $a' \in \mathcal{N}G$ and consider the map $a \mapsto t(aa') - t(a'a)$, $a \in \mathcal{N}G$. This latter map is WOT-continuous on bounded sets and vanishes on $L(\mathbf{C}G)$, as we have just proved. Hence, using the same argument as above, we conclude that $t(aa') = t(a'a)$ for all $a \in \mathcal{N}G$.

(v) We know that any operator $a \in \mathcal{Z}G$ is the WOT-limit of a bounded sequence of operators in $Z(L(\mathbf{C}G))$; therefore, the equality $t(a) = a$ is an immediate consequence of Proposition 5.34(iii), in view of the WOT-continuity of t on bounded sets.

(vi) We fix an operator $a \in Z(L(\mathbf{C}G))$ and consider the map $a' \mapsto t(aa') - at(a')$, $a' \in \mathcal{N}G$. This map is WOT-continuous on bounded sets and vanishes on $L(\mathbf{C}G)$ (cf. Proposition 5.34(iv)). Approximating any operator of $\mathcal{N}G$ by a bounded sequence in $L(\mathbf{C}G)$, we conclude that $t(aa') = at(a')$ for all $a' \in \mathcal{N}G$. We now fix an element $a' \in \mathcal{N}G$ and consider the map $a \mapsto t(aa') - at(a')$, $a \in \mathcal{Z}G$. This map is WOT-continuous on bounded sets and vanishes on $Z(L(\mathbf{C}G))$, as we have just proved. Hence, approximating any operator of $\mathcal{Z}G$ by a bounded sequence in $Z(L(\mathbf{C}G))$, it follows that $t(aa') = at(a')$ for all $a \in \mathcal{Z}G$.

(vii) Since the trace τ is WOT-continuous, the equality $\tau = \tau' \circ t$ follows from the WOT-continuity of t on bounded sets, combined with Proposition 5.34(v), by approximating any operator $a \in \mathcal{N}G$ by a bounded sequence of operators in $L(\mathbf{C}G)$. $\qquad\square$

Before proceeding any further, we explicit the center-valued trace t defined above, in the special case where the group G is finite.

Remarks 5.39 (i) Let $R = \mathbf{M}_n(\mathbf{C})$ be the algebra of $n \times n$ matrices with entries in $\mathbf{C}$. Then, the center $Z(R) \subseteq R$ consists of the scalar multiples of the identity matrix I_n. We note that the matrix units E_{ij} satisfy the following equalities:

- $E_{ij} = E_{ii}E_{ij} - E_{ij}E_{ii}$ if $i \neq j$,
- $E_{ii} - E_{jj} = E_{ij}E_{ji} - E_{ji}E_{ij}$ for all i, j and
- $\sum_{i=1}^{n} E_{ii} = I_n$.

It follows that there is a unique **C**-linear trace $t : R \longrightarrow Z(R)$, which is the identity on $Z(R)$; t is given by letting $t(A) = \frac{\mathrm{tr}(A)}{n} I_n$ for all matrices $A \in R$. (Here, we denote by tr the usual trace of a matrix.)

(ii) Assume that the group G is finite. Then, the von Neumann algebra $\mathcal{N}G$ is isomorphic with the complex group algebra $\mathbf{C}G$. We assume that G has r mutually non-isomorphic irreducible complex representations $V_1, \ldots, V_r$ and let $\chi_1, \ldots, \chi_r$ be the corresponding characters. We consider the dimensions $n_i = \dim V_i = \chi_i(1)$, $i = 1, \ldots, r$, of these representations and the Wedderburn decomposition

$$\mathbf{C}G \simeq \prod_{i=1}^{r} \mathbf{M}_{n_i}(\mathbf{C}) \ .$$

This decomposition identifies any element $a \in \mathbf{C}G$ with the r-tuple of matrices $(\varrho_1(a), \ldots, \varrho_r(a))$, where $\varrho_i(a) \in \mathbf{M}_{n_i}(\mathbf{C})$ is the matrix describing the action of a on $V_i = \mathbf{C}^{n_i}$ for all $i = 1, \ldots, r$; in particular, $\mathrm{tr}(\varrho_i(a)) = \chi_i(a)$ for all $i = 1, \ldots, r$. The center

$$Z(\mathbf{C}G) \simeq \prod_{i=1}^{r} Z(\mathbf{M}_{n_i}(\mathbf{C})) = \prod_{i=1}^{r} \mathbf{C} \cdot I_{n_i}$$

is the direct product of r copies of $\mathbf{C}$. Therefore, in view of (i) above, the center-valued trace $t : \mathcal{N}G \longrightarrow \mathcal{Z}G$ is identified with the map

$$t : \mathbf{C}G \longrightarrow \prod_{i=1}^{r} \mathbf{C} \cdot I_{n_i} \ ,$$

which is defined by letting $t(a) = \left(\frac{\chi_1(a)}{n_1} I_{n_1}, \ldots, \frac{\chi_r(a)}{n_r} I_{n_r} \right)$ for all $a \in \mathbf{C}G$.

5.2.2 Matrices with Entries in $\mathcal{N}G$

In this subsection, we fix a countable group G, the associated von Neumann algebra $\mathcal{N}G$ and its center $\mathcal{Z}G$. For any positive integer n we consider the algebra $\mathbf{M}_n(\mathcal{N}G)$ of $n \times n$ matrices with entries in $\mathcal{N}G$. Then, $\mathbf{M}_n(\mathcal{N}G)$ is a von Neumann algebra of operators acting on the n-fold direct sum $\ell^2 G \oplus \cdots \oplus \ell^2 G$, with center $Z(\mathbf{M}_n(\mathcal{N}G)) = \{aI_n; a \in \mathcal{Z}G\}$ (cf. Exercise 1.3.5(i)). The center-valued trace t, defined in Theorem 5.38, induces the trace

$$t_n : \mathbf{M}_n(\mathcal{N}G) \longrightarrow \mathcal{Z}G \ ,$$

which maps any matrix $A = (a_{ij})_{i,j} \in \mathbf{M}_n(\mathcal{N}G)$ onto $\sum_{i=1}^{n} t(a_{ii}) \in \mathcal{Z}G$ (cf. Proposition 1.39(i)).

Proposition 5.40 *For any integer $n > 0$ the trace $t_n : \mathbf{M}_n(\mathcal{N}G) \longrightarrow \mathcal{Z}G$ defined above has the following properties:*

(i) $t_n(aI_n) = na$ for all $a \in \mathcal{Z}G$,

(ii) $t_n(aA) = at_n(A)$ for all $a \in \mathcal{Z}G$ and $A \in \mathbf{M}_n(\mathcal{N}G)$ (i.e. t_n is $\mathcal{Z}G$-linear) and

(iii) The trace functional τ_n on the matrix algebra $\mathbf{M}_n(\mathcal{N}G)$, which is induced by the canonical trace τ on $\mathcal{N}G$, factors as the composition

$$\mathbf{M}_n(\mathcal{N}G) \xrightarrow{t_n} \mathcal{Z}G \xrightarrow{\tau'} \mathbf{C} \;,$$

where τ' is the restriction of τ on $\mathcal{Z}G$.

Proof. (i) It follows from Theorem 5.38(v) that $t_n(A) = \operatorname{tr}(A)$ for all matrices $A \in \mathbf{M}_n(\mathcal{Z}G)$. In particular, $t_n(aI_n) = na$ for all $a \in \mathcal{Z}G$.

(ii) Let us fix an element $a \in \mathcal{Z}G$ and a matrix $A = (a_{ij})_{i,j} \in \mathbf{M}_n(\mathcal{N}G)$. Then, $aA = (aa_{ij})_{i,j}$ and hence

$$t_n(aA) = \sum\nolimits_{i=1}^{n} t(aa_{ii}) = \sum\nolimits_{i=1}^{n} at(a_{ii}) = a\sum\nolimits_{i=1}^{n} t(a_{ii}) = at_n(A) \;,$$

where the second equality follows from Theorem 5.38(vi).

(iii) For any matrix $A = (a_{ij})_{i,j} \in \mathbf{M}_n(\mathcal{N}G)$ we compute

$$\tau_n(A) = \sum\nolimits_{i=1}^{n} \tau(a_{ii}) = \sum\nolimits_{i=1}^{n} \tau(t(a_{ii})) = \tau\Big(\sum\nolimits_{i=1}^{n} t(a_{ii})\Big) = \tau(t_n(A)) \;,$$

where the second equality follows from Theorem 5.38(vii). $\qquad\square$

Our goal is to demonstrate the usefulness of the t_n's in the study of idempotent matrices with entries in $\mathcal{N}G$. As a first result in that direction, we may reformulate Kaplansky's positivity theorem (Theorem 3.12), as follows:

Corollary 5.41 *Let n be a positive integer. Then, the following conditions are equivalent for an idempotent matrix $E \in \mathbf{M}_n(\mathcal{N}G)$:*

(i) $E = 0$,

(ii) $t_n(E) = 0$, where t_n is the trace induced on the algebra $\mathbf{M}_n(\mathcal{N}G)$ by the center-valued trace on $\mathcal{N}G$.

Proof. It is clear that (i) implies (ii). On the other hand, the implication (ii)→(i) is an immediate consequence of Theorem 3.12, in view of Proposition 5.40(iii). $\qquad\square$

At this point we have to introduce some general concepts. We recall that a bounded operator u on a Hilbert space $\mathcal{H}$ is called a partial isometry if there are closed subspaces $V, V' \subseteq \mathcal{H}$, such that u maps V isometrically onto V' and vanishes on the orthogonal complement $V^{\perp}$. In that case, the adjoint u^* maps V' isometrically onto V and vanishes on the orthogonal complement $V'^{\perp}$. Therefore, $u^*u = p_V$ and $uu^* = p_{V'}$ are the orthogonal projections onto the subspaces V and V' respectively and hence

$$uu^*u = u \quad \text{and} \quad u^*uu^* = u^* \, . \qquad (5.13)$$

If e, f are two projections on $\mathcal{H}$, then $e \le f$ if $ef = fe = e$ (cf. §1.1.1.II); in that case, the (closed) subspace $\operatorname{im} e$ is contained in $\operatorname{im} f$, whereas the operator $f - e$ is the orthogonal projection onto the subspace $(\operatorname{im} e)^{\perp} \cap \operatorname{im} f$. We write $e < f$ if $e \le f$ and $e \ne f$.

Definition 5.42 Let $\mathcal{N}$ be a von Neumann algebra of operators acting on a Hilbert space $\mathcal{H}$ and consider two projections $e, f \in \mathcal{N}$.

(i) The projections e, f are called equivalent (more precisely, equivalent rel $\mathcal{N}$) if there is a partial isometry $u \in \mathcal{N}$, such that $e = u^*u$ and $f = uu^*$; in that case, we write $e \sim f$.

(ii) The projection e is called weaker (more precisely, weaker rel $\mathcal{N}$) than f if there is a projection $e' \in \mathcal{N}$, such that $e \sim e'$ and $e' \le f$; in that case, we write $e \precsim f$.

(iii) The projection e is called strictly weaker (more precisely, strictly weaker rel $\mathcal{N}$) than f if $e \precsim f$ and $e \not\sim f$; in that case, we write $e \prec f$.

In particular, for any positive integer n we may apply the concepts of equivalence and weak ordering of projections to the special case of the von Neumann algebra $\mathbf{M}_n(\mathcal{N}G)$ of $n \times n$ matrices with entries in $\mathcal{N}G$. The following result describes the behavior of the $\mathcal{Z}G$-valued trace t_n with respect to these concepts.

Proposition 5.43 *Let n be a positive integer and E, F two projections in the von Neumann algebra $\mathbf{M}_n(\mathcal{N}G)$ of $n \times n$ matrices with entries in $\mathcal{N}G$.*
(i) If $E \sim F$ then $t_n(E) = t_n(F)$.
(ii) If $E < F$ then $t_n(E) \ne t_n(F)$.
(iii) If $E \prec F$ then $t_n(E) \ne t_n(F)$.

Proof. (i) Let $U \in \mathbf{M}_n(\mathcal{N}G)$ be a partial isometry, such that $E = U^*U$ and $F = UU^*$. Then,

$$t_n(E) = t_n(U^*U) = t_n(UU^*) = t_n(F) \, ,$$

where the second equality follows since t_n is a trace.

(ii) In view of the additivity of t_n, it suffices to show that $t_n(F - E) \ne 0$. Since $E < F$, the operator $F - E$ is a non-zero projection and hence the result follows from Corollary 5.41.

(iii) Let $E' \in \mathbf{M}_n(\mathcal{N}G)$ be a projection, such that $E \sim E'$ and $E' \le F$. Then, we must have $E' < F$ and hence $t_n(E) = t_n(E') \ne t_n(F)$, in view of (i) and (ii) above. $\qquad \square$

Using the comparison theory of projections in the algebra of matrices with entries in $\mathcal{N}G$ (cf. Appendix E), we may complement the assertion of Proposition 5.43(i), as follows:

Proposition 5.44 *Let n a positive integer. Then, the following conditions are equivalent for two projections E, F in the von Neumann algebra $\mathbf{M}_n(\mathcal{N}G)$ of $n \times n$ matrices with entries in $\mathcal{N}G$:*

(i) $E \sim F$,

(ii) $t_n(E) = t_n(F) \in \mathcal{Z}G$, where t_n is the trace on $\mathbf{M}_n(\mathcal{N}G)$ induced by the center-valued trace on $\mathcal{N}G$.

Proof. (i)→(ii): This is precisely the assertion of Proposition 5.43(i).

(ii)→(i): Assume on the contrary that the projections $E, F \in \mathbf{M}_n(\mathcal{N}G)$ are not equivalent, whereas $t_n(E) = t_n(F)$. Then, we may invoke Corollary E.8 of Appendix E, in order to find a central projection $C \in Z(\mathbf{M}_n(\mathcal{N}G))$, such that $CE \prec CF$ or $CF \prec CE$. In view of Proposition 5.43(iii), we conclude that in either case we have

$$t_n(CE) \neq t_n(CF) \, . \tag{5.14}$$

Taking into account the form of the matrices in the center of the algebra $\mathbf{M}_n(\mathcal{N}G)$, it follows that $C = aI_n$ for some $a \in \mathcal{Z}G$ (in fact, a must be itself a projection). Then, Proposition 5.40(ii) implies that

$$t_n(CE) = t_n(aE) = at_n(E) = at_n(F) = t_n(aF) = t_n(CF) \, .$$

This contradicts (5.14), proving therefore that the projections E, F must be equivalent if $t_n(E) = t_n(F)$. $\qquad\qquad\square$

In order to reformulate the previous results in K-theoretic terms, we note that the center-valued trace t induces an additive map

$$t_* : K_0(\mathcal{N}G) \longrightarrow \mathcal{Z}G \, ,$$

which maps the class of any idempotent $n \times n$ matrix E with entries in $\mathcal{N}G$ onto $t_n(E) \in \mathcal{Z}G$ (cf. Proposition 1.39(ii)). Using the results obtained above, we shall prove that t_* is injective, thereby justifying the use of the center-valued trace in order to represent K-theory classes in $K_0(\mathcal{N}G)$.

Remark 5.45 Let $\mathcal{N}$ be a von Neumann algebra and P a finitely generated projective $\mathcal{N}$-module. Then, there is a positive integer n and an idempotent endomorphism α of the free $\mathcal{N}$-module $\mathcal{N}^n$, such that $P = \operatorname{im} \alpha$. Viewing $\mathcal{N}^n$ as the set of row-vectors of size n, we can find an idempotent $n \times n$ matrix E with entries in $\mathcal{N}$, such that $\alpha = \widetilde{E}$ is given by right multiplication with E. Since the matrix E^* is idempotent as well, we may invoke Lemma 3.13, in order to find a projection $F \in \mathbf{M}_n(\mathcal{N})$, such that $E^*F = F$ and $FE^* = E^*$. Taking adjoints, we conclude that $FE = F$ and $EF = E$. It follows that $\widetilde{E}\widetilde{F} = \widetilde{F}$ and $\widetilde{F}\widetilde{E} = \widetilde{E}$ and hence $\operatorname{im} \widetilde{E} = \operatorname{im} \widetilde{F}$. Therefore, P is the image of the idempotent endomorphism $\widetilde{F} : \mathcal{N}^n \longrightarrow \mathcal{N}^n$, which is given by right multiplication with F.

Theorem 5.46 *Let* $t_* : K_0(\mathcal{N}G) \longrightarrow \mathcal{Z}G$ *be the additive map induced by the center-valued trace* t.

(i) If P, Q *are two finitely generated projective* $\mathcal{N}G$*-modules, such that* $t_*[P] = t_*[Q] \in \mathcal{Z}G$, *then* P *and* Q *are isomorphic.*

(ii) t_* *is injective.*

Proof. (i) In view of Remark 5.45, there is a positive integer n and two projections $E, F \in \mathbf{M}_n(\mathcal{N}G)$, such that $P = \mathrm{im}\,\widetilde{E}$ and $Q = \mathrm{im}\,\widetilde{F}$, where $\widetilde{E}, \widetilde{F} : (\mathcal{N}G)^n \longrightarrow (\mathcal{N}G)^n$ are the maps given by right multiplication with E and F respectively. Since $t_*[P] = t_*[Q]$, we conclude that

$$t_n(E) = t_*[E] = t_*[P] = t_*[Q] = t_*[F] = t_n(F) \in \mathcal{Z}G .$$

Then, Proposition 5.44 implies the existence of a partial isometry $U \in \mathbf{M}_n(\mathcal{N}G)$ with $E = U^*U$ and $F = UU^*$. Using (5.13) for U, it follows that the $2n \times 2n$ matrix $V = \begin{pmatrix} U^* & I_n - E \\ I_n - F & U \end{pmatrix}$ is invertible with inverse $V^{-1} = \begin{pmatrix} U & I_n - F \\ I_n - E & U^* \end{pmatrix}$, whereas $V \cdot \begin{pmatrix} F & 0 \\ 0 & 0 \end{pmatrix} \cdot V^{-1} = \begin{pmatrix} E & 0 \\ 0 & 0 \end{pmatrix}$. Therefore, Proposition 1.29 implies that the finitely generated projective $\mathcal{N}G$-modules P and Q are isomorphic.

(ii) This is an immediate consequence of (i). □

Example 5.47 Assume that the group G is finite and has r mutually non-isomorphic irreducible complex representations $V_1, \ldots, V_r$ with corresponding characters $\chi_1, \ldots, \chi_r$. We consider the dimensions $n_i = \dim V_i = \chi_i(1)$, $i = 1, \ldots, r$, of these representations and the Wedderburn decomposition

$$\mathbf{C}G \simeq \prod_{i=1}^r \mathbf{M}_{n_i}(\mathbf{C}) .$$

Let $e_i \in \mathbf{C}G$ be the central idempotent which acts as the identity on V_i and induces the zero map on V_j for all $j \neq i$.[5] Then, there is an isomorphism of left $\mathbf{C}G$-modules $\mathbf{C}G \cdot e_i \simeq V_i^{n_i}$ for all $i = 1, \ldots, r$. Since $\chi_i(e_i) = n_i$ for all i, whereas $\chi_j(e_i) = 0$ if $j \neq i$, we conclude that

$$n_i\, t_*[V_i] = t_*[V_i^{n_i}] = t_*[e_i] = t(e_i) = (0, \ldots, 0, I_{n_i}, 0, \ldots, 0)$$

(cf. Remark 5.39(ii)) and hence $t_*[V_i] = \left(0, \ldots, 0, \frac{1}{n_i} I_{n_i}, 0, \ldots, 0\right)$ for all $i = 1, \ldots, r$. If V is any finite dimensional complex representation of G, then $V \simeq \bigoplus_{i=1}^r V_i^{l_i}$ for suitable non-negative integers $l_1, \ldots, l_r$. Taking into account the additivity of t_*, we conclude that $t_*[V] = \left(\frac{l_1}{n_1} I_{n_1}, \ldots, \frac{l_r}{n_r} I_{n_r}\right)$. In this way, applying Theorem 5.46(i) to the special case of a finite group, we recover the Krull-Schmidt theorem, which asserts that two finite dimensional

---

[5] Under the Wedderburn decomposition, e_i corresponds to the r-tuple of matrices $(0, \ldots, 0, I_{n_i}, 0, \ldots, 0)$.

complex representations $V \simeq \bigoplus_{i=1}^{r} V_i^{l_i}$ and $V' \simeq \bigoplus_{i=1}^{r} V_i^{l_i'}$ are isomorphic if and only if the r-tuples of integers $(l_1, \ldots, l_r)$ and $(l_1', \ldots, l_r')$ coincide (cf. [41, Theorem 19.21]).

We conclude this section by proving Eckmann's generalization of Swan's theorem on induced representations (Theorem 2.38) for arbitrary groups. We note that the group algebra $\mathbf{C}G$ is viewed as a subalgebra of the von Neumann algebra $\mathcal{N}G$, by means of the algebra homomorphism L. For any positive integer n we consider the induced algebra homomorphism

$$ L_n : \mathbf{M}_n(\mathbf{C}G) \longrightarrow \mathbf{M}_n(\mathcal{N}G) \, . $$

We also consider the Hattori-Stallings traces

$$ r^{\mathbf{C}G} : \mathbf{M}_n(\mathbf{C}G) \longrightarrow T(\mathbf{C}G) \quad \text{and} \quad r^{\mathcal{N}G} : \mathbf{M}_n(\mathcal{N}G) \longrightarrow T(\mathcal{N}G) \, . $$

Lemma 5.48 *(i) The composition*

$$ T(\mathbf{C}G) \xrightarrow{T(L)} T(\mathcal{N}G) \xrightarrow{\bar{t}} \mathcal{Z}G \, , $$

where $T(L)$ is the map induced from the algebra homomorphism L by passage to the quotients and $\bar{t}$ is that induced by the center-valued trace t, is given by

$$ [g] \mapsto \begin{cases} \frac{1}{[G:C_g]} L_{\zeta_{[g]}} & \text{if } [g] \in \mathcal{C}_f(G) \\ 0 & \text{if } [g] \notin \mathcal{C}_f(G) \end{cases} $$

for all $[g] \in \mathcal{C}(G)$. Here, $\mathcal{C}_f(G)$ is the subset of $\mathcal{C}(G)$ consisting of the finite conjugacy classes and $\zeta_{[g]} = \sum \{x : x \in [g]\} \in \mathbf{C}G$ for all $[g] \in \mathcal{C}_f(G)$.

(ii) There is a commutative diagram

$$ \begin{array}{ccccc} \mathbf{M}_n(\mathbf{C}G) & \xrightarrow{L_n} & \mathbf{M}_n(\mathcal{N}G) & \xrightarrow{t_n} & \mathcal{Z}G \\ {\scriptstyle r^{\mathbf{C}G}} \downarrow & & {\scriptstyle r^{\mathcal{N}G}} \downarrow & & \| \\ T(\mathbf{C}G) & \xrightarrow{T(L)} & T(\mathcal{N}G) & \xrightarrow{\bar{t}} & \mathcal{Z}G \end{array} $$

for all $n \geq 1$, where t_n is the trace on $\mathbf{M}_n(\mathcal{N}G)$, which is induced by the center-valued trace t.

Proof. (i) This is an immediate consequence of Theorem 5.38(i), which implies that the composition

$$ \mathbf{C}G \xrightarrow{L} \mathcal{N}G \xrightarrow{t} \mathcal{Z}G \, , $$

maps any element $g \in G$ onto $\frac{1}{[G:C_g]} L_{\zeta_{[g]}}$ (resp. onto 0) if $[g] \in \mathcal{C}_f(G)$ (resp. if $[g] \notin \mathcal{C}_f(G)$).

(ii) The commutativity of the square on the left follows from the functoriality of the Hattori-Stallings trace (cf. Lemma 1.38), whereas the commutativity of the square on the right follows from the definition of t_n. $\qquad\square$

Corollary 5.49 *Let P be a finitely generated projective $\mathbf{C}G$-module with Hattori-Stallings rank $r_{HS}(P) = \sum_{[g]\in\mathcal{C}(G)} r_g(P)[g] \in T(\mathbf{C}G)$ and consider the finitely generated projective $\mathcal{N}G$-module $P' = \mathcal{N}G \otimes_{\mathbf{C}G} P$. Then,*

$$t_*[P'] = \sum_{[g]\in\mathcal{C}_{f,tor}(G)} \frac{r_g(P)}{[G:C_g]} L_{\zeta_{[g]}} \in \mathcal{Z}G,$$

where $\mathcal{C}_{f,tor}(G)$ is the subset of $\mathcal{C}(G)$ consisting of the finite conjugacy classes of elements of finite order.

Proof. Let n be a positive integer and E an idempotent $n \times n$ matrix with entries in $\mathbf{C}G$, such that P is the image of the endomorphism $\widetilde{E}$ of the free left $\mathbf{C}G$-module $(\mathbf{C}G)^n$, which is associated with E. Then,

$$r^{\mathbf{C}G}(E) = r_{HS}(P) = \sum_{[g]\in\mathcal{C}(G)} r_g(P)[g] . \tag{5.15}$$

If $E' = L_n(E) \in \mathbf{M}_n(\mathcal{N}G)$, then P' is the image of the endomorphism $\widetilde{E'}$ of the free left $\mathcal{N}G$-module $(\mathcal{N}G)^n$, which is associated with E'. We now compute

$$\begin{aligned}
t_*[P'] &= t_*[E'] \\
&= t_n(E') \\
&= (t_n \circ L_n)(E) \\
&= \left(\bar{t} \circ T(L) \circ r^{\mathbf{C}G}\right)(E) \\
&= \left(\bar{t} \circ T(L)\right)\left(\sum_{[g]\in\mathcal{C}(G)} r_g(P)[g]\right) \\
&= \sum_{[g]\in\mathcal{C}(G)} r_g(P)\left(\bar{t} \circ T(L)\right)[g] \\
&= \sum_{[g]\in\mathcal{C}_f(G)} \frac{r_g(P)}{[G:C_g]} L_{\zeta_{[g]}} .
\end{aligned}$$

In the above chain of equalities, the fourth one follows from Lemma 5.48(ii), the fifth one from (5.15) and the last one from Lemma 5.48(i). On the other hand, Bass' theorem implies that any element $g \in G$ with $r_g(P) \neq 0$ is conjugate to infinitely many powers of it (cf. Theorem 3.26(i)). Therefore, if $g \in G_f$ is an element of infinite order then $r_g(P) = 0$. □

We are now ready to state and prove Eckmann's generalization of Swan's theorem. We note that if k is a subring of the field $\mathbf{C}$ of complex numbers then the group algebra kG, being a subalgebra of $\mathbf{C}G$, may be viewed as a subalgebra of the von Neumann algebra $\mathcal{N}G$.

Theorem 5.50 *(Eckmann) Let k be a subring of the field $\mathbf{C}$ of complex numbers with $k \cap \mathbf{Q} = \mathbf{Z}$ and kG the corresponding group algebra. If P is a finitely generated projective kG-module, then the left $\mathcal{N}G$-module $\mathcal{N}G \otimes_{kG} P$ is free.*

Proof. Let $r_{HS}(P) = \sum_{[g]\in\mathcal{C}(G)} r_g(P)[g] \in T(kG)$ be the Hattori-Stallings rank of P and consider the finitely generated projective $\mathbf{C}G$-module $P_0 = \mathbf{C}G \otimes_{kG} P$. Then, $r_g(P_0) = r_g(P)$ for all $g \in G$ (cf. Proposition 1.46) and hence $r_{HS}(P_0) = \sum_{[g] \in \mathcal{C}(G)} r_g(P)[g] \in T(\mathbf{C}G)$. Invoking Corollary 5.49, we conclude that

$$t_*[\mathcal{N}G \otimes_{kG} P] = t_*[\mathcal{N}G \otimes_{\mathbf{C}G} P_0] = \sum_{[g] \in \mathcal{C}_{f,tor}(G)} \frac{r_g(P)}{[G : C_g]} L_{\zeta_{[g]}} \,,$$

where $\mathcal{C}_{f,tor}(G)$ is the subset of $\mathcal{C}(G)$ consisting of the finite conjugacy classes of elements of finite order. In view of Linnell's theorem, $r_g(P) = 0$ for any element $g \in G \setminus \{1\}$ of finite order (cf. Theorem 3.32(ii)); hence,

$$t_*[\mathcal{N}G \otimes_{kG} P] = \frac{r_1(P)}{[G : C_1]} L_{\zeta_{[1]}} = r_1(P) \cdot 1 \in \mathcal{Z}G \,.$$

The complex number $r_1(P)$ is rational, in view of Zaleskii's theorem (Theorem 3.19), and non-negative, in view of Kaplansky's positivity theorem (Corollary 3.14(i)). Since $r_1(P) \in k$ and $k \cap \mathbf{Q} = \mathbf{Z}$, it follows that $n = r_1(P)$ is a non-negative integer. We note that $t_*[(\mathcal{N}G)^n] = t_*[I_n] = t_n(I_n) = n \cdot 1 \in \mathcal{Z}G$ (cf. Proposition 5.40(i)) and hence $t_*[\mathcal{N}G \otimes_{kG} P] = t_*[(\mathcal{N}G)^n] \in \mathcal{Z}G$. Therefore, we may invoke Theorem 5.46(i) in order to conclude that the $\mathcal{N}G$-modules $\mathcal{N}G \otimes_{kG} P$ and $(\mathcal{N}G)^n$ are isomorphic.   □

Remarks 5.51 (i) The assertion of Theorem 5.50 can be reformulated in K-theoretic terms, as follows: For any subring k of the field $\mathbf{C}$ of complex numbers with $k \cap \mathbf{Q} = \mathbf{Z}$, the additive map

$$\widetilde{K}_0(kG) \longrightarrow \widetilde{K}_0(\mathcal{N}G)$$

between the reduced K-theory groups (cf. Remark 1.27(ii)), which is induced by the composition $kG \hookrightarrow \mathbf{C}G \xrightarrow{L} \mathcal{N}G$, is the zero map.

(ii) The arithmetic condition on the coefficient ring k can't be omitted from the statement of Theorem 5.50. Indeed, if G is a group with torsion, then there are always finitely generated projective $\mathbf{C}G$-modules, such that the induced $\mathcal{N}G$-modules are not free (cf. Exercise 5.3.10 below).

5.3 Exercises

1. Show that the dual group $\widehat{\mathbf{Q}}$ of the additive group $(\mathbf{Q}, +)$ of rational numbers can be identified with the subgroup of the countable direct product $\prod_{n=0}^{\infty} S^1$ of copies of S^1, which consists of those sequences $(z_n)_n$ that satisfy the equalities $z_{n+1}^{n+1} = z_n$ for all $n \geq 0$.
2. Let Γ be a compact topological group with Haar measure μ.
   (i) Show that $\mu(U) > 0$ for any non-empty open subset $U \subseteq \Gamma$.
   (ii) Let $f \in C(\Gamma)$ be a continuous function on Γ and M_f the associated multiplication operator on $L^2(\Gamma)$. Show that the operator norm of M_f is equal to the supremum norm of f.

(*Hint:* In order to prove that $\| M_f \| \geq \| f \|_\infty$, consider a point $\gamma_0 \in \Gamma$ with $| f(\gamma_0) | = \| f \|_\infty$. For any $\varepsilon > 0$ choose an open neighborhood U of γ_0, such that $| f(\gamma_0) |^2 - | f(\gamma) |^2 < \varepsilon$ for all $\gamma \in U$, and test with the function $\chi_U \in L^2(\Gamma)$.)

3. Show that a subgroup $A \subseteq S^1$ is trivial if and only if $| z - 1 | < \sqrt{3}$ for all $z \in A$.

4. Let $\mathcal{H}$ be a Hilbert space, $\mathcal{A} \subseteq \mathcal{B}(\mathcal{H})$ a unital self-adjoint subalgebra and $\mathcal{N} = \mathcal{A}''$ its WOT-closure.

   (i) Show that the center $Z(\mathcal{N})$ of $\mathcal{N}$ contains the WOT-closure of the center $Z(\mathcal{A})$ of $\mathcal{A}$.

   In contrast to the situation described in Proposition 5.37, the inclusion $Z(\mathcal{A})'' \subseteq Z(\mathcal{N})$ may be proper. In fact, it is the goal of this Exercise to provide an example where $Z(\mathcal{A})'' \neq Z(\mathcal{N})$.[6] To that end, we let $\mathcal{H}_0$ be an infinite dimensional Hilbert space and consider the Hilbert space $\mathcal{H} = \mathcal{H}_0 \oplus \mathbf{C}$.

   (ii) For any $a \in \mathcal{B}(\mathcal{H}_0)$ and any scalar $\lambda \in \mathbf{C}$ we consider the linear map $T(a, \lambda) : \mathcal{H} \longrightarrow \mathcal{H}$, which maps any element $(v, z) \in \mathcal{H}$ onto $(a(v) + \lambda v, \lambda z)$. Show that $T(a, \lambda) \in \mathcal{B}(\mathcal{H})$.

   (iii) Consider the ideal $\mathcal{F} \subseteq \mathcal{B}(\mathcal{H}_0)$ of finite rank operators and let $\mathcal{A} = \{T(a, \lambda) : a \in \mathcal{F}, \lambda \in \mathbf{C}\}$, in the notation of (ii) above. Show that $\mathcal{A}$ is a unital self-adjoint subalgebra of $\mathcal{B}(\mathcal{H})$, whose center $Z(\mathcal{A})$ consists of the scalar multiples of the identity.

   (iv) Let $\mathcal{A} \subseteq \mathcal{B}(\mathcal{H})$ be the subalgebra defined in (iii) above. Show that the center $Z(\mathcal{A}'')$ of the bicommutant $\mathcal{A}''$ is 2-dimensional and conclude that the inclusion $Z(\mathcal{A})'' \subseteq Z(\mathcal{A}'')$ is proper.

5. Let G be a countable group, $\mathcal{N}G$ the associated von Neumann algebra and $\mathcal{Z}G$ its center. We consider a $\mathbf{C}$-linear trace $t' : \mathcal{N}G \longrightarrow \mathcal{Z}G$, which is WOT-continuous on bounded sets and maps $\mathcal{Z}G$ identically onto itself. The goal of this Exercise is to show that t' coincides with the center-valued trace t constructed in Theorem 5.38.

   (i) Let $g \in G$ be an element with finitely many conjugates and C_g its centralizer in G. Show that $t'(L_g) = \frac{1}{[G:C_g]} L_{\zeta_{[g]}} \in \mathcal{Z}G$.

   (ii) Let $(g_n)_n$ be a sequence of distinct elements of G. Show that the sequence of operators $(L_{g_n})_n$ in $\mathcal{B}(\ell^2 G)$ is WOT-convergent to 0.

   (iii) Let $g \in G$ be an element with infinitely many conjugates. Show that $t'(L_g) = 0$.

   (iv) Show that $t' = t$.

6. A group G is called i.c.c. (infinite conjugacy class) if any element $g \in G \setminus \{1\}$ has infinitely many conjugates. Equivalently, G is an i.c.c. group if the subgroup $G_f \subseteq G$, defined in the paragraph before Lemma 5.27, is trivial.

   (i) Show that the free group on $n \geq 2$ generators is i.c.c.

---

[6] The following example was communicated to me by E. Katsoulis.

(ii) Let S_∞ be the group consisting of those permutations σ of the set $\mathbf{N} = \{0, 1, 2, \ldots\}$, for which there is an integer $n_0 = n_0(\sigma)$, such that $\sigma(n) = n$ for all $n \geq n_0$. Show that S_∞ is an i.c.c. group.

7. Let G be a countable i.c.c. group (cf. Exercise 6 above.)
   (i) Show that the center $\mathcal{Z}G$ of the von Neumann algebra $\mathcal{N}G$ consists of the scalar multiples of the identity operator $1 \in \mathcal{B}(\ell^2 G)$.
   (ii) Show that the center-valued trace t on $\mathcal{N}G$ is such that $t(a) = \tau(a) \cdot 1$, where $\tau : \mathcal{N}G \longrightarrow \mathbf{C}$ is the canonical trace and 1 the identity operator on $\ell^2 G$.
   (iii) Show that the additive map $\tau_* : K_0(\mathcal{N}G) \longrightarrow \mathbf{C}$, which is induced by the canonical trace τ, is injective.

8. Let G be a countable group and $e \in \mathbf{C}G$ an idempotent.
   (i) Consider the operator $L_e \in \mathcal{N}G$ and show that $\| t(L_e) \| \leq 1$, where t is the center-value trace on the von Neumann algebra $\mathcal{N}G$.
   (*Hint:* Use Lemma 3.13.)
   (ii) Show that $\sum_{[g] \in \mathcal{C}_f(G)} \frac{|r_g(e)|^2}{[G : C_g]} \leq 1$, where we denote by C_g the centralizer of any element $g \in G$ and let $\mathcal{C}_f(G)$ be the subset of $\mathcal{C}(G)$ consisting of the finite conjugacy classes.[7]

9. Let G be a countable group and consider the complex group algebra $\mathbf{C}G$, the associated von Neumann algebra $\mathcal{N}G$ and its center $\mathcal{Z}G$.
   (i) Let t be the center-valued trace on $\mathcal{N}G$. Show that $t(a^*) = t(a)^*$ for all $a \in \mathcal{N}G$.
   (ii) Let n be a positive integer and consider the trace t_n, which is induced by the center-valued trace t on the matrix algebra $\mathbf{M}_n(\mathcal{N}G)$. Show that $t_n(A^*) = t_n(A)^* \in \mathcal{Z}G$ for all $A \in \mathbf{M}_n(\mathcal{N}G)$.
   (iii) Let $t_* : K_0(\mathcal{N}G) \longrightarrow \mathcal{Z}G$ be the additive map induced by the center-valued trace t. Show that the subgroup $\operatorname{im} t_*$ consists of self-adjoint operators.
   (*Hint:* Use Lemma 3.13.)
   (iv) Let P be a finitely generated projective $\mathbf{C}G$-module. Show that $r_{g^{-1}}(P) = \overline{r_g(P)}$ for all $g \in G_f$.

10. Let G be a countable group. In this Exercise we examine the extent to which the assertion of Theorem 5.50 (in its equivalent formulation stated in Remark 5.51(i)) is valid for finitely generated projective modules over the complex group algebra $\mathbf{C}G$. To that end, we consider the additive map in reduced K-theory

$$\widetilde{K_0}(L) : \widetilde{K_0}(\mathbf{C}G) \longrightarrow \widetilde{K_0}(\mathcal{N}G) \,,$$

which is induced by the algebra homomorphism L. Our goal is to show that the image $\operatorname{im} \widetilde{K_0}(L)$ is a torsion group if and only if the group G_f is torsion-free.[8]

---

[7] A stronger version of this inequality was obtained by I.B.S. Passi and D. Passman in [54].

[8] This result is essentially due to B. Eckmann [21].

(i) If the group G_f has non-trivial torsion elements show that the subgroup $\operatorname{im} \widetilde{K_0}(L) \subseteq \widetilde{K_0}(\mathcal{N}G)$ is not a torsion group.

(*Hint:* Assume that $g \in G_f$ is an element of finite order $n > 1$ and consider the idempotent $e = \frac{1}{n} \sum_{i=0}^{n-1} g^i \in \mathbf{C}G$.)

(ii) If the group G_f is torsion-free show that the subgroup $\operatorname{im} \widetilde{K_0}(L) \subseteq \widetilde{K_0}(\mathcal{N}G)$ is a torsion group.

11. Let $\mathcal{N}$ be a von Neumann algebra of operators acting on a Hilbert space $\mathcal{H}$ and consider an abelian group V and an additive V-valued trace τ on $\mathcal{N}$. It is the goal of this Exercise to show that τ vanishes on nilpotent elements of $\mathcal{N}$.[9] To that end, we consider a fixed element $a \in \mathcal{N}$.

(i) For any non-negative integer n we let $p_n \in \mathcal{B}(\mathcal{H})$ be the orthogonal projection onto the closed subspace $\overline{\operatorname{im} a^n} \subseteq \mathcal{H}$. Show that the sequence $(p_n)_n$ is a decreasing sequence of projections in $\mathcal{N}$, such that $p_{n+1}ap_n = ap_n$ for all $n \geq 0$.

(ii) For any positive integer n we consider the complementary projection $e_n = p_{n-1} - p_n \in \mathcal{N}$. Show that $e_ip_j = 0$ if $0 < i \leq j$ and hence conclude that $e_nae_n = 0$ for all $n \geq 1$.

(iii) Show that $\tau(ae_n) = 0$ for any positive integer n.

(*Hint:* Use the trace property of τ.)

(iv) If a is nilpotent, then show that $\tau(a) = 0$.

(*Hint:* In the notation above, if $p_n = 0$ then $e_1 + \cdots + e_n = 1$.)

12. Let G be a group and A a nilpotent matrix with entries in the complex group algebra $\mathbf{C}G$. Show that $r_g(A) = 0$ for any element $g \in G_f$.[10]

(*Hint:* Assuming that G is countable, compute the operator $t_n(A') \in \mathcal{Z}G$, where n is the size of A and $A' = L_n(A)$. Then, use Exercise 11(iv) above.)

[9] The following argument was communicated to me by A. Katavolos.
[10] This result is proved by I.B.S. Passi and D. Passman in [54].

Notes and Comments on Chap. 5. The relevance of the integrality of the trace conjecture in the study of idempotents in the reduced C^*-algebra of a torsion-free group G was first noted by A. Connes [15], J. Cuntz [17], M. Pimsner and D. Voiculescu [57]. In the case where the group G has non-trivial torsion elements, P. Baum and A. Connes had conjectured in [6] that the image of τ_* is the subgroup of **Q** generated by the inverses of the orders of the finite subgroups of G. This latter conjecture was disproved by R. Roy [59]. Subsequently, W. Lück [46] formulated a modified version of that conjecture, according to which the image of τ_* is contained in the subring of **Q** generated by the inverses of the orders of the finite subgroups of G. The identification of the reduced C^*-algebra of an abelian group G with the algebra of continuous functions on the dual group $\widehat{G}$ and the relation between the existence of non-trivial torsion elements in G to the connectedness of $\widehat{G}$ are standard themes in Harmonic Analysis (cf. [33]). The proof of the integrality of the trace conjecture for free groups, which was presented in §5.1.3, is due to A. Connes [15]. In that direction, it should be noted that M. Pimsner and D. Voiculescu have proved in [57] that the additive map τ_* induces an isomorphism $K_0(C_r^*G) \simeq \mathbf{Z}$, in the case where the group G is free. A center-valued trace t, such as the one defined on $\mathcal{N}G$ in §5.2.1, can be defined for any finite von Neumann algebra; the construction of t in that generality can be found in Chap. 8 of [36]. Eckmann's theorem (Theorem 5.50) was proved in [20] (see also [64]). The relevant induction map (cf. Remark 5.51(i)) is studied in [25] for any coefficient ring $k \subseteq \mathbf{C}$. Further applications of the use of $\mathcal{N}G$ in the study of the Hattori-Stallings rank of idempotent matrices with entries in group algebras can be found in [26].

A

Tools from Commutative Algebra

In this Appendix, we present those results from Commutative Algebra that were used in Chaps. 2 and 3. For more details on the subject, the reader is referred to the specialized books [3] and [52] or the encyclopedic treatise [8].

A.1 Localization and Local Rings

I. LOCALIZATION. Let R be a commutative ring and $S \subseteq R$ a multiplicatively closed subset containing 1. The localization $R[S^{-1}]$ of R at S is the ring obtained from R by formally inverting the elements of S. More precisely, $R[S^{-1}]$ consists of the equivalence classes of pairs of the form (r, s), where $r \in R$ and $s \in S$, under the equivalence relation defined by the rule

$$(r, s) \sim (r', s') \text{ if and only if there exists } s_0 \in S \text{ such that } s_0 s r' = s_0 s' r$$

for $r, r' \in R$ and $s, s' \in S$. We denote the equivalence class of a pair (r, s), where $r \in R$ and $s \in S$, by r/s. Addition and multiplication in $R[S^{-1}]$ are defined by the rules

$$r/s + r'/s' = (s'r + sr')/ss' \quad \text{and} \quad (r/s) \cdot (r'/s') = (rr')/(ss')$$

for $r, r' \in R$ and $s, s' \in S$. It is straightforward to verify that these operations are well-defined and endow $R[S^{-1}]$ with the structure of a commutative ring.

Remark A.1 Let S be a multiplicatively closed subset of a commutative ring R containing 1. Then, the group of units $U(R[S^{-1}])$ of the localization $R[S^{-1}]$ consists of those formal fractions $x = r/s \in R[S^{-1}]$, where $r \in R$ and $s \in S$ are such that there exists $t \in R$ with $rt \in S$. Indeed, it is clear that an element x of that form is invertible with inverse $st/rt \in R[S^{-1}]$. Conversely, if $x = r/s \in U(R[S^{-1}])$ and $x^{-1} = r'/s' \in R[S^{-1}]$, for some $r' \in R$ and $s' \in S$, then $rr'/ss' = 1/1 \in R[S^{-1}]$ and hence there exists $s'' \in S$ with $rr's'' = ss's'' \in R$. Then, the element $t = r's'' \in R$ is such that $rt = rr's'' = ss's'' \in S$, as needed. In particular, assume that the multiplicatively closed subset S has the following property: If $r \in R$ is such that there exists $t \in R$ with $rt \in S$, then $r \in S$.[1] In that case, the group of units of the ring $R[S^{-1}]$ consists of the fractions r/s with $r, s \in S$.

We now consider the map

$$\lambda = \lambda_{R,S} : R \longrightarrow R[S^{-1}],$$

which is given by $r \mapsto r/1$, $r \in R$. It is clear that λ is a ring homomorphism; as such, it endows the localization $R[S^{-1}]$ with the structure of a commutative R-algebra. The following Proposition characterizes $R[S^{-1}]$ as the universal commutative R-algebra in which the elements of S (more precisely, their canonical images) are invertible.

Proposition A.2 *Let R be a commutative ring, $S \subseteq R$ a multiplicatively closed subset containing 1, $R[S^{-1}]$ the corresponding localization and λ the ring homomorphism defined above. Then, for any commutative ring A and any ring homomorphism $\varphi : R \longrightarrow A$ with the property that $\varphi(S) \subseteq U(A)$, there exists a unique ring homomorphism $\Phi : R[S^{-1}] \longrightarrow A$ with $\Phi \circ \lambda = \varphi$.*

Proof. Given a pair (A, φ), we may define Φ by letting $\Phi(r/s) = \varphi(r)\varphi(s)^{-1}$ for all $r/s \in R[S^{-1}]$. It is easily seen that Φ is a well-defined ring homomorphism and $\Phi \circ \lambda = \varphi$. The uniqueness of Φ satisfying that condition follows since $r/s = \lambda(r)\lambda(s)^{-1} \in R[S^{-1}]$ for all $r \in R$, $s \in S$. $\square$

Corollary A.3 *(i) Let R be a commutative ring and $T \subseteq S \subseteq R$ two multiplicatively closed subsets containing 1. If S' is the multiplicatively closed subset of $R' = R[T^{-1}]$ consisting of the elements of the form s/t, where $s \in S$ and $t \in T$, then there is a unique homomorphism of R-algebras*

$$R'[S'^{-1}] \longrightarrow R[S^{-1}].$$

Moreover, this homomorphism is bijective and identifies $(r/t)/(s/t')$ with rt'/st for any $r \in R$, $s \in S$ and $t, t' \in T$.

(ii) Let R be a commutative ring and $s, t \in R$ two elements with $u = st$. If S, T and U are the multiplicatively closed subsets of R generated by s, t and u respectively[2] and T' is the image of T in $R' = R[S^{-1}]$, then there is a unique R-algebra homomorphism

$$R'[T'^{-1}] \longrightarrow R[U^{-1}].$$

Moreover, this homomorphism is bijective and identifies $(r/s^n)/(t^m/1)$ with $rs^m t^n/u^{n+m}$ for any $r \in R$ and $n, m \geq 0$.

---

[1] This property can be rephrased by saying that the complement $I = R \setminus S$ is such that $IR \subseteq I$.

[2] By this, we mean that S consists of the powers s^n, $n \geq 0$, of s and similarly for T and U.

Proof. Assertion (i) follows since $R'[S'^{-1}]$ is the universal commutative R-algebra in which the image of S consists of invertible elements. Similarly, the claim in (ii) is a consequence of the observation that the universal commutative R-algebra in which the images of s and t are invertible is precisely the universal commutative R-algebra in which the image of their product u is invertible. $\square$

Our next goal is to prove the flatness of localization. To that end, we let S be a multiplicatively closed subset of a commutative ring R containing 1 and fix an R-module M. We then consider the $R[S^{-1}]$-module $M[S^{-1}]$, which is defined as follows: As a set, $M[S^{-1}]$ consists of the equivalence classes of pairs of the form (m, s), where $m \in M$ and $s \in S$, under the equivalence relation defined by the rule

$$(m, s) \sim (m', s') \text{ if and only if there exists } s_0 \in S \text{ such that } s_0 s m' = s_0 s' m$$

for $m, m' \in M$ and $s, s' \in S$. We denote the equivalence class of a pair (m, s), where $m \in M$ and $s \in S$, by m/s. Addition and the $R[S^{-1}]$-action on $M[S^{-1}]$ are defined by the rules

$$m/s + m'/s' = (s'm + sm')/ss' \quad \text{and} \quad (r/s) \cdot (m'/s') = (rm')/(ss')$$

for $m, m' \in M$, $r \in R$ and $s, s' \in S$. It is straightforward to verify that these operations are well-defined and endow $M[S^{-1}]$ with the structure of an $R[S^{-1}]$-module.

Pursuing further the analogy with the construction of the ring $R[S^{-1}]$, we consider the map

$$\lambda = \lambda_{M,S} : M \longrightarrow M[S^{-1}],$$

which is given by $m \mapsto m/1$, $m \in M$. Then, λ is R-linear and has the universal property described in the following result.

Proposition A.4 *Let R be a commutative ring, $S \subseteq R$ a multiplicatively closed subset containing 1, M an R-module, $M[S^{-1}]$ the $R[S^{-1}]$-module defined above and $\lambda : M \longrightarrow M[S^{-1}]$ the corresponding R-linear map. Then, for any $R[S^{-1}]$-module N and any R-linear map $\varphi : M \longrightarrow N'$ there exists a unique $R[S^{-1}]$-linear map $\Phi : M[S^{-1}] \longrightarrow N$ with $\Phi \circ \lambda = \varphi$. (Here, we denote by N' the R-module obtained from N by restricting the scalars along the natural homomorphism $R \longrightarrow R[S^{-1}]$.)*

Proof. Given a pair (N, φ), we may define Φ by letting $\Phi(m/s) = (1/s) \cdot \varphi(m)$ for all $m/s \in M[S^{-1}]$. Then, Φ is a well-defined $R[S^{-1}]$-linear map such that $\Phi \circ \lambda = \varphi$. The uniqueness of Φ satisfying that condition follows since $m/s = (1/s) \cdot \lambda(m) \in M[S^{-1}]$ for all $m \in M$, $s \in S$. $\square$

Corollary A.5 *Let R be a commutative ring, $S \subseteq R$ a multiplicatively closed subset containing 1 and M an R-module.*
(i) There is an $R[S^{-1}]$-module isomorphism $M \otimes_R R[S^{-1}] \simeq M[S^{-1}]$, which identifies $m \otimes 1/s$ with m/s for all $m \in M$ and $s \in S$.

(ii) If there exists an element $s \in S$ such that $sM = 0$, then the $R[S^{-1}]$-module $M \otimes_R R[S^{-1}]$ vanishes.

(iii) If the R-module M is finitely generated and the $R[S^{-1}]$-module $M \otimes_R R[S^{-1}]$ vanishes, then there exists $s \in S$ such that $sM = 0$.

Proof. (i) This follows from Proposition A.4, in view of the universal property of the $R[S^{-1}]$-module $M \otimes_R R[S^{-1}]$.

(ii) If $s \in S$ and $m \in M$ are such that $sm = 0$, then $m/1 = 0/1 \in M[S^{-1}]$. Therefore, if $sM = 0$ then $M \otimes_R R[S^{-1}] \simeq M[S^{-1}] = 0$.

(iii) Assume that M is generated over R by $m_1, \ldots, m_n$. Since the module $M[S^{-1}] \simeq M \otimes_R R[S^{-1}]$ vanishes, we have $m_i/1 = 0/1 \in M[S^{-1}]$ for all $i = 1, \ldots, n$. But then there exists for each i an element $s_i \in S$, such that $s_i m_i = 0$. If $s = \prod_{i=1}^{n} s_i \in S$ then s annihilates all generators $m_1, \ldots, m_n$ of M and hence $sM = 0$.   □

Corollary A.6 *Let R be a commutative ring and $S \subseteq R$ a multiplicatively closed subset containing 1. Then, the R-module $R[S^{-1}]$ is flat.*

Proof. In view of Corollary A.5(i), we have to verify that the functor $M \mapsto M[S^{-1}]$, M an R-module, is left exact. In other words, we have to verify that for any submodule N of an R-module M the map $N[S^{-1}] \longrightarrow M[S^{-1}]$, which maps $n/s \in N[S^{-1}]$ onto $n/s \in M[S^{-1}]$ is injective. But this is clear from the definition of equality in the modules $M[S^{-1}]$ and $N[S^{-1}]$.   □

II. LOCALIZATION AT A PRIME IDEAL. An important class of multiplicatively closed subsets of a commutative ring R is that arising from prime ideals. If $\wp \subseteq R$ is a prime ideal (i.e. if $\wp$ is a proper ideal of R such that the quotient ring $R/\wp$ is an integral domain) then the complement $S = R \setminus \wp$ is multiplicatively closed and contains 1. The corresponding localization is denoted by $R_\wp$ and referred to (by an obvious abuse of language) as the localization of R at $\wp$. For example, if R is an integral domain then 0 is a prime ideal and the localization R_0 is the field of fractions of R.

The following result describes a property of localization that was used in a crucial way in §2.1.1, in the proof of the continuity of the geometric rank function associated with a finitely generated projective module.

Proposition A.7 *Let R be a commutative ring, M (resp. P) a finitely generated (resp. finitely generated projective) R-module and $\varphi : M \longrightarrow P$ an R-linear map. Assume that $\wp \subseteq R$ is a prime ideal, such that φ becomes an isomorphism when localized at $\wp$. Then, there exists an element $u \in R \setminus \wp$, such that φ becomes an isomorphism after inverting u (i.e. when localizing at the multiplicatively closed subset $U \subseteq R$ generated by u).*

Proof. The localized map $\varphi \otimes 1 : M \otimes_R R_\wp \longrightarrow P \otimes_R R_\wp$ being bijective, the R-flatness of $R_\wp$ (cf. Corollary A.6) implies that both $\ker \varphi \otimes_R R_\wp$ and $\operatorname{coker} \varphi \otimes_R R_\wp$ vanish. Since the R-module $\operatorname{coker} \varphi$ is finitely generated, Corollary A.5(iii) implies that there exists $s \in R \setminus \wp$ such that $s \cdot \operatorname{coker} \varphi = 0$. Let S be the multiplicatively closed subset of R generated by s; we note that $S \subseteq R \setminus \wp$. If $R' = R[S^{-1}]$, $\varphi' = \varphi \otimes 1$ and $\wp' = \wp R[S^{-1}]$, then the R-flatness of R' and the vanishing of coker $\varphi \otimes_R R'$ (cf. Corollary A.5(ii)) show that there is a short exact sequence of R'-modules

$$0 \longrightarrow \ker \varphi \otimes_R R' \longrightarrow M \otimes_R R' \xrightarrow{\varphi'} P \otimes_R R' \longrightarrow 0 .$$

Since the R'-module $P \otimes_R R'$ is projective, this short exact sequence splits; in particular, the R'-module $\ker \varphi' = \ker \varphi \otimes_R R'$ is finitely generated. We note that

$$\ker \varphi' \otimes_{R'} R'_{\wp'} = \ker \varphi \otimes_R R'_{\wp'} = \ker \varphi \otimes_R R_\wp = 0 ,$$

where the second equality follows using Corollary A.3(i). Hence, we may invoke Corollary A.5(iii) once again in order to find an element $t' \in R' \setminus \wp'$ with $t' \cdot \ker \varphi' = 0$. Without loss of generality, we assume that $t' = t/1$ for some $t \in R \setminus \wp$. Let U (resp. T') be the multiplicatively closed subset of R (resp. R') generated by $u = st \in R \setminus \wp$ (resp. by t'). Then,

$$\ker \varphi \otimes_R R[U^{-1}] = \ker \varphi \otimes_R R'[T'^{-1}] = \ker \varphi' \otimes_{R'} R'[T'^{-1}] = 0 ,$$

where the first (resp. third) equality follows from Corollary A.3(ii) (resp. Corollary A.5(ii)). Since $u = st$ annihilates coker φ, Corollary A.5(ii) implies that coker $\varphi \otimes_R R[U^{-1}] = 0$. It follows that the map

$$\varphi \otimes 1 : M \otimes_R R[U^{-1}] \longrightarrow P \otimes_R R[U^{-1}]$$

is an isomorphism and this finishes the proof.   □

III. LOCAL RINGS. A commutative ring is called local if it has a unique maximal ideal. For example, any field is local, whereas the ring $\mathbf{Z}$ is not.

Remarks A.8 (i) It is clear that any proper ideal in a commutative ring consists of singular (i.e. non-invertible) elements. In fact, if R is a local ring then its maximal ideal $\mathbf{m}$ is precisely the set of singular elements. Indeed, if $r \in R$ is not invertible then the ideal Rr is proper and hence contained in the (unique) maximal ideal $\mathbf{m}$, i.e. $r \in \mathbf{m}$.

(ii) As a converse to (i), we note that if the set of singular elements of a commutative ring R forms an ideal I, then R is a local ring with maximal ideal $\mathbf{m} = I$. This is clear since any proper ideal J of R consists of singular elements and is therefore contained in I.

(iii) If R is a commutative ring and $\wp \subseteq R$ a prime ideal, then the localization $R_\wp$ is a local ring with maximal ideal

$$\wp R_\wp = \{r/s : r \in \wp, s \notin \wp\} \subseteq R_\wp .$$

This is an immediate consequence of (ii) above, as soon as one notices that an element of the localization $R_\wp$ is invertible if and only if it is of the form r/s, with $r, s \notin \wp$ (cf. Remark A.1).

Our next goal is to show that any finitely generated projective module over a local ring is free,[3] a fact used in the very definition of the geometric rank function associated with a finitely generated projective module in §2.1.1. To that end, we need the following lemma.

Lemma A.9 *Let R be a commutative ring, $I \subseteq R$ a proper ideal and M a finitely generated R-module satisfying the equality $IM = M$. Then, there exists an element $r \in I$ such that $(1+r)M = 0$. In particular, M is trivial (i.e. $M = 0$) if either one of the following two conditions is satisfied:*
(i) (Nakayama) R is a local ring or
(ii) R is an integral domain and the R-module M is torsion-free.

Proof. If $M = \sum_{i=1}^{n} Rm_i$ then $IM = \sum_{i=1}^{n} Im_i$ and hence we can write $m_i = \sum_{j=1}^{n} r_{ij}m_j$, for suitable elements $r_{ij} \in I$. It follows that the matrix

$$
A = \begin{bmatrix}
1 - r_{11} & -r_{12} & \cdots & -r_{1n} \\
-r_{21} & 1 - r_{22} & \cdots & -r_{2n} \\
\vdots & \vdots & & \vdots \\
-r_{n1} & -r_{n2} & \cdots & 1 - r_{nn}
\end{bmatrix} \in \mathbf{M}_n(R)
$$

annihilates the $n \times 1$ column-vector

$$
\vec{m} = \begin{bmatrix}
m_1 \\
m_2 \\
\vdots \\
m_n
\end{bmatrix}
$$

Multiplying the equation $A \cdot \vec{m} = \mathbf{0}$ to the left by the matrix $\mathrm{adj}\, A$, we deduce that the element $\det A \in R$ annihilates all of the m_i's. Since these elements generate M, we conclude that $(\det A)M = 0$. It is immediate by the form of the matrix A that $\det A = 1 + r$, for a suitable element $r \in I$.

(i) Assume that the ring R is local with maximal ideal $\mathbf{m}$. Since the ideal I is proper, we have $I \subseteq \mathbf{m}$ and hence $1 + r \notin \mathbf{m}$. In view of Remark A.8(i), we conclude that $1 + r$ is invertible in R. The equation $(1+r)M = 0$ then shows that $M = 0$.

(ii) Assume that R is an integral domain, whereas M is torsion-free and non-zero. Then, the equation $(1+r)M = 0$ implies that $1 + r = 0$. But then $1 = -r \in I$, contradicting our assumption that I is proper. $\square$

Proposition A.10 *Let R be a local ring and P a finitely generated projective R-module. Then, P is free.*

Proof. Let $\mathbf{m}$ be the maximal ideal of R and $k = R/\mathbf{m}$ the residue field. Then, $P/\mathbf{m}P = P \otimes_R k$ is a finite dimensional k-vector space; hence, it has a basis

---

[3] In fact, any projective module (not necessarily finitely generated) over a local ring is free; this result is due to I. Kaplansky [37].

$\overline{x_1}, \ldots, \overline{x_n}$ for suitable elements $x_1, \ldots, x_n \in P$. We shall prove that the x_i's form a basis of P by showing that the R-linear map $\varphi : R^n \longrightarrow P$, which maps $(r_1, \ldots, r_n) \in R^n$ onto $r_1 x_1 + \ldots + r_n x_n \in P$ is an isomorphism. The exact sequence of R-modules

$$R^n \xrightarrow{\varphi} P \longrightarrow \operatorname{coker} \varphi \longrightarrow 0$$

induces the exact sequence of k-vector spaces

$$k^n \xrightarrow{\varphi \otimes 1} P \otimes_R k \longrightarrow \operatorname{coker} \varphi \otimes_R k \longrightarrow 0 \, .$$

Since $\varphi \otimes 1$ is bijective, the k-vector space $\operatorname{coker} \varphi \otimes_R k$ vanishes. Therefore, Nakayama's lemma implies that the finitely generated R-module $\operatorname{coker} \varphi$ is actually zero and hence φ is onto. We note that the short exact sequence of R-modules

$$0 \longrightarrow \ker \varphi \longrightarrow R^n \xrightarrow{\varphi} P \longrightarrow 0$$

splits, since P is projective. It follows that the R-module $\ker \varphi$ is finitely generated, whereas there is an induced (split) short exact sequence

$$0 \longrightarrow \ker \varphi \otimes_R k \longrightarrow k^n \xrightarrow{\varphi \otimes 1} P \otimes_R k \longrightarrow 0 \, .$$

As before, Nakayama's lemma implies that $\ker \varphi = 0$ and hence φ is 1-1. $\quad\square$

A.2 Integral Dependence

Let us consider a commutative ring R and a commutative R-algebra T. An element $t \in T$ is called integral over R if there is a monic polynomial $f(X) \in R[X]$, such that $f(t) = 0 \in T$. It is clear that elements in the canonical image of R in T are integral over R; if these are the only elements of T that are integral over R, then R is said to be integrally closed in T. On the other extreme, the algebra T is called integral over R (or an integral extension of R) if any element $t \in T$ is integral over R. The following result establishes a few equivalent formulations of the integrality condition.

Proposition A.11 Let R be a commutative ring and T a commutative R-algebra. Then, the following conditions are equivalent for an element $t \in T$:

(i) t is integral over R.

(ii) The R-subalgebra $R[t] \subseteq T$ is finitely generated as an R-module.

(iii) There is a finitely generated R-submodule $M \subseteq T$ containing 1, such that $tM \subseteq M$.

Proof. (i)→(ii): Assume that t is integral over R. Then, there is an integer $n \geq 1$ and elements $r_0, \ldots, r_{n-1} \in R$, such that

$$t^n + r_{n-1} t^{n-1} + \cdots + r_1 t + r_0 = 0 \in T \, .$$

It follows that t^n is an R-linear combination of $1, t, \ldots, t^{n-1}$ and hence $R[t] = \sum_{i=0}^{n-1} Rt^i$ is a finitely generated R-module.

(ii)$\to$(iii): This is clear, since we may choose $M = R[t]$.

(iii)$\to$(i): Let $M = \sum_{i=1}^{n} Rm_i \subseteq T$ be an R-submodule of T that contains 1 and is closed under multiplication by t. Then, there are equations of the form $tm_i = \sum_{j=1}^{n} r_{ij}m_j$, $i = 1, \ldots, n$, for suitable elements $r_{ij} \in R$. It follows that the $n \times n$ matrix

$$A = \begin{bmatrix} t - r_{11} & -r_{12} & \cdots & -r_{1n} \\ -r_{21} & t - r_{22} & \cdots & -r_{2n} \\ \vdots & \vdots & & \vdots \\ -r_{n1} & -r_{n2} & \cdots & t - r_{nn} \end{bmatrix} \in \mathbf{M}_n(T)$$

annihilates the $n \times 1$ column-vector

$$\vec{m} = \begin{bmatrix} m_1 \\ m_2 \\ \vdots \\ m_n \end{bmatrix}$$

Multiplying the equation $A \cdot \vec{m} = 0$ to the left by the matrix adj A, we deduce that the element $\det A \in T$ annihilates all of the m_i's. Since these elements generate the R-module M, it follows that $\det A$ annihilates M; in particular, $\det A = (\det A)1 = 0$. Then, the expansion of $\det A$ provides us with an equation that shows t to be integral over R. □

Corollary A.12 *Let R be a commutative ring, T a commutative R-algebra and U a commutative T-algebra.*

(i) If $t_1, \ldots, t_n \in T$ are integral over R, then the finitely generated R-subalgebra $R[t_1, \ldots, t_n] \subseteq T$ is finitely generated as an R-module.

(ii) If $u \in U$ is integral over T and T is integral over R, then u is integral over R. In particular, if U is integral over T and T integral over R, then U is integral over R.

Proof. (i) We use induction on n, the case $n = 1$ following from Proposition A.11. Since t_n is integral over R, it is also integral over $R[t_1, \ldots, t_{n-1}]$. Therefore, Proposition A.11 implies that $R[t_1, \ldots, t_n] = R[t_1, \ldots, t_{n-1}][t_n]$ is finitely generated as an $R[t_1, \ldots, t_{n-1}]$-module. By the induction hypothesis, $R[t_1, \ldots, t_{n-1}]$ is finitely generated as an R-module and hence $R[t_1, \ldots, t_n]$ is finitely generated as an R-module (cf. Lemma 1.1(i)).

(ii) Since u is integral over T, there is an integer $n \geq 1$ and elements $t_0, \ldots, t_{n-1} \in T$, such that

$$u^n + t_{n-1}u^{n-1} + \cdots + t_1 u + t_0 = 0 \in U .$$

This equation shows that u is integral over the ring $R' = R[t_0, \ldots, t_{n-1}]$. It follows from Proposition A.11 that $R'[u]$ is finitely generated as an R'-module. But R' is finitely generated as an R-module, in view of (i) above, and hence $R'[u]$ is finitely generated as an R-module (cf. Lemma 1.1(i)). Using Proposition A.11 once again, we deduce that u is integral over R.   □

Corollary A.13 *Let R be a commutative ring, T a commutative R-algebra and $R' = \{t \in T : t$ is integral over $R\}$. Then:*
(i) R' is an R-subalgebra of T and
(ii) R' is integrally closed in T.
The R-algebra R' is called the integral closure of R in T.

Proof. (i) For any $t_1, t_2 \in R'$ the R-algebra $R[t_1, t_2]$ is finitely generated as an R-module, in view of Corollary A.12(i). Since $R[t_1, t_2]$ contains 1 and is closed under multiplication by $t_1 \pm t_2$ and $t_1 t_2$, Proposition A.11 implies that $t_1 \pm t_2, t_1 t_2 \in R'$. It follows that R' is a subring of T. Since $r \cdot 1 \in T$ is obviously integral over R for any $r \in R$, it follows that R' is an R-submodule of T (and hence an R-subalgebra of it).

(ii) This is an immediate consequence of Corollary A.12(ii).   □

The next result describes some basic properties of the integral closure of a domain in an extension of its field of fractions.

Lemma A.14 *Let R be an integral domain, K its field of fractions, L an algebraic extension field of K and T the integral closure of R in L.*
(i) For any $x \in L$ there exists $r \in R \setminus \{0\}$, such that $rx \in T$. In particular, L is the field of fractions of T.
(ii) Any field automorphism σ of L over K restricts to an automorphism of T over R.

Proof. (i) Any element $x \in L$ is algebraic over K and hence satisfies an equation of the form

$$r_n x^n + r_{n-1} x^{n-1} + \cdots + r_1 x + r_0 = 0 \,,$$

for an integer $n \geq 1$ and suitable elements $r_0, \ldots, r_n \in R$, with $r_n \neq 0$. Multiplying that equation by r_n^{n-1}, we obtain the equation

$$(r_n x)^n + r_{n-1} (r_n x)^{n-1} + \cdots + r_n^{n-2} r_1 (r_n x) + r_n^{n-1} r_0 = 0 \,.$$

Hence, $r_n x$ is integral over R, i.e. $r_n x \in T$.
(ii) Any element $t \in T$ satisfies an equation of the form

$$t^n + r_{n-1} t^{n-1} + \cdots + r_1 t + r_0 = 0 \,,$$

for an integer $n \geq 1$ and suitable elements $r_0, \ldots, r_{n-1} \in R$. If σ is an automorphism of L over K, then the element $\sigma(t) \in L$ satisfies the equation

$$\sigma(t)^n + r_{n-1} \sigma(t)^{n-1} + \cdots + r_1 \sigma(t) + r_0 = 0 \,.$$

Hence, $\sigma(t)$ is integral over R, i.e. $\sigma(t) \in T$. Considering the automorphism σ^{-1} of L over K, it follows that σ maps T bijectively onto itself.   □

Having the ring $\mathbf{Z}$ of integers in mind, we examine the case where R is a principal ideal domain or, more generally, a unique factorization domain.

Lemma A.15 *A unique factorization domain is integrally closed in its field of fractions.*

Proof. Let R be a unique factorization domain and K its field of fractions. An element $x \in K$ is not contained in R if it can be expressed as a quotient of the form a/b, where the elements $a, b \in R \setminus \{0\}$ are relatively prime and b is not a unit. If such an element x is integral over R, it satisfies an equation of the form

$$x^n + r_{n-1}x^{n-1} + \cdots + r_1 x + r_0 = 0 \in K,$$

for an integer $n \geq 1$ and suitable elements $r_0, \ldots, r_{n-1} \in R$. Multiplying that equation by b^n, we obtain the equality

$$a^n + r_{n-1}ba^{n-1} + \cdots + r_1 b^{n-1}a + r_0 b^n = 0 \in R. \tag{A.1}$$

Since b is not a unit, it has a prime divisor p. Then, p divides the sum

$$r_{n-1}ba^{n-1} + \cdots + r_1 b^{n-1}a + r_0 b^n = b(r_{n-1}a^{n-1} + \cdots + r_1 b^{n-2}a + r_0 b^{n-1})$$

and hence (A.1) implies that p divides a^n. But the prime p does not divide a, since a and b are relatively prime, and this is a contradiction.   □

Corollary A.16 *Let R be a unique factorization domain, K its field of fractions, L a Galois extension field of K and T the integral closure of R in L. Let $x \in T$ and consider its Galois conjugates $x_1 (= x), x_2, \ldots, x_k$. Then, the polynomial $f(X) = \prod_{i=1}^{k}(X - x_i)$ has coefficients in R.*

Proof. First of all, we note that the x_i's are contained in T, in view of Lemma A.14(ii); therefore, $f(X)$ is a polynomial in $T[X]$. Being a symmetric polynomial in the x_i's, any coefficient $y \in T$ of $f(X)$ is invariant under the action of the Galois group of L over K and hence $y \in K$. Since $T \cap K = R$, in view of Lemma A.15, we conclude that $y \in R$.   □

Corollary A.17 *Let R be a principal ideal domain contained in an integral domain T and assume that T is integral over R. Then, for any prime element $p \in R$ there exists a maximal ideal $\mathbf{m} \subseteq T$, such that $R \cap \mathbf{m} = pR$; in that case, the field $T/\mathbf{m}$ is an extension of R/pR.*

Proof. First of all, we note that elements of R that are invertible in T must be already invertible in R; this follows from Lemma A.15, since T is integral over R, whereas the principal ideal domain R is a unique factorization domain.[4] In particular, a prime element $p \in R$ is not invertible in T and hence the ideal $pT \subseteq T$ is proper. If $\mathbf{m} \subseteq T$ is a maximal ideal containing pT, then the

---

[4] In fact, one doesn't need the assumption that R is a principal ideal domain; cf. Exercise A.5.1.

contraction $R \cap \mathbf{m}$ is a prime (and hence proper) ideal of R containing pR. But pR is a maximal ideal of R and hence $R \cap \mathbf{m} = pR$.   □

We now specialize the above discussion and consider the subring $\mathcal{R}$ of $\overline{\mathbf{Q}}$ consisting of those algebraic numbers that are integral over $\mathbf{Z}$; this is the ring of algebraic integers. The following three properties of $\mathcal{R}$ are immediate consequences of the general results established above.

(AI1) For any algebraic number $x \in \overline{\mathbf{Q}}$ there is a non-zero integer n, such that $nx \in \mathcal{R}$.

(AI2) Let x be an algebraic integer and $x_1(= x), x_2, \ldots, x_k$ its Galois conjugates. Then, all x_i's are algebraic integers, whereas the polynomial $\prod_{i=1}^{k}(X - x_i)$ has coefficients in $\mathbf{Z}$.

(AI3) For any prime number $p \in \mathbf{Z}$ there exists a maximal ideal $\mathcal{M} \subseteq \mathcal{R}$, such that $\mathbf{Z} \cap \mathcal{M} = p\mathbf{Z}$; then, the field $\mathcal{R}/\mathcal{M}$ has characteristic p.

A.3 Noether Normalization

I. FINITELY GENERATED ALGEBRAS OVER FIELDS. In order to obtain some information on the structure of finitely generated commutative algebras over a field, we begin with a few simple observations on polynomials.

Lemma A.18 *Let R be an integral domain of characteristic 0 and consider m distinct polynomials $f_1(X), \ldots, f_m(X) \in R[X]$. Then, there exists $t_0 \in \mathbf{N}$ such that for all $t \in \mathbf{N}$ with $t > t_0$ the m elements $f_1(t), \ldots, f_m(t) \in R$ are distinct.*

Proof. For any $i \neq j$, the equation $f_i(X) = f_j(X)$ has finitely many roots in R. Since $\mathbf{N} \subseteq \mathbf{Z}$ is contained in R, the set $\Lambda_{ij} = \{t \in \mathbf{N} : f_i(t) = f_j(t)\}$ is finite. Being a finite union of finite sets, the set $\Lambda = \bigcup_{i \neq j} \Lambda_{ij}$ is finite as well. We now let $t_0 = \max \Lambda$ and note that if $t \in \mathbf{N}$ exceeds t_0, then $t \notin \Lambda_{ij}$ (i.e. $f_i(t) \neq f_j(t)$) for all $i \neq j$.   □

Corollary A.19 *Let $\left(k_j^{(1)}\right)_j, \ldots, \left(k_j^{(m)}\right)_j \in \mathbf{N}^n$ be m distinct n-tuples of non-negative integers. Then, there are positive integers $t_1, \ldots, t_n$, with $t_n = 1$, such that the m integers $\sum_{j=1}^{n} t_j k_j^{(1)}, \ldots, \sum_{j=1}^{n} t_j k_j^{(m)} \in \mathbf{N}$ are distinct.*

Proof. For all $i = 1, \ldots, m$, we consider the polynomial

$$f_i(X) = \sum_{j=1}^{n} k_j^{(i)} X^{n-j} \in \mathbf{Z}[X] \,.$$

In view of our assumption, the polynomials $f_1(X), \ldots, f_m(X)$ are distinct. Hence, we may invoke Lemma A.18 in order to find $t \in \mathbf{N}$, $t > 0$, such that the m integers $\sum_{j=1}^{n} k_j^{(1)} t^{n-j}, \ldots, \sum_{j=1}^{n} k_j^{(m)} t^{n-j} \in \mathbf{Z}$ are distinct. The proof is finished by letting $t_j = t^{n-j}$ for all $j = 1, \ldots, n$.   □

Let k be a commutative ring, R a commutative k-algebra and $k[X_1,\ldots,X_n]$ the polynomial k-algebra in n variables. Then, for any n-tuple of elements $(r_1,\ldots,r_n) \in R^n$ there is a unique k-algebra homomorphism

$$\mathbf{r} : k[X_1,\ldots,X_n] \longrightarrow R \,,$$

which maps X_i onto r_i for all $i = 1,\ldots,n$. The image of a polynomial $f(X_1,\ldots,X_n) \in k[X_1,\ldots,X_n]$ under $\mathbf{r}$ is denoted by $f(r_1,\ldots,r_n)$ and referred to as the evaluation of f at the n-tuple $(r_1,\ldots,r_n)$.

In particular, let us consider n other variables $Y_1,\ldots,Y_{n-1},X$ and the corresponding polynomial algebra $R = k[Y_1,\ldots,Y_{n-1},X]$. If $t_1,\ldots,t_{n-1}$ are positive integers, then the n-tuple $(Y_1 + X^{t_1},\ldots,Y_{n-1} + X^{t_{n-1}},X) \in R^n$ induces a homomorphism of k-algebras

$$k[X_1,\ldots,X_n] \longrightarrow R = k[Y_1,\ldots,Y_{n-1},X] \,,$$

which maps X_i onto $Y_i + X^{t_i}$ for all $i = 1,\ldots,n-1$ and X_n onto X. For any polynomial $f(X_1,\ldots,X_n) \in k[X_1,\ldots,X_n]$ the element

$$g(Y_1,\ldots,Y_{n-1},X) = f(Y_1 + X^{t_1},\ldots,Y_{n-1} + X^{t_{n-1}},X) \in R$$

may be viewed as a polynomial in X with coefficients in $k[Y_1,\ldots,Y_{n-1}]$. If the degree of g in X is m, then we can write

$$f(Y_1 + X^{t_1},\ldots,Y_{n-1} + X^{t_{n-1}},X) = \sum_{i=0}^{m} g_i(Y_1,\ldots,Y_{n-1})X^i, \qquad (A.2)$$

where $g_i(Y_1,\ldots,Y_{n-1}) \in k[Y_1,\ldots,Y_{n-1}]$ for all i and $g_m(Y_1,\ldots,Y_{n-1}) \neq 0$.

Corollary A.20 Let k be a commutative ring, $X_1,\ldots,X_n,Y_1,\ldots,Y_{n-1},X$ independent indeterminates and $f(X_1,\ldots,X_n) \in k[X_1,\ldots,X_n]$ a non-zero polynomial. Then, there are positive integers $t_1,\ldots,t_{n-1}$, such that the leading coefficient $g_m(Y_1,\ldots,Y_{n-1}) \in k[Y_1,\ldots,Y_{n-1}]$ of X in the expression (A.2) of

$$g(Y_1,\ldots,Y_{n-1},X) = f(Y_1 + X^{t_1},\ldots,Y_{n-1} + X^{t_{n-1}},X) \in k[Y_1,\ldots,Y_{n-1},X]$$

is a non-zero element of k.

Proof. Let $aX_1^{k_1} \cdots X_n^{k_n}$ be a monomial of the polynomial $f(X_1,\ldots,X_n)$, where $a \in k$ is non-zero and the k_i's are non-negative integers. For any choice of positive integers $t_1,\ldots,t_{n-1}$, it is clear that the monomial in $Y_1,\ldots,Y_{n-1},X$ of the summand $a(Y_1 + X^{t_1})^{k_1} \cdots (Y_{n-1} + X^{t_{n-1}})^{k_{n-1}} X^{k_n}$ of $g(Y_1,\ldots,Y_{n-1},X)$ with the highest degree in X is the product

$$aX^{t_1 k_1} \cdots X^{t_{n-1} k_{n-1}} X^{k_n} = aX^{t_1 k_1 + \cdots + t_{n-1} k_{n-1} + k_n} \,,$$

whereas all other monomials of $a(Y_1 + X^{t_1})^{k_1} \cdots (Y_{n-1} + X^{t_{n-1}})^{k_{n-1}} X^{k_n}$ have degree in X strictly less than $t_1 k_1 + \cdots + t_{n-1} k_{n-1} + k_n$. In view of Corollary A.19, we may choose the positive numbers $t_1,\ldots,t_{n-1}$, in such a way that the exponents $t_1 k_1 + \cdots + t_{n-1} k_{n-1} + k_n$ that result from the various monomials of $f(X_1, \ldots, X_n)$ are distinct. Then, the largest of these exponents is the degree m of g in X, whereas the leading term $g_m X^m$ is of the form $a' X^m$ for some non-zero element $a' \in k$. □

We are now ready to state and prove the normalization lemma.

Theorem A.21 *(Noether normalization lemma) Let k be a field and R a finitely generated commutative k-algebra. Then, there exists a k-subalgebra $R_0 \subseteq R$, such that:*
   (i) R_0 is a polynomial k-algebra and
   (ii) R is an integral extension of R_0.

Proof. Let $R = k[r_1, \ldots, r_n]$ for suitable elements $r_1, \ldots, r_n \in R$. We shall use induction on n.

If $n = 1$, then either r_1 is algebraically independent over k, in which case we may let $R_0 = R$, or else r_1 satisfies a (monic) polynomial equation with coefficients in k, in which case $\dim_k R < \infty$ and we let $R_0 = k$.

Assume that $n > 1$ and the result has been proved for k-algebras generated by $n - 1$ elements. If the elements $r_1, \ldots, r_n$ are algebraically independent over k, we may let $R_0 = R$. If not, there exists a non-zero polynomial $f(X_1, \ldots, X_n) \in k[X_1, \ldots, X_n]$, such that $f(r_1, \ldots, r_n) = 0$. In view of Corollary A.20, we can find positive integers $t_1, \ldots, t_{n-1}$ such that the leading coefficient $g_m(Y_1, \ldots, Y_{n-1}) \in k[Y_1, \ldots, Y_{n-1}]$ of X in the expression (A.2) of the polynomial

$$g(Y_1, \ldots, Y_{n-1}, X) = f(Y_1 + X^{t_1}, \ldots, Y_{n-1} + X^{t_{n-1}}, X) \in k[Y_1, \ldots, Y_{n-1}, X]$$

is a non-zero element $a \in k$. Replacing the polynomial $f(X_1, \ldots, X_n)$ by $a^{-1} f(X_1, \ldots, X_n)$, we may assume that $g_m(Y_1, \ldots, Y_{n-1}) = 1$. Then, g may be viewed as a monic polynomial in X with coefficients in $k[Y_1, \ldots, Y_{n-1}]$. Let $r_1' = r_1 - r_n^{t_1}, \ldots r_{n-1}' = r_{n-1} - r_n^{t_{n-1}}$ and consider the k-subalgebra $R' = k[r_1', \ldots, r_{n-1}'] \subseteq R$. It is clear that $R = k[r_1, \ldots, r_n]$ can be generated by $r_1', \ldots, r_{n-1}', r_n$ and hence $R = k[r_1', \ldots, r_{n-1}', r_n] = R'[r_n]$. Since

$$\begin{aligned} g(r_1', \ldots, r_{n-1}', r_n) &= f(r_1' + r_n^{t_1}, \ldots, r_{n-1}' + r_n^{t_{n-1}}, r_n) \\ &= f(r_1, \ldots, r_{n-1}, r_n) \\ &= 0, \end{aligned}$$

r_n is a root of the monic polynomial $g(r_1', \ldots, r_{n-1}', X) \in R'[X]$. Therefore, $R = R'[r_n]$ is a finitely generated R'-module and hence R is integral over R' (cf. Proposition A.11). In view of the induction hypothesis, there exists a k-subalgebra $R_0 \subseteq R'$, such that R_0 is a polynomial k-algebra and R' is integral over it. Since R is integral over R', we may invoke Corollary A.12(ii) and conclude that R is integral over R_0 as well. □

Often, we apply Noether's normalization lemma in the form of one of the following corollaries.

Corollary A.22 *Let k be a subring of a field K.*

(i) If K is integral over k, then k is a field.

(ii) If k is a field and K is finitely generated as a k-algebra, then K is a finite algebraic extension of k.

Proof. (i) If a is a non-zero element of k, then $a^{-1} \in K$ is integral over k and hence

$$a^{-n} + b_{n-1}a^{-n+1} + \cdots + b_1 a^{-1} + b_0 = 0 \, ,$$

for an integer $n \geq 1$ and suitable elements $b_0, \ldots, b_{n-1} \in k$. Multiplying that equation by a^{n-1}, we conclude that

$$a^{-1} + b_{n-1} + \cdots + b_1 a^{n-2} + b_0 a^{n-1} = 0$$

and hence $a^{-1} \in k$.

(ii) In view of Noether's normalization lemma, we can find a polynomial k-algebra K_0 contained in K, such that K is integral over it. Using the result of part (i), we conclude that K_0 is a field. Being a polynomial algebra over k, K_0 can be a field only if $K_0 = k$. Therefore, K is an integral (i.e. algebraic) extension of k. □

Corollary A.23 *Let k be an algebraically closed field and R a finitely generated commutative k-algebra. Then, the set $\operatorname{Hom}_{k-Alg}(R, k)$ is non-empty.*

Proof. Let $\mathbf{m} \subseteq R$ be a maximal ideal. Then, the field $K = R/\mathbf{m}$ is finitely generated as a k-algebra and hence Corollary A.22(ii) implies that K is an algebraic extension of k. Being algebraically closed, k has no proper algebraic extensions; therefore, $K = k$. It follows that the quotient map $R \longrightarrow R/\mathbf{m}$ is a homomorphism of the type we are looking for. □

II. FINITELY GENERATED ALGEBRAS OVER UFD'S. We now turn our attention to finitely generated commutative algebras B over a unique factorization domain A. Our goal is to examine the prime elements of A that are invertible in B.

In general, if A is an integral domain and $a \in A$ a non-zero element, we denote by $A[a^{-1}]$ the A-subalgebra of the field of fractions of A generated by a^{-1}. If M is an A-module then $M[a^{-1}]$ denotes the $A[a^{-1}]$-module $M \otimes_A A[a^{-1}]$.

We begin with a generalization of Noether's normalization lemma.

Lemma A.24 *Let A be an integral domain and B a finitely generated commutative A-algebra, which is torsion-free as an A-module. Then, there is a non-zero element $a \in A$ and an $A[a^{-1}]$-subalgebra $B_0 \subseteq B[a^{-1}]$, such that:*

(i) B_0 is a polynomial $A[a^{-1}]$-algebra and

(ii) $B[a^{-1}]$ is integral over B_0.

Proof. Let $B = A[x_1, \ldots, x_n]$ for suitable elements $x_1, \ldots, x_n \in B$ and consider the field of fractions K of A. Since B is torsion-free as an A-module, it can be regarded as a subring of the finitely generated commutative K-algebra $R = K \otimes_A B = K[x_1, \ldots, x_n]$. In view of Noether's normalization lemma, there is a polynomial K-subalgebra $R_0 = K[Y_1, \ldots, Y_m] \subseteq R$, such that R is integral over it. Replacing, if necessary, Y_i by $s_i Y_i$, for a suitable $s_i \in A \setminus \{0\}$, we may assume that $Y_i \in B$ for all $i = 1, \ldots, m$. Then, for any A-subalgebra $T \subseteq K$ the T-subalgebra $T[Y_1, \ldots, Y_m] \subseteq R_0$ is contained in $T \otimes_A B \subseteq R$.

Since $x_i \in R$ is integral over R_0, there is a monic polynomial $f_i(X) \in R_0[X]$, such that $f_i(x_i) = 0$ for all $i = 1, \ldots, n$. The polynomial $f_i(X)$ involves finitely many elements of $R_0 = K[Y_1, \ldots, Y_m]$, whereas each one of these involves finitely many elements of K. It follows that there is a non-zero element $a_i \in A$, such that $f_i(X)$ is a polynomial in X with coefficients in $A[a_i^{-1}, Y_1, \ldots, Y_m] \subseteq R_0$; hence, x_i is integral over $A[a_i^{-1}, Y_1, \ldots, Y_m]$. Letting $a = \prod_{i=1}^n a_i \in A$, we conclude that all of the x_i's are integral over $B_0 = A[a^{-1}, Y_1, \ldots, Y_m]$. Since the integral closure of B_0 in R is a B_0-subalgebra and hence an $A[a^{-1}]$-subalgebra of R (cf. Corollary A.13(i)), it follows that $A[a^{-1}, x_1, \ldots, x_n]$ is integral over B_0. This finishes the proof, since $A[a^{-1}, x_1, \ldots, x_n] = A[a^{-1}] \otimes_A B = B[a^{-1}]$. □

Proposition A.25 *Let A be a unique factorization domain and B a finitely generated commutative A-algebra, containing A as a subring. Then, up to associates, there are only finitely many prime elements $p \in A$ that are invertible in B.*

Proof. The torsion submodule B_t of the A-module B is easily seen to be an ideal of B. Since A is a subring of B, the unit element $1 \in B$ is not contained in B_t. Therefore, B_t is a proper ideal of B and hence the quotient $\overline{B} = B/B_t$ is a *non-zero* A-algebra. Since a prime element $p \in A$ that is invertible in B is also invertible in $\overline{B}$, we may replace B by its quotient $\overline{B}$ and reduce to the case where the finitely generated commutative A-algebra B is torsion-free as an A-module.

Let K be the field of fractions of A. Being torsion-free as an A-module, B can be regarded as a subring of the finitely generated commutative K-algebra $R = K \otimes_A B$. In view of Lemma A.24, there is a non-zero element $a \in A$ and a polynomial $A[a^{-1}]$-subalgebra $B_0 \subseteq B[a^{-1}]$, such that $B[a^{-1}]$ is integral over it. Let $R_0 = K \otimes_{A[a^{-1}]} B_0 \subseteq R$. Being a localization of A, the subring $A[a^{-1}] \subseteq K$ is also a unique factorization domain (cf. Exercise A.5.2). Invoking Gauss' lemma, we conclude that the same is true for the polynomial $A[a^{-1}]$-algebra B_0. Then, Lemma A.15 shows that B_0 is integrally closed in its field of fractions and, a fortiori, in R_0. We now consider the commutative diagram

$$
\begin{array}{ccccc}
K & \longrightarrow & R_0 & \longrightarrow & R \\
\uparrow & & \uparrow & & \uparrow \\
A[a^{-1}] & \longrightarrow & B_0 & \longrightarrow & B[a^{-1}] \longleftarrow B
\end{array}
$$

where all arrows are inclusions. Since any element of $B[a^{-1}]$ is integral over B_0, which is itself integrally closed in R_0, we conclude that the intersection $B[a^{-1}] \cap R_0$ is equal to B_0. Therefore, it follows that

$$B \cap K \subseteq B[a^{-1}] \cap K = B[a^{-1}] \cap R_0 \cap K = B_0 \cap K = A[a^{-1}],$$

where the last equality is a consequence of the fact that B_0 is a polynomial $A[a^{-1}]$-algebra. Hence, if a prime element $p \in A$ is invertible in B then $p^{-1} \in B \cap K \subseteq A[a^{-1}]$. It is easily seen that this can happen only if the prime p divides a. The proof is finished, since, up to associates, there are only finitely many such primes.   □

Corollary A.26 *Let A be a unique factorization domain and B a finitely generated commutative A-algebra. Consider an element $x \in B$ and assume that no power x^n, $n \geq 1$, is torsion as an element of the A-module B. Then, up to associates, there are only finitely many prime elements $p \in A$ for which $x \in pB$.*

Proof. Consider the localization $B' = B[S^{-1}]$ of B at the multiplicatively closed subset S consisting of the powers of x. Then, $B' = B[Y]/(xY - 1)$ is a finitely generated commutative A-algebra, which, in view of the assumption made on x, contains A as a subring. Hence, Proposition A.25 implies that, up to associates, the set of prime elements $p \in A$ that are invertible in B' is finite. This finishes the proof, since any prime element $p \in A$ for which $x \in pB$ is necessarily invertible in B'. Indeed, if $x = pb$ for some $b \in B$, then $p^{-1} = b/x \in B'$.   □

In particular, letting $A = \mathbf{Z}$, we obtain the following corollary.

Corollary A.27 *Let k be a finitely generated commutative ring of characteristic 0. Then, there are only finitely many prime numbers that are invertible in k. If, in addition, k is an integral domain, then for any $x \in k \setminus \{0\}$ there are only finitely many prime numbers $p \in \mathbf{Z}$ for which $x \in pk$.*   □

A.4 The Krull Intersection Theorem

I. THE ASCENDING CHAIN CONDITION. We begin by developing a few basic properties of Noetherian rings and modules.

Proposition A.28 *Let R be a commutative ring. Then, the following conditions are equivalent for an R-module M:*
   (i) Any submodule $N \subseteq M$ is finitely generated.
   (ii) Any ascending chain of submodules of M has a maximum element.
If these conditions hold, then M is said to be a Noetherian module.

Proof. (i)→(ii): Let $(N_i)_i$ be an ascending chain of submodules of M and consider the union $N = \bigcup_i N_i$. By assumption, there are elements $x_1, \ldots, x_n$ that generate the R-module N. Since there are only finitely many of them, all of the x_i's are contained in N_{i_0} for some index i_0. Therefore, $N = \sum_{i=1}^{n} Rx_i \subseteq N_{i_0}$ and hence $N = N_{i_0}$ is the maximum element of the given chain of submodules.

(ii)→(i): If a submodule $N \subseteq M$ is not finitely generated, then we may use an inductive argument in order to construct a sequence of elements $(x_i)_i$ of N, such that the sequence of submodules $(N_i)_i$, where $N_i = \sum_{t=1}^{i} Rx_t$ for all $i \geq 0$, is strictly increasing. But the existence of such a sequence contradicts condition (ii). $\qquad\square$

In particular, a commutative ring R if said to be Noetherian if the regular module R is Noetherian.

Lemma A.29 *Let R be a commutative ring.*

(i) Any submodule and any quotient module of a Noetherian module is Noetherian.

(ii) If $0 \longrightarrow M' \longrightarrow M \longrightarrow M'' \longrightarrow 0$ is a short exact sequence of R-modules and M', M'' are Noetherian, then M is Noetherian as well.

(iii) If R is a Noetherian ring then any finitely generated R-module is Noetherian.

Proof. (i) If N is a submodule (resp. a quotient module) of a Noetherian module M then any submodule of N, being also a submodule (resp. a quotient module of a submodule) of M, is finitely generated.

(ii) This follows since any submodule of M, being an extension of a submodule of M'' by a submodule of M', is finitely generated.

(iii) By an iterated application of (ii), it follows that any finitely generated free R-module is Noetherian. Since any finitely generated R-module is a quotient of such a free module, the result follows from (i). $\qquad\square$

Lemma A.30 *Let R be a commutative ring.*

(i) If R is a principal ideal domain, then R is Noetherian.

(ii) If R is a quotient of a commutative Noetherian ring, then R is Noetherian as well.

Proof. (i) This is immediate, since any ideal of R is principal (and hence finitely generated).

(ii) If R is a quotient of a commutative ring T, then any ideal I of R is a quotient of an ideal J of T. If T is Noetherian, then J is finitely generated as a T-module and hence I is finitely generated as an R-module. $\qquad\square$

In order to obtain non-trivial examples of Noetherian rings, we need the following result of Hilbert.

Theorem A.31 *(Hilbert basis theorem) If R is a commutative Noetherian ring, then so is the polynomial ring $R[X]$.*

Proof. Let I be an ideal of $R[X]$ and consider the increasing sequence of ideals $(J_n)_n$ of R, where $J_n \subseteq R$ consists of the leading coefficients of all polynomials in I that have degree $\leq n$. We also consider the ideal $J_\infty = \bigcup_n J_n$. Since R is Noetherian, the ideal J_n is finitely generated for all $n \in \mathbf{N} \cup \{\infty\}$. Hence, for all n there is a finite set of polynomials $F_n \subseteq I$, having degree $\leq n$, whose leading coefficients generate J_n.

We claim that if $n_0 = \max\{\deg g : g \in F_\infty\}$, then I is generated by the (finite) set $F = F_0 \cup F_1 \cup \cdots \cup F_{n_0-1} \cup F_\infty$. Indeed, let us consider a non-zero polynomial $f \in I$ and show that $f \in \sum_{g \in F} R[X]g$, by using induction on $n = \deg f$. Assume that elements of I having degree $< n$ are contained in $\sum_{g \in F} R[X]g$ and suppose that $n < n_0$. Then, the leading coefficient of f is an R-linear combination of the leading coefficients of the polynomials in F_n and hence the polynomial $f - \sum_{g \in F_n} r_g X^{n-\deg g} g \in I$ has degree $< n$, for suitable elements $r_g \in R$, $g \in F_n$. In view of the induction hypothesis, we conclude that $f - \sum_{g \in F_n} r_g X^{n-\deg g} g \in \sum_{g \in F} R[X]g$; since $F_n \subseteq F$, it follows that $f \in \sum_{g \in F} R[X]g$. Now suppose that $n \geq n_0$. Since the leading coefficient of f is an R-linear combination of the leading coefficients of the polynomials in F_∞, the polynomial $f - \sum_{g \in F_\infty} r_g X^{n-\deg g} g \in I$ has degree $< n$, for suitable elements $r_g \in R$, $g \in F_\infty$. In view of the induction hypothesis, we conclude that $f - \sum_{g \in F_\infty} r_g X^{n-\deg g} g \in \sum_{g \in F} R[X]g$; since $F_\infty \subseteq F$, it follows that $f \in \sum_{g \in F} R[X]g$ in this case as well. $\qquad \square$

Corollary A.32 *Any finitely generated commutative ring is Noetherian.*

Proof. Being a principal ideal domain, the ring $\mathbf{Z}$ is Noetherian (Lemma A.30(i)). Therefore, an iterated application of the Hilbert basis theorem shows that the polynomial ring $\mathbf{Z}[X_1, \ldots, X_n]$ is Noetherian for all n. Since any finitely generated commutative ring is a quotient of such a polynomial ring, the result follows from Lemma A.30(ii). $\qquad \square$

II. The intersection of the powers of an ideal. We now study the intersection of the powers of an ideal in a commutative Noetherian ring.

Proposition A.33 *Let R be a commutative ring, $I \subseteq R$ an ideal and M a Noetherian R-module.*

(i) Assume that N, L are two submodules of M, such that N is maximal with respect to the property $N \cap L = IL$. Then, for any $r \in I$ there exists $n \in \mathbf{N}$, such that $r^n M \subseteq N$. In particular, if I is finitely generated, then $I^t M \subseteq N$ for $t \gg 0$.

(ii) Assume that the ideal I is finitely generated and consider a submodule L of M, such that $L \subseteq \bigcap_n I^n M$. Then, $IL = L$.

Proof. (i) We fix an element $r \in I$ and consider for any non-negative integer i the submodule $M_i \subseteq M$ consisting of those elements $x \in M$ for which $r^i x \in N$. Since the R-module M is Noetherian, the increasing sequence $(M_i)_i$ of submodules of M must be eventually constant; hence, there exists an integer $n \in \mathbf{N}$ such that $M_n = M_{n+1}$. We claim that

$$(r^n M + N) \cap L = N \cap L . \tag{A.3}$$

Of course, the $\supseteq$-inclusion is clear. Conversely, let $z \in (r^n M + N) \cap L$. Then, there are elements $x \in M$ and $y \in N$, such that $z = r^n x + y \in L$. Since $rz \in IL = N \cap L \subseteq N$, we have $r^{n+1}x = rz - ry \in N$ and hence $x \in M_{n+1} = M_n$. Therefore, $r^n x \in N$ and hence $z = r^n x + y \in N$, i.e. $z \in N \cap L$. Having established (A.3), the maximality of N implies that $r^n M + N = N$ and hence $r^n M \subseteq N$.

If the ideal I is generated by elements $r_1, \ldots, r_k$ and $n_1, \ldots, n_k$ are positive integers, then $I^t \subseteq \sum_{i=1}^{k} r_i^{n_i} R$, where $t = 1 + \sum_{i=1}^{k}(n_i - 1)$. Indeed, it is easily seen that any summand in the expansion of a typical product

$$\prod_{l=1}^{t} \sum_{i=1}^{k} r_i s_i^{(l)} = \left(\sum_{i=1}^{k} r_i s_i^{(1)} \right) \cdot \left(\sum_{i=1}^{k} r_i s_i^{(2)} \right) \cdots \left(\sum_{i=1}^{k} r_i s_i^{(t)} \right),$$

where all $s_i^{(l)}$'s are elements of R, is divisible by $r_i^{n_i}$ for at least one i. In particular, it follows that $I^t M \subseteq \sum_{i=1}^{k} r_i^{n_i} M$. Having chosen the n_i's in such a way that $r_i^{n_i} M \subseteq N$ for all i, it follows that $I^t M \subseteq N$.

(ii) Consider the class consisting of those submodules $N \subseteq M$ that satisfy the condition $N \cap L = IL$. This class is non-empty, since it contains IL. Applying Zorn's lemma, we may choose a submodule $N \subseteq M$ maximal in that class. By part (i), we have $I^t M \subseteq N$ for some $t \gg 0$ and hence $L \subseteq \bigcap_n I^n M \subseteq I^t M \subseteq N$; therefore, $L = N \cap L = IL$. $\square$

Theorem A.34 *(Krull intersection theorem) Let $I \subseteq R$ be a proper ideal of a commutative Noetherian ring R and consider a finitely generated R-module M. Then, the submodule $L = \bigcap_n I^n M$ is trivial (i.e. $L = 0$) if either one of the following two conditions is satisfied:*

(i) R is a local ring or

(ii) R is an integral domain and M a torsion-free R-module.

Proof. Since R is a Noetherian ring, the ideal I is finitely generated, whereas Lemma A.29(iii) shows that the R-module M is Noetherian. Therefore, Proposition A.33(ii) implies that $IL = L$. Since M is Noetherian, L is finitely generated; hence, the proof is finished by invoking Lemma A.9. $\square$

Corollary A.35 *Let k be a finitely generated integral domain and $I \subseteq k$ a proper ideal. Then, $\bigcap_n I^n = 0$.*

Proof. This follows from Theorem A.34, in view of Corollary A.32. $\square$

A.5 Exercises

1. Let T be a commutative ring, which is integral over a subring $R \subseteq T$. Then, show that $U(R) = R \cap U(T)$.

2. Let R be a unique factorization domain and $S \subseteq R$ a multiplicatively closed subset containing 1. Show that the localization $R[S^{-1}]$ is a unique factorization domain as well.

3. Let k be a field and R a finitely generated commutative k-algebra.
   (i) Give an example showing that the set $\operatorname{Hom}_{k-Alg}(R, k)$ may be empty if k is not algebraically closed.
   (ii) If $\overline{k}$ is the algebraic closure of k, show that $\operatorname{Hom}_{k-Alg}(R, \overline{k}) \neq \emptyset$.

4. Show that the conclusion of Proposition A.25 may be false if either
   (i) the commutative A-algebra B is not finitely generated or
   (ii) the structural homomorphism $A \longrightarrow B$ is not injective.

B

Discrete Ring-Valued Integrals

Let X be a compact space and $C(X)$ the associated algebra of continuous complex-valued functions, endowed with the supremum norm. A regular Borel measure μ on X induces a continuous linear functional $\mathcal{I}_\mu$ on $C(X)$, namely the functional $f \mapsto \int f \, d\mu$, $f \in C(X)$. Moreover, the Riesz representation theorem (cf. [61, Theorem 6.19]) asserts that any continuous linear functional on $C(X)$ arises from a unique regular Borel measure μ in this way.

In this Appendix, we consider a discrete version of the above process and consider measures (more precisely, premeasures) which have values in an abelian group A. Since A will have, in general, no topological structure, we suitably restrict the class of functions that can be integrated. In fact, we consider only the locally constant integer-valued functions on X; these are precisely the continuous functions from X to the discrete space $\mathbf{Z}$. It will turn out that the algebra (in the measure-theoretic sense) of subsets of X on which the measure has to be defined is that consisting of the clopen subsets of X. The special case where $A = R$ is a commutative ring and the measure of any clopen subset of X is an idempotent therein is of particular importance for the applications we have in mind.

B.1 Discrete Group-Valued Integrals

Let us fix a compact topological space X and consider the set $L(X)$ consisting of the clopen subsets $Y \subseteq X$. We note that $L(X)$ is a subalgebra of the Boolean algebra $\mathcal{P}(X)$ of all subsets of X (cf. Examples 1.3(ii),(iii)). Since X is compact, any continuous function f on X with values in $\mathbf{Z}$ takes only finitely many values, say $a_1, \ldots, a_n$. We assume that the a_i's are distinct and note that the inverse image of the singleton $\{a_i\} \subseteq \mathbf{Z}$ under f is a clopen subset X_i of X for all $i = 1, \ldots, n$. Then, $f = \sum_{i=1}^{n} a_i \chi_{X_i}$; we refer to that equation as the canonical decomposition of f. It is clear that a decomposition $f = \sum_{j=1}^{m} b_j \chi_{Y_j}$, where $b_j \in \mathbf{Z}$ and $Y_j \in L(X)$ for all $j = 1, \ldots, m$, coincides with the canonical one if the following two conditions are satisfied:

(i) The integers $b_1, \ldots, b_m$ are distinct and

(ii) The clopen subsets $Y_1, \ldots, Y_m$ of X are non-empty, mutually disjoint and cover X.

The set of all locally constant integer-valued functions on X is a commutative ring with operations defined pointwise; we denote this ring by $[X, \mathbf{Z}]$. We note that the map $Y \mapsto \chi_Y$, $Y \in L(X)$, is an isomorphism of Boolean algebras

$$\chi : L(X) \longrightarrow \mathrm{Idem}([X, \mathbf{Z}]) . \tag{B.1}$$

Indeed, $\chi_\emptyset$ and χ_X are the constant functions with value 0 and 1 respectively, whereas for any two clopen subsets $Y, Y' \subseteq X$ we have

$$\chi_{Y \cap Y'} = \chi_Y \chi_{Y'} = \chi_Y \wedge \chi_{Y'}$$

and

$$\chi_{Y \cup Y'} = \chi_Y + \chi_{Y'} - \chi_{Y \cap Y'} = \chi_Y + \chi_{Y'} - \chi_Y \chi_{Y'} = \chi_Y \vee \chi_{Y'} .$$

We now let $\mathcal{L}$ be a Boolean algebra of subsets of a set Ω and A an abelian group. An A-valued premeasure μ on $\mathcal{L}$ is a function $\mu : \mathcal{L} \longrightarrow A$ satisfying the following two conditions:

(μ1) $\mu(\emptyset) = 0$ and

(μ2) if $Y_1, \ldots, Y_n \in \mathcal{L}$ are disjoint, then $\mu(\bigcup_{i=1}^n Y_i) = \sum_{i=1}^n \mu(Y_i)$.

Given an A-valued premeasure μ on the Boolean algebra $L(X)$ of clopen subsets of X, we may define for any function $f \in [X, \mathbf{Z}]$ with canonical decomposition $f = \sum_{i=1}^n a_i \chi_{X_i}$ the element $\mathcal{I}_\mu(f) = \sum_{i=1}^n a_i \mu(X_i) \in A$.

Definition B.1 *The element $\mathcal{I}_\mu(f) \in A$ defined above is called the discrete A-valued integral of f with respect to the premeasure μ.*

Lemma B.2 *Let X be a compact space, A an abelian group and μ an A-valued premeasure on $L(X)$.*

(i) If $a_1, \ldots, a_n$ are integers, $X_1, \ldots, X_n$ mutually disjoint clopen subsets covering X and $f = \sum_{i=1}^n a_i \chi_{X_i}$, then $\mathcal{I}_\mu(f) = \sum_{i=1}^n a_i \mu(X_i)$.

(ii) $\mathcal{I}_\mu(f + g) = \mathcal{I}_\mu(f) + \mathcal{I}_\mu(g)$ for all $f, g \in [X, \mathbf{Z}]$.

Proof. (i) Let $f = \sum_{j=1}^m b_j \chi_{Y_j}$ be the canonical decomposition of f; then, the Y_j's are mutually disjoint and cover X. It follows that X_i is the disjoint union of the family $(X_i \cap Y_j)_j$ for all $i = 1, \ldots, n$ and hence

$$f = \sum_i a_i \chi_{X_i} = \sum_i a_i \left(\sum_j \chi_{X_i \cap Y_j} \right) = \sum_{i,j} a_i \chi_{X_i \cap Y_j} . \tag{B.2}$$

Similarly, the X_i's being mutually disjoint and covering X, Y_j is the disjoint union of the family $(X_i \cap Y_j)_i$ for all $j = 1, \ldots, m$; hence,

$$f = \sum_j b_j \chi_{Y_j} = \sum_j b_j \left(\sum_i \chi_{X_i \cap Y_j} \right) = \sum_{i,j} b_j \chi_{X_i \cap Y_j} . \tag{B.3}$$

Since the family $(X_i \cap Y_j)_{i,j}$ consists of mutually disjoint sets, we may compare the decompositions of f in (B.2) and (B.3) in order to conclude that whenever $X_i \cap Y_j \neq \emptyset$ we have $a_i = b_j$. Since $\mu(\emptyset) = 0$, it follows that $a_i\mu(X_i \cap Y_j) = b_j\mu(X_i \cap Y_j)$ for all i, j. Therefore, using the additivity of μ (property $(\mu 2)$), we have

$$
\begin{aligned}
\mathcal{I}_\mu(f) &= \sum_j b_j\mu(Y_j) \\
&= \sum_j b_j\left(\sum_i \mu(X_i \cap Y_j)\right) \\
&= \sum_{i,j} b_j\mu(X_i \cap Y_j) \\
&= \sum_{i,j} a_i\mu(X_i \cap Y_j) \\
&= \sum_i a_i\left(\sum_j \mu(X_i \cap Y_j)\right) \\
&= \sum_i a_i\mu(X_i)\,.
\end{aligned}
$$

(ii) Let $f = \sum_{i=1}^n a_i\chi_{X_i}$ and $g = \sum_{j=1}^m b_j\chi_{Y_j}$ be the canonical decompositions of f and g respectively; then, $f + g = \sum_{i,j}(a_i + b_j)\chi_{X_i \cap Y_j}$. Since the family $(X_i \cap Y_j)_{i,j}$ consists of disjoint clopen subsets covering X, we may invoke (i) above and the additivity of μ in order to conclude that

$$
\begin{aligned}
\mathcal{I}_\mu(f + g) &= \sum_{i,j}(a_i + b_j)\mu(X_i \cap Y_j) \\
&= \sum_{i,j} a_i\mu(X_i \cap Y_j) + \sum_{i,j} b_j\mu(X_i \cap Y_j) \\
&= \sum_i a_i\left(\sum_j \mu(X_i \cap Y_j)\right) + \sum_j b_j\left(\sum_i \mu(X_i \cap Y_j)\right) \\
&= \sum_i a_i\mu(X_i) + \sum_j b_j\mu(Y_j) \\
&= \mathcal{I}_\mu(f) + \mathcal{I}_\mu(g)\,,
\end{aligned}
$$

as needed.   □

With the notation established above, we can state the following discrete version of the Riesz representation theorem.

Proposition B.3 *Let X be a compact space and A an abelian group.*

(i) For any A-valued premeasure μ on $L(X)$ the associated discrete A-valued integral $\mathcal{I}_\mu : [X, \mathbf{Z}] \longrightarrow A$ is a homomorphism of abelian groups.

(ii) Conversely, for any group homomorphism $\varphi : [X, \mathbf{Z}] \longrightarrow A$ there is a unique A-valued premeasure μ on $L(X)$, such that $\varphi = \mathcal{I}_\mu$.

Proof. (i) This is precisely Lemma B.2(ii).

(ii) Given a homomorphism $\varphi : [X, \mathbf{Z}] \longrightarrow A$, define an A-valued map μ on $L(X)$ by letting $\mu(Y) = \varphi(\chi_Y)$ for all $Y \in L(X)$. Since $\chi_\emptyset$ is the constant function with value 0, it follows that $\mu(\emptyset) = 0$. If $Y_1, \dots, Y_n$ are disjoint clopen subsets of X and $Y = \bigcup_{i=1}^n Y_i$, then $\chi_Y = \sum_{i=1}^n \chi_{Y_i}$; therefore,

$$
\mu(Y) = \varphi(\chi_Y) = \sum_{i=1}^n \varphi(\chi_{Y_i}) = \sum_{i=1}^n \mu(Y_i)
$$

and μ is indeed a premeasure. By the very definition of μ, it follows that the group homomorphisms $\mathcal{I}_\mu$ and φ coincide on the set $\{\chi_Y : Y \in L(X)\}$. Since this set generates the group $[X, \mathbf{Z}]$, we conclude that $\mathcal{I}_\mu = \varphi$. The uniqueness of μ satisfying that condition is clear. $\qquad\square$

For a topological space X and an abelian group A, we denote the set of A-valued premeasures on $L(X)$ by $\mathcal{M}(L(X); A)$. We note that $\mathcal{M}(L(X); A)$ has the structure of an abelian group, where the sum $\mu + \mu'$ of two premeasures $\mu, \mu' \in \mathcal{M}(L(X); A)$ is defined by letting $(\mu + \mu')(Y) = \mu(Y) + \mu'(Y)$ for all $Y \in L(X)$.

Corollary B.4 *Let X be a compact space and A an abelian group. Then, the map $\mu \mapsto \mathcal{I}_\mu$, $\mu \in \mathcal{M}(L(X); A)$, is an isomorphism of groups*

$$\mathcal{I} : \mathcal{M}(L(X); A) \longrightarrow Hom([X, \mathbf{Z}], A) .$$

Proof. In view of Proposition B.3, it only remains to show that $\mathcal{I}_{\mu+\mu'} = \mathcal{I}_\mu + \mathcal{I}_{\mu'}$ for all $\mu, \mu' \in \mathcal{M}(L(X); A)$. For any $Y \in L(X)$ we have

$$\mathcal{I}_{\mu+\mu'}(\chi_Y) = (\mu + \mu')(Y) = \mu(Y) + \mu'(Y) = \mathcal{I}_\mu(\chi_Y) + \mathcal{I}_{\mu'}(\chi_Y) .$$

Since the group $[X, \mathbf{Z}]$ is generated by the characteristic functions of clopen subsets $Y \in L(X)$, it follows that the homomorphisms $\mathcal{I}_{\mu+\mu'}$ and $\mathcal{I}_\mu + \mathcal{I}_{\mu'}$ are equal. $\qquad\square$

B.2 Idempotent-Valued Premeasures

In this section, we examine the special properties that are enjoyed by the integrals considered above, in the case where the abelian group A is endowed with the structure of a commutative ring.

Let $\mathcal{L}$ be a Boolean algebra of subsets of a set Ω, R a commutative ring and $\nu : \mathcal{L} \longrightarrow Idem(R)$ a morphism of Boolean algebras. Then, ν associates with any $Y \in \mathcal{L}$ an idempotent $\nu(Y) \in R$, in such a way that the following properties are satisfied:

(ν1) $\nu(\emptyset) = 0$ and $\nu(\Omega) = 1$,
(ν2) $\nu(Y \cup Y') = \nu(Y) + \nu(Y') - \nu(Y)\nu(Y')$ for all $Y, Y' \in \mathcal{L}$ and
(ν3) $\nu(Y \cap Y') = \nu(Y)\nu(Y')$ for all $Y, Y' \in \mathcal{L}$.

Lemma B.5 *Let $\mathcal{L}$ be a Boolean algebra of subsets of a set Ω and consider a commutative ring R.*

(i) If $\nu : \mathcal{L} \longrightarrow Idem(R)$ is a Boolean algebra morphism then ν is a premeasure on $\mathcal{L}$, such that $\nu(\Omega) = 1$.

(ii) Conversely, assume that ν is a premeasure on $\mathcal{L}$ with values in the set of idempotents of R, such that $\nu(\Omega) = 1$. If $2 \in R$ is not a zero-divisor then $\nu : \mathcal{L} \longrightarrow Idem(R)$ is a Boolean algebra morphism.

Proof. (i) It suffices to verify that ν satisfies property (μ2). To that end, let us consider two disjoint subsets $Y, Y' \in \mathcal{L}$. Since ν is $\wedge$-preserving (property (ν3))

and $\nu(\emptyset) = 0$, we have $\nu(Y)\nu(Y') = \nu(Y \cap Y') = 0$. But ν is $\vee$-preserving as
well (property $(\nu2)$) and hence $\nu(Y \cup Y') = \nu(Y) + \nu(Y') - \nu(Y)\nu(Y') = \nu(Y) +$
$\nu(Y')$. Using induction on n, one can prove that whenever $Y_1, \ldots, Y_n \in \mathcal{L}$ are
mutually disjoint and $Y = \bigcup_{i=1}^{n} Y_i$, then $\nu(Y) = \sum_{i=1}^{n} \nu(Y_i)$.

(ii) By assumption, ν satisfies property $(\nu1)$. Since property $(\nu2)$ is a
consequence of properties $(\mu2)$ and $(\nu3)$,[1] it suffices to prove that ν satisfies
property $(\nu3)$. To that end, we note that whenever $Y_1, Y_2 \in \mathcal{L}$ are disjoint, the
idempotents $e_1 = \nu(Y_1)$ and $e_2 = \nu(Y_2)$ are orthogonal. Indeed, in that case,
$e_1 + e_2 = \nu(Y_1) + \nu(Y_2) = \nu(Y_1 \cup Y_2)$ is an idempotent and hence

$$e_1 + e_2 = (e_1 + e_2)^2 = e_1^2 + e_2^2 + 2e_1e_2 = e_1 + e_2 + 2e_1e_2 .$$

Therefore, $2e_1e_2 = 0$ and hence $e_1e_2 = 0$. Now let $Y, Y' \in \mathcal{L}$ and $e = \nu(Y \cap Y')$.
Then, $Y \setminus Y' \in \mathcal{L}$ is disjoint from $Y \cap Y'$ and $\nu(Y \setminus Y') = \nu(Y) - e$ (in view
of property $(\mu2)$). Therefore, $e(\nu(Y) - e) = 0$ and hence $e = e^2 = e\nu(Y)$.
Similarly, one can show that $e = e\nu(Y')$. Finally, since the subsets $Y \setminus Y'$ and
$Y' \setminus Y$ are disjoint, we have

$$\begin{aligned}
0 &= \nu(Y \setminus Y')\nu(Y' \setminus Y) \\
&= (\nu(Y) - e)(\nu(Y') - e) \\
&= \nu(Y)\nu(Y') - e\nu(Y) - e\nu(Y') + e^2 \\
&= \nu(Y)\nu(Y') - e - e + e \\
&= \nu(Y)\nu(Y') - e
\end{aligned}$$

and hence $\nu(Y \cap Y') = e = \nu(Y)\nu(Y')$. □

We now consider a compact topological space X, a commutative ring R and
study the discrete R-valued integral $\mathcal{I}_\nu$, associated with a Boolean algebra
morphism ν defined on the algebra $\mathcal{L} = L(X)$ of clopen subsets of X with
values in $\mathrm{Idem}(R)$.

Lemma B.6 *Let X be a compact space, R a commutative ring and consider
a Boolean algebra morphism $\nu : L(X) \longrightarrow \mathrm{Idem}(R)$. Then:*
(i) $\mathcal{I}_\nu(fg) = \mathcal{I}_\nu(f)\mathcal{I}_\nu(g)$ for all $f, g \in [X, \mathbf{Z}]$.
(ii) If $\mathbf{1}$ is the constant function on X with value $1 \in \mathbf{Z}$, then $\mathcal{I}_\nu(\mathbf{1}) = 1$.

Proof. (i) Let $f = \sum_{i=1}^{n} a_i \chi_{X_i}$ and $g = \sum_{j=1}^{m} b_j \chi_{Y_j}$ be the canonical decompositions of f and g respectively; then, $fg = \sum_{i,j} a_i b_j \chi_{X_i \cap Y_j}$. Using the
additivity of $\mathcal{I}_\nu$ and property $(\nu3)$, we conclude that

$$\begin{aligned}
\mathcal{I}_\nu(fg) &= \sum_{i,j} a_i b_j \nu(X_i \cap Y_j) \\
&= \sum_{i,j} a_i b_j \nu(X_i)\nu(Y_j) \\
&= \left(\sum_i a_i \nu(X_i)\right)\left(\sum_j b_j \nu(Y_j)\right) \\
&= \mathcal{I}_\nu(f)\mathcal{I}_\nu(g) .
\end{aligned}$$

(ii) This is clear, since $\mathbf{1} = \chi_X$ and $\nu(X) = 1$. □

---

[1] Indeed, property $(\mu2)$ for ν is easily seen to imply that $\nu(Y \cup Y') = \nu(Y) + \nu(Y') - \nu(Y \cap Y')$ for all $Y, Y' \in \mathcal{L}$.

Proposition B.7 *Let X be a compact space and R a commutative ring.*

(i) For any Boolean algebra morphism $\nu : L(X) \longrightarrow Idem(R)$ the associated discrete R-valued integral $\mathcal{I}_\nu : [X, \mathbf{Z}] \longrightarrow R$ is a ring homomorphism.

(ii) Conversely, for any ring homomorphism $\varphi : [X, \mathbf{Z}] \longrightarrow R$ there is a unique Boolean algebra morphism $\nu : L(X) \longrightarrow Idem(R)$, such that $\varphi = \mathcal{I}_\nu$.

Proof. (i) This follows from Lemmas B.2(ii) and B.6.

(ii) Given a ring homomorphism $\varphi : [X, \mathbf{Z}] \longrightarrow R$, we may consider the premeasure $\nu : L(X) \longrightarrow R$ defined in the proof of Proposition B.3; recall that $\nu(Y) = \varphi(\chi_Y)$ for any $Y \in L(X)$. Then, ν is the unique premeasure satisfying $\mathcal{I}_\nu = \varphi$; we have to show that ν takes values in the set $Idem(R)$ and is a Boolean algebra morphism with values therein. Since χ_Y is an idempotent in the function ring $[X, \mathbf{Z}]$ and φ is multiplicative, it is clear that $\nu(Y) \in Idem(R)$ for all $Y \in L(X)$. But then ν is the composition

$$L(X) \xrightarrow{\chi} Idem([X, \mathbf{Z}]) \xrightarrow{Idem(\varphi)} Idem(R) \, ,$$

where χ is the isomorphism (B.1) and $Idem(\varphi)$ the morphism of Boolean algebras induced by φ. In particular, it follows that ν is a morphism of Boolean algebras, as needed.   □

Remarks B.8 (i) Let X be a compact space and R a commutative ring. Then, we can reformulate Proposition B.7 as the assertion that the map

$$\mathcal{I} : \mathrm{Hom}_{Boole}(L(X), Idem(R)) \longrightarrow \mathrm{Hom}_{Ring}([X, \mathbf{Z}], R) \, ,$$

which is given by $\nu \mapsto \mathcal{I}_\nu$, $\nu \in \mathrm{Hom}_{Boole}(L(X), Idem(R))$, is bijective.

(ii) The discussion about discrete ring-valued integrals in this Appendix was motivated by the various constructions associated with the geometric rank in §2.1. Having that special case in mind, we chose to consider only locally constant integer-valued functions. In fact, we could have replaced the ring $\mathbf{Z}$ by an arbitrary commutative ring k and insisted that the abelian group A (resp. the commutative ring R) be a k-module (resp. a commutative k-algebra). In exactly the same way as above, one can define for any A-valued premeasure (resp. for any $Idem(R)$-valued Boolean algebra morphism) on $L(X)$ a corresponding A-valued (resp. R-valued) integral defined on the class of locally constant k-valued functions on the compact space X and obtain identifications

$$\mathcal{M}(L(X); A) \simeq \mathrm{Hom}_k([X, k], A)$$

and

$$\mathrm{Hom}_{Boole}(L(X), Idem(R)) \simeq \mathrm{Hom}_{k-Alg}([X, k], R) \, .$$

B.3 Exercises

1. Let X, X' be compact topological spaces and $f : X \longrightarrow X'$ a continuous map. We also consider the Boolean algebra morphism $L(f) : L(X') \longrightarrow$

$L(X)$, which is given by $Y' \mapsto f^{-1}(Y')$, $Y' \in L(X')$ (cf. Example 1.5(iii)).
(i) Let A be an abelian group and μ an A-valued premeasure on $L(X)$.
Show that $\mu' = \mu \circ L(f)$ is an A-valued premeasure on $L(X')$ and

$$\mathcal{I}_{\mu'} = \mathcal{I}_\mu \circ [f, \mathbf{Z}] : [X', \mathbf{Z}] \longrightarrow A \, ,$$

where $[f, \mathbf{Z}] : [X', \mathbf{Z}] \longrightarrow [X, \mathbf{Z}]$ is the map $g \mapsto g \circ f$, $g \in [X', \mathbf{Z}]$.[2]
(ii) (naturality of $\mathcal{I}$ with respect to the topological space) Let A be an abelian group. Show that the following diagram is commutative

$$\begin{array}{ccc} \mathcal{M}(L(X); A) & \xrightarrow{\mathcal{I}} & \mathrm{Hom}([X, \mathbf{Z}], A) \\ {\scriptstyle L(f)^* \downarrow} & & {\scriptstyle \downarrow [f, \mathbf{Z}]^*} \\ \mathcal{M}(L(X'); A) & \xrightarrow{\mathcal{I}} & \mathrm{Hom}([X', \mathbf{Z}], A) \end{array}$$

where $L(f)^*$ is the map $\mu \mapsto \mu \circ L(f)$, $\mu \in \mathcal{M}(L(X); A)$, and $[f, \mathbf{Z}]^*$ the map $\varphi \mapsto \varphi \circ [f, \mathbf{Z}]$, $\varphi \in \mathrm{Hom}([X, \mathbf{Z}], A)$.
(iii) Let R be a commutative ring. Show that the following diagram is commutative

$$\begin{array}{ccc} \mathrm{Hom}_{Boole}(L(X), \mathrm{Idem}(R)) & \xrightarrow{\mathcal{I}} & \mathrm{Hom}_{Ring}([X, \mathbf{Z}], R) \\ {\scriptstyle L(f)^* \downarrow} & & {\scriptstyle \downarrow [f, \mathbf{Z}]^*} \\ \mathrm{Hom}_{Boole}(L(X'), \mathrm{Idem}(R)) & \xrightarrow{\mathcal{I}} & \mathrm{Hom}_{Ring}([X', \mathbf{Z}], R) \end{array}$$

where $L(f)^*$ and $[f, \mathbf{Z}]^*$ are the restrictions of the corresponding maps in (ii) above.
2. Let A, A' be abelian groups, $\sigma : A \longrightarrow A'$ a group homomorphism and X a compact topological space.
(i) Show that if μ is an A-valued premeasure on $L(X)$, then $\mu' = \sigma \circ \mu$ is an A'-valued premeasure on $L(X)$ and

$$\mathcal{I}_{\mu'} = \sigma \circ \mathcal{I}_\mu : [X, \mathbf{Z}] \longrightarrow A' \, .$$

(ii) (naturality of $\mathcal{I}$ with respect to the abelian group) Show that the following diagram is commutative

$$\begin{array}{ccc} \mathcal{M}(L(X); A) & \xrightarrow{\mathcal{I}} & \mathrm{Hom}([X, \mathbf{Z}], A) \\ {\scriptstyle \sigma_* \downarrow} & & {\scriptstyle \downarrow \sigma_*} \\ \mathcal{M}(L(X); A') & \xrightarrow{\mathcal{I}} & \mathrm{Hom}([X, \mathbf{Z}], A') \end{array}$$

where we denote by σ_* both maps $\mu \mapsto \sigma \circ \mu$, $\mu \in \mathcal{M}(L(X); A)$, and $\varphi \mapsto \sigma \circ \varphi$, $\varphi \in \mathrm{Hom}([X, \mathbf{Z}], A)$.

---

[2] If we denote the values of $\mathcal{I}_\mu$ and $\mathcal{I}_{\mu'}$ using the standard $\int$-sign, the equality $\mathcal{I}_{\mu'} = \mathcal{I}_\mu \circ [f, \mathbf{Z}]$ takes the familiar form of the *change of variables formula*: $\int_{X'} g(z) \, d\mu'(z) = \int_X g(f(t)) \, d\mu(t)$ for all $g \in [X', \mathbf{Z}]$.

(iii) Let R, R' be commutative rings and $\tau : R \longrightarrow R'$ a ring homomorphism. Show that the following diagram is commutative

$$
\begin{array}{ccc}
\mathrm{Hom}_{Boole}(L(X), \mathrm{Idem}(R)) & \overset{\mathcal{I}}{\longrightarrow} & \mathrm{Hom}_{Ring}([X, \mathbf{Z}], R) \\
{\scriptstyle Idem(\tau)_*} \downarrow & & \downarrow {\scriptstyle \tau_*} \\
\mathrm{Hom}_{Boole}(L(X), \mathrm{Idem}(R')) & \overset{\mathcal{I}}{\longrightarrow} & \mathrm{Hom}_{Ring}([X, \mathbf{Z}], R')
\end{array}
$$

where $\mathrm{Idem}(\tau)$ is the Boolean algebra morphism induced by τ and $\mathrm{Idem}(\tau)_*$ (resp. τ_*) the map given by composing to the left with $\mathrm{Idem}(\tau)$ (resp. with τ).

3. Let $\mathcal{L}$ be a Boolean algebra of subsets of a set Ω and R a commutative ring. Prove the following strengthening of Lemma B.5:
   (i) Let $\nu : \mathcal{L} \longrightarrow \mathrm{Idem}(R)$ be a function satisfying properties $(\nu 2)$ and $(\nu 3)$. If $\nu(\emptyset) = 0$, then ν is a premeasure.
   (ii) Conversely, assume that ν is a premeasure on $\mathcal{L}$ with values in the set of idempotents of R and let $\nu(\Omega) = e \in \mathrm{Idem}(R)$. If $2 \in R$ is not a zero-divisor then $\nu(Y) \in Re$ for all $Y \in \mathcal{L}$ and ν is a Boolean algebra morphism from $\mathcal{L}$ to the algebra of idempotents $\mathrm{Idem}(Re)$ of the commutative ring Re.

4. The goal of this Exercise is to show that the regularity hypothesis about $2 \in R$ can't be omitted in Lemma B.5(ii). To that end, let Ω be a finite set and $\mathcal{L} = \mathcal{P}(\Omega)$ its power set. Consider the commutative ring $R = \mathbf{Z}/2\mathbf{Z} = \{0, 1\}$ and define an R-valued map ν on $\mathcal{L}$, by letting $\nu(Y) = 0$ (resp. 1) if the subset $Y \subseteq \Omega$ has an even (resp. odd) number of elements. Show that:
   (i) ν is a premeasure on $\mathcal{L}$ with values in $\mathrm{Idem}(R)$.
   (ii) If Ω has an odd number of elements then $\nu(\Omega) = 1$.
   (iii) If Ω has more than one elements, then $\nu : \mathcal{L} \longrightarrow \mathrm{Idem}(R)$ is not a Boolean algebra morphism.

C

Frobenius' Density Theorem

Let $f(X) \in \mathbf{Z}[X]$ be a monic polynomial of degree n without multiple roots in $\mathbf{C}$. For any prime number p we may reduce $f(X)$ modulo p and obtain a polynomial $f_p(X) \in \mathbf{F}_p[X]$. Even if the original polynomial is irreducible in $\mathbf{Z}[X]$, it may very well happen that its reduction modulo p is reducible in $\mathbf{F}_p[X]$. Moreover, the partition of n induced by the degrees of the irreducible factors of $f_p(X)$ in $\mathbf{F}_p[X]$ may vary with p. We also consider the roots $a_1, \ldots, a_n$ of $f(X)$ in $\mathbf{C}$ and the corresponding splitting field $K = \mathbf{Q}(a_1, \ldots, a_n)$. The Galois group Γ of K over $\mathbf{Q}$ may be viewed as a subgroup of the group S_n of permutations on n letters, by restricting its action to the a_i's. Counting the lengths of the cycles in the cycle decomposition of an element $\gamma \in \Gamma \subseteq S_n$, we obtain a partition of n. In this way, we obtain partitions of n by two different methods:

    (i) by factoring $f(X)$ modulo prime numbers and

    (ii) by viewing elements of the Galois group Γ as permutations of the roots of $f(X)$.

It turns out that these two methods are related to each other; it is the goal of the present Appendix to describe this relationship. As a consequence, we prove that an irreducible monic polynomial $f(X) \in \mathbf{Z}[X]$ decomposes into the product of linear factors modulo p for all but finitely many prime numbers p only if $\deg f(X) = 1$. This fact played an important role in the arguments that were used in §3.1.2, in the proof of Zaleskii's theorem.

    Frobenius' theorem is related to Dirichlet's theorem on primes in arithmetic progressions; for further details on these results, the reader may consult the lucid exposition [67].

C.1 The Density Theorem

Let us fix a monic polynomial $f(X) \in \mathbf{Z}[X]$ of degree n. We assume that $f(X)$ has n distinct roots $a_1, \ldots, a_n$ in $\mathbf{C}$. For any prime number p we consider the quotient map $\mathbf{Z} \longrightarrow \mathbf{F}_p$ and the induced map between the polynomial rings

$\mathbf{Z}[X] \longrightarrow \mathbf{F}_p[X]$. In this way, $f(X)$ induces a monic polynomial $f_p(X) \in \mathbf{F}_p[X]$ of degree n, its reduction modulo p. The polynomial ring $\mathbf{F}_p[X]$ being a unique factorization domain, we may write

$$f_p(X) = \prod_{i=1}^{s_p} g_i^{(p)}(X),$$

where s_p is a positive integer and $g_i^{(p)}(X)$ a monic irreducible polynomial in $\mathbf{F}_p[X]$ for all $i = 1, \ldots, s_p$. Moreover, this decomposition is unique up to the ordering of the factors. We let $n_i^{(p)} = \deg g_i^{(p)}(X)$ for all i and note that $n = \sum_{i=1}^{s_p} n_i^{(p)}$. Therefore, assuming that the ordering is such that $n_i^{(p)} \geq n_j^{(p)}$ whenever $i \leq j$, we obtain a partition of n, in the sense of the following definition.

Definition C.1 *Let n be a positive integer. A partition δ of n of length s is a sequence $(n_1, \ldots, n_s)$ of positive integers, such that $n = \sum_{i=1}^{s} n_i$ and $n_i \geq n_j$ for all $i \leq j$.*

Let Π be the set of all prime numbers and $\mathbf{\Delta}_n$ the set of all partitions of n. The above considerations show that the monic polynomial $f(X)$ induces a map

$$\delta_f : \Pi \longrightarrow \mathbf{\Delta}_n,$$

where $\delta_f(p) = \left(n_1^{(p)}, \ldots, n_{s_p}^{(p)} \right)$ is the partition of n associated with the degrees of the irreducible factors of the reduction $f_p(X) \in \mathbf{F}_p[X]$ of $f(X)$ modulo p for all prime numbers p. We say that $\delta_f(p)$ is the decomposition length type of the polynomial $f(X)$ modulo p. For example, if $f(X)$ is irreducible modulo p, then its decomposition length type modulo p is (n). On the other hand, if $f(X)$ decomposes into the product of linear factors modulo p, then $\delta_f(p) = (1, \ldots, 1)$.

We now consider the splitting field $K = \mathbf{Q}(a_1, \ldots, a_n)$ of $f(X)$ and the corresponding Galois group Γ. Any element $\gamma \in \Gamma$ permutes the roots $a_1, \ldots, a_n$ and hence defines an element $\widetilde{\gamma}$ in the group S_n of permutations on n letters.[1] We may decompose $\widetilde{\gamma}$ into the product of disjoint cycles

$$\widetilde{\gamma} = \prod_{i=1}^{s_\gamma} c_i^{(\gamma)},$$

in such a way that each one of the n letters appears once and only once in the decomposition. (Hence, cycles of length 1 are allowed.) Moreover, this decomposition is unique up to the ordering of the $c_i^{(\gamma)}$'s. We let $n_i^{(\gamma)}$ be the length of the cycle $c_i^{(\gamma)}$ for all $i = 1, \ldots, s_\gamma$ and assume that the ordering of the cycles is such that $n_i^{(\gamma)} \geq n_j^{(\gamma)}$ for all $i \leq j$. Since $n = \sum_{i=1}^{s_\gamma} n_i^{(\gamma)}$, the sequence $\left(n_1^{(\gamma)}, \ldots, n_{s_\gamma}^{(\gamma)} \right)$ is a partition of n. In this way, we obtain a map

---

[1] The map $\gamma \mapsto \widetilde{\gamma}$ depends on the given parametrization of the roots of $f(X)$. If we relabel the roots, the permutation $\widetilde{\gamma}$ will change by an inner automorphism of S_n.

$$\delta'_f : \Gamma \longrightarrow \mathbf{\Delta}_n \,,$$

which is defined by letting $\delta'_f(\gamma) = \left(n_1^{(\gamma)}, \dots, n_{s_\gamma}^{(\gamma)}\right) \in \mathbf{\Delta}_n$ for all $\gamma \in \Gamma$.[2] We say that $\delta'_f(\gamma)$ is the cycle pattern type of γ. For example, the cycle pattern type of the identity element $1 \in \Gamma$ is $(1, \dots, 1)$. In fact, the map $\gamma \mapsto \tilde{\gamma}$ being an embedding of Γ into S_n, 1 is the only element $\gamma \in \Gamma$ with $\delta'_f(\gamma) = (1, \dots, 1)$.

The relationship between the maps δ_f and δ'_f is described by Frobenius' density theorem. In order to state the result, we need the notion of density of a set of prime numbers.

Definition C.2 *Let Π be the set of all prime numbers. Then, a subset $\Pi_0 \subseteq \Pi$ is said to have density d if the limit*

$$\lim_n \frac{card\,(\Pi_0 \cap [0,n])}{card\,(\Pi \cap [0,n])} = \lim_n \frac{card\{p : p \in \Pi_0, p \le n\}}{card\{p : p\ prime, p \le n\}}$$

exists and equals d.

For example, a finite set of prime numbers has density 0. It follows that the density of a set Π_0 of prime numbers is 1 if the complement $\Pi \setminus \Pi_0$ is finite.

We are now ready to state Frobenius' density theorem.

Theorem C.3 *Let $f(X) \in \mathbf{Z}[X]$ be a monic polynomial of degree n without multiple roots in $\mathbf{C}$. We consider the Galois group Γ of $f(X)$ and the maps δ_f and δ'_f defined above. We fix a partition $\delta \in \mathbf{\Delta}_n$ and let $\Pi_\delta = \delta_f^{-1}(\delta)$ and $\Gamma_\delta = \delta_f'^{-1}(\delta)$. Then, Π_δ has a density which is equal to $\dfrac{card\,(\Gamma_\delta)}{card\,(\Gamma)}$.* □

The following consequence of the density theorem played an important role in the argumentation of §3.1.2.

Corollary C.4 *Let $f(X) \in \mathbf{Z}[X]$ be a monic irreducible polynomial, which splits completely into the product of linear factors modulo p for all but finitely many prime numbers p. Then, the polynomial $f(X)$ is linear.*

Proof. Being irreducible in $\mathbf{Z}[X]$, the polynomial $f(X)$ has distinct roots in $\mathbf{C}$; let Γ be its Galois group. Then, the order N of Γ is equal to the degree of the splitting field K of $f(X)$ over $\mathbf{Q}$. We apply Frobenius' density theorem to the special case of the partition $\delta = (1, \dots, 1)$ of $n = \deg f(X)$. In view of our hypothesis, the set Π_δ in Theorem C.3 consists of all but finitely many prime numbers and hence its density is 1. Since the set Γ_δ is the singleton $\{1\}$, we have $1 = \frac{1}{N}$ and hence $N = 1$. It follows that $K = \mathbf{Q}$, in which case the roots $a_1, \dots, a_n$ of $f(X)$ are rational numbers. The ring $\mathbf{Z}$ being integrally closed in $\mathbf{Q}$, in view of Lemma A.15, we conclude that $a_i \in \mathbf{Z}$ for all $i = 1, \dots, n$ and hence $f(X) = \prod_{i=1}^n (X - a_i)$ in $\mathbf{Z}[X]$. Since the polynomial $f(X)$ is irreducible in $\mathbf{Z}[X]$, we must have $n = 1$. □

---

[2] Since the sequences of lengths of the cycles in the cycle decompositions of two conjugate permutations are the same, the map δ'_f does not depend on the labelling of the roots of $f(X)$ (cf. footnote (1)).

C.2 Exercises

1. Let Γ be a finite group acting on the non-empty finite set X. For any $\gamma \in \Gamma$ we consider the fixed set $\mathrm{Fix}(\gamma) = \{x \in X : \gamma x = x\}$ and denote by ν_γ its cardinality. Similarly, for any $x \in X$ we consider the stabilizer $\mathrm{Stab}(x) = \{\gamma \in \Gamma : \gamma x = x\}$ and denote by μ_x its order.
   (i) Show that $\sum_{\gamma \in \Gamma} \nu_\gamma = \sum_{x \in X} \mu_x$.
   (ii) If the action is transitive, then show that $\sum_{\gamma \in \Gamma} \nu_\gamma = |\Gamma|$.
   (iii) Assume that the action is transitive. If $\mathrm{Fix}(\gamma) \neq \emptyset$ for all $\gamma \in \Gamma$, then show that X is a singleton.

2. Let $f(X) \in \mathbf{Z}[X]$ be a monic irreducible polynomial, whose reduction $f_p(X) \in \mathbf{F}_p[X]$ has a root in $\mathbf{F}_p$ for all but finitely many prime numbers p. The goal of this Exercise is to prove that $f(X)$ is linear (generalizing thereby Corollary C.4).
   (i) Let Γ be the Galois group of the polynomial $f(X)$. Show that any element of Γ fixes at least one root of $f(X)$ in $\mathbf{C}$.
   (ii) Show that the polynomial $f(X)$ has only one root in $\mathbf{C}$ and hence conclude that $\deg f(X) = 1$.
   (*Hint:* Use Exercise 1(iii) above.)

3. Let a be an integer with $\sqrt{a} \notin \mathbf{Z}$. Show that there are infinitely many prime numbers p for which a is the square of an integer modulo p and infinitely many prime numbers p for which a is not the square of any integer modulo p.

D

Homological Techniques

In this Appendix, we collect the basic results from Homological Algebra that were used in Chap. 4. Our goal is not to present a complete treatment on group homology, but rather to state the results needed in the book and place them in the perspective of the general theory. Consequently, we shall give no proofs of the statements to be made and refer instead the interested reader to specialized books on the subject, such as [9, 34] and [48].

D.1 Complexes and Homology

D.1.1 Chain Complexes

We fix a ring R. A chain complex of left R-modules is a pair (C, d), where $C = \bigoplus_{i \geq 0} C_i$ is a graded left R-module and $d = (d_n)_n$ a homogeneous R-linear endomorphism of C of degree -1, satisfying the equality $d^2 = 0$ (d is the differential of the complex). The chain complex (C, d) is presented pictorially as

$$C_0 \xleftarrow{d_1} C_1 \xleftarrow{d_2} \cdots \xleftarrow{d_n} C_n \xleftarrow{d_{n+1}} \cdots$$

The elements of the submodule $\ker d_n$ (resp. $\operatorname{im} d_{n+1}$) of C_n are referred to as n-cycles (resp. n-boundaries). The homology $H(C, d)$ of (C, d) is the graded R-module, which is given in degree n by $H_n(C, d) = \ker d_n / \operatorname{im} d_{n+1}$. The complex (C, d) is called acyclic if $H_n(C, d) = 0$ for all n.

Let (C, d) and (C', d') be two chain complexes. Then, a chain map

$$\varphi : (C, d) \longrightarrow (C', d')$$

is a homogeneous R-linear map of degree 0 from C to C', such that $\varphi d = d' \varphi$. A chain map φ as above induces R-linear maps

$$\varphi_n : H_n(C, d) \longrightarrow H_n(C', d')$$

for all n. If these latter maps are isomorphisms, the chain map φ is called a quasi-isomorphism. In particular, an isomorphism of chain complexes (defined in the obvious way) is a quasi-isomorphism. Two chain maps

$$\varphi, \psi : (C, d) \longrightarrow (C', d')$$

are called homotopic if there exists a homogeneous map $\Sigma : C \longrightarrow C'$ of degree $+1$, such that $\Sigma d + d'\Sigma = \varphi - \psi$. In that case, the maps induced in homology by φ and ψ are equal, i.e.

$$\varphi_n = \psi_n : H_n(C, d) \longrightarrow H_n(C', d')$$

for all n. A chain complex (C, d) is called contractible if the chain endomorphisms id_C and 0 of (C, d) are homotopic. It follows that a contractible chain complex is acyclic. A chain map $\varphi : (C, d) \longrightarrow (C', d')$ is called a homotopy equivalence if there exists a chain map $\psi : (C', d') \longrightarrow (C, d)$, such that the compositions $\varphi\psi$ and $\psi\varphi$ are homotopic with the identity maps $\mathrm{id}_{C'}$ and id_C respectively.

If $\varphi : (C, d) \longrightarrow (C', d')$ is a chain map then the kernel of the R-linear map $\varphi : C \longrightarrow C'$ is a graded d-invariant R-submodule of C; therefore, $(\ker \varphi, d)$ is a chain complex. Similarly, $(\mathrm{im}\, \varphi, d')$ is a chain subcomplex of (C', d'). In this way, one extends the notion of exactness to the category of chain complexes and maps. In particular, one may consider a short exact sequence of chain complexes

$$0 \longrightarrow (C', d') \overset{i}{\longrightarrow} (C, d) \overset{p}{\longrightarrow} (C'', d'') \longrightarrow 0 \,.$$

A short exact sequence as above induces a long exact sequence of R-modules

$$\cdots \longrightarrow H_n(C', d') \overset{i_n}{\longrightarrow} H_n(C, d) \overset{p_n}{\longrightarrow} H_n(C'', d'') \overset{\partial_n}{\longrightarrow} H_{n-1}(C', d') \longrightarrow \cdots$$

The notions of cochain complexes, cochain maps and cohomology can be defined in the same way, by considering differentials of degree $+1$.

D.1.2 Double Complexes

A double chain complex (or chain bicomplex) of left R-modules consists of a bigraded left R-module $C = \bigoplus_{i,j \geq 0} C_{ij}$ together with R-linear maps d_h and d_v, which are homogeneous of degrees $(-1, 0)$ and $(0, -1)$ respectively and satisfy the equalities $d_h^2 = 0$, $d_v^2 = 0$ and $d_h d_v + d_v d_h = 0$. The map d_h (resp. d_v) is referred to as the horizontal (resp. vertical) differential of the double complex.

If (C, d_h, d_v) and (C', d_h', d_v') are chain bicomplexes, then a chain bicomplex map

$$\varphi : (C, d_h, d_v) \longrightarrow (C', d_h', d_v')$$

is a homogeneous R-linear map of degree $(0, 0)$ from C to C', such that $\varphi d_h = d_h'\varphi$ and $\varphi d_v = d_v'\varphi$. It is clear that a chain bicomplex map φ as above restricts in the horizontal direction to a chain map between the rows (C_{*j}, d_h) and (C'_{*j}, d'_h) for all j. Similarly, φ restricts in the vertical direction to a chain map between the columns (C_{i*}, d_v) and (C'_{i*}, d'_v) for all i.

For any double complex (C, d_h, d_v) there is an associated (total) chain complex $(\text{Tot } C, d)$, which is defined by letting $(\text{Tot } C)_n = \bigoplus_{i+j=n} C_{ij}$ for all n and $d = d_h + d_v$. Moreover, any chain bicomplex map

$$\varphi : (C, d_h, d_v) \longrightarrow (C', d'_h, d'_v)$$

induces a chain map

$$\text{Tot } \varphi : (\text{Tot } C, d) \longrightarrow (\text{Tot } C', d') .$$

We say that φ is a quasi-isomorphism if this is the case for $\text{Tot } \varphi$.

Proposition D.1 *Let* $\varphi : (C, d_h, d_v) \longrightarrow (C', d'_h, d'_v)$ *be a chain bicomplex map. Then,* φ *is a quasi-isomorphism if either one of the following two conditions is satisfied:*

(i) φ *restricts to a quasi-isomorphism between the columns* (C_{i*}, d_v) *and* (C'_{i*}, d'_v) *for all* i.

(ii) φ *restricts to a quasi-isomorphism between the rows* (C_{*j}, d_h) *and* (C'_{*j}, d'_h) *for all* j.   □

Corollary D.2 *Let* (C, d_h, d_v) *be a double complex.*

(i) If the rows (C_{*j}, d_h) *are acyclic in positive degrees for all* j*, then the chain complexes* $(\text{Tot } C, d)$ *and* $\left(\bigoplus_j C_{0j}/d_h C_{1j}, \overline{d_v} \right)$ *are quasi-isomorphic.*

(ii) If the columns (C_{i*}, d_v) *are acyclic for all* $i > 0$*, then the complex* $(\text{Tot } C, d)$ *is quasi-isomorphic with the 0-th column* (C_{0*}, d_v).   □

The notion of a double cochain complex can be defined in the same way, by considering differentials of degrees $(+1, 0)$ and $(0, +1)$.

Examples D.3 (i) Let R be a ring and (C, d) (resp. (C', d')) a chain complex of right (resp. left) R-modules. Then, there is a chain bicomplex of abelian groups $(C \otimes_R C', d \otimes 1, \pm 1 \otimes d')$, consisting of $C_i \otimes_R C'_j$ in degree (i, j); the signs $+$ and $-$ in the vertical differentials alternate in order for the operator $d_h d_v + d_v d_h$ to vanish. We note that there are natural maps (called Künneth maps)

$$K : H_i(C) \otimes_R H_j(C') \longrightarrow H_{i+j}(\text{Tot}(C \otimes_R C'))$$

for all $i, j \geq 0$, which are given by letting $[x_i] \otimes [x'_j] \mapsto [x_i \otimes x'_j]$ for any homology classes $[x_i] \in H_i(C)$ and $[x'_j] \in H_j(C')$. Similar remarks apply to cochain complexes and cohomology.

(ii) Let R be a ring, (C, d) a chain complex and (C', d') a cochain complex of left R-modules. Then, there is a double cochain complex of abelian groups $(\text{Hom}_R(C, C'), d^*, \pm d'_*)$, which consists of $\text{Hom}_R(C_i, C'^j)$ in degree (i, j). Here,

d^* and d'_* denote the additive maps between the Hom-groups induced by d and d' respectively. We note that there are natural maps

$$H^{i+j}(\operatorname{Tot}\operatorname{Hom}_R(C,C')) \longrightarrow \operatorname{Hom}_R\big(H_i(C),H^j(C')\big)$$

for all $i,j \geq 0$, which are given by mapping a cohomology class $[f] \in H^{i+j}(\operatorname{Tot}\operatorname{Hom}_R(C,C'))$ onto the R-linear map from $H_i(C)$ to $H^j(C')$ induced by the component $f_{ij} \in \operatorname{Hom}_R(C_i,C'^j)$ of f.

D.1.3 Tor and Ext

As an example of the notions introduced above, we define the functors Tor and Ext. To that end, let us fix a right R-module M and a left R-module N. For any R-projective resolutions

$$0 \longleftarrow M \xleftarrow{\varepsilon} P_* \quad \text{and} \quad 0 \longleftarrow N \xleftarrow{\eta} Q_*$$

we may consider the chain complexes of abelian groups $P_* \otimes_R N$, $M \otimes_R Q_*$ and the double complex $P_* \otimes_R Q_*$ (cf. Example D.3(i)). Then, there are quasi-isomorphisms

$$P_* \otimes_R N \xleftarrow{1 \otimes \eta} \operatorname{Tot}(P_* \otimes_R Q_*) \xrightarrow{\varepsilon \otimes 1} M \otimes_R Q_* \,.$$

In particular, we may define the Tor-groups by letting

$$\operatorname{Tor}_n^R(M,N) = H_n(P_* \otimes_R N) \simeq H_n(\operatorname{Tot}(P_* \otimes_R Q_*)) \simeq H_n(M \otimes_R Q_*)$$

for all n. If

$$0 \longleftarrow M \xleftarrow{\varepsilon'} P'_* \quad \text{and} \quad 0 \longleftarrow N \xleftarrow{\eta'} Q'_*$$

are also R-projective resolutions of M and N, then there are chain maps

$$\varphi : P_* \longrightarrow P'_* \quad \text{and} \quad \psi : Q_* \longrightarrow Q'_*$$

that satisfy the equalities $\varepsilon = \varepsilon'\varphi_0$ and $\eta = \eta'\psi_0$ and are unique up to homotopy with that property. These chain maps are homotopy equivalences and hence there is a commutative diagram of quasi-isomorphisms

$$
\begin{array}{ccccc}
P_* \otimes_R N & \xleftarrow{1 \otimes \eta} & \operatorname{Tot}(P_* \otimes_R Q_*) & \xrightarrow{\varepsilon \otimes 1} & M \otimes_R Q_* \\
{\scriptstyle \varphi \otimes 1} \downarrow & & {\scriptstyle Tot(\varphi \otimes \psi)} \downarrow & & {\scriptstyle 1 \otimes \psi} \downarrow \\
P'_* \otimes_R N & \xleftarrow{1 \otimes \eta'} & \operatorname{Tot}(P'_* \otimes_R Q'_*) & \xrightarrow{\varepsilon' \otimes 1} & M \otimes_R Q'_*
\end{array}
$$

It follows that the definition of the Tor-groups is independent of the chosen resolutions, in the sense that the vertical quasi-isomorphisms in the diagram above induce identifications in homology.

The behavior of the Tor-groups with respect to direct sums is described in the next result.

Proposition D.4 *(i) Let R be a ring, $(M_i)_i$ a family of right R-modules and $(N_j)_j$ a family of left R-modules. If $M = \bigoplus_i M_i$ and $N = \bigoplus_j N_j$, then the inclusions $M_i \hookrightarrow M$ and $N_j \hookrightarrow N$ induce an isomorphism of abelian groups $\operatorname{Tor}_n^R(M, N) \simeq \bigoplus_{i,j} \operatorname{Tor}_n^R(M_i, N_j)$ for all n.*

(ii) Let $R_1, \ldots, R_s$ be rings and $R = \prod_{i=1}^s R_i$ their direct product. For each $i = 1, \ldots, s$ we consider a right R_i-module M_i and a left R_i-module N_i. Then, the abelian group $M = \bigoplus_{i=1}^s M_i$ (resp. $N = \bigoplus_{i=1}^s N_i$) has the structure of a right (resp. left) R-module and there is a natural isomorphism $\operatorname{Tor}_n^R(M, N) \simeq \bigoplus_{i=1}^s \operatorname{Tor}_n^{R_i}(M_i, N_i)$ for all n. □

In order to define the Ext-groups of two left R-modules M and N, one considers a projective resolution

$$0 \longleftarrow M \longleftarrow P_*$$

of M and an injective resolution

$$0 \longrightarrow N \longrightarrow I^*$$

of N. Then, there are induced quasi-isomorphisms of cochain complexes

$$\operatorname{Hom}_R(P_*, N) \longrightarrow \operatorname{Tot} \operatorname{Hom}_R(P_*, I^*) \longleftarrow \operatorname{Hom}_R(M, I^*)$$

(cf. Example D.3(ii)) and one defines for any n

$$\begin{aligned}
\operatorname{Ext}_R^n(M, N) &= H^n(\operatorname{Hom}_R(P_*, N)) \\
&\simeq H^n(\operatorname{Tot} \operatorname{Hom}_R(P_*, I^*)) \\
&\simeq H^n(\operatorname{Hom}_R(M, I^*)) \, .
\end{aligned}$$

Remarks D.5 (i) Let R be a ring, M a right R-module and N a left R-module. In degree 0, there is an identification $\operatorname{Tor}_0^R(M, N) \simeq M \otimes_R N$. The groups $\operatorname{Tor}_n^R(M, N)$ vanish for all $n > 0$ if either M or N is projective (or, more generally, flat) as an R-module.

(ii) Let R be a ring and M, N two left R-modules. In degree 0, there is an identification $\operatorname{Ext}_R^0(M, N) \simeq \operatorname{Hom}_R(M, N)$. The groups $\operatorname{Ext}_R^n(M, N)$ vanish for all $n > 0$ if either M is projective or N is injective as an R-module.

(iii) Let (C, d) and (C', d') be two chain complexes of abelian groups, such that either one of them is free. Then, the homology of the double complex $(C \otimes C', d \otimes 1, \pm 1 \otimes d')$ fits into a natural short exact sequence

$$0 \to (H_*(C) \otimes H_*(C'))_n \xrightarrow{K} H_n(\operatorname{Tot}(C \otimes C')) \to (\operatorname{Tor}(H_*(C), H_*(C')))_{n-1} \to 0 \, ,$$

where

$$(H_*(C) \otimes H_*(C'))_n = \bigoplus_{i+j=n} H_i(C) \otimes H_j(C') \, ,$$

K is the Künneth map (cf. Example D.3(i)) and

$$(\operatorname{Tor}(H_*(C), H_*(C')))_{n-1} = \bigoplus_{i+j=n-1} \operatorname{Tor}_1^{\mathbf{Z}}(H_i(C), H_j(C'))$$

for all n. In particular, if the complexes (C, d) and (C', d') are free resolutions of the abelian groups M and M' respectively, then the chain complex $\operatorname{Tot}(C \otimes C')$ is a free resolution of $M \otimes M'$, provided that $\operatorname{Tor}_1^{\mathbf{Z}}(M, M') = 0$.

D.2 Group Homology and Cohomology

D.2.1 Basic Definitions

We now consider a commutative ring k, a group G and specialize the above discussion to the case where R is the group algebra kG. For any left kG-module M we define the homology groups $H_n(G, M)$ and the cohomology groups $H^n(G, M)$ of G with coefficients in M, by letting

$$H_n(G, M) = \operatorname{Tor}_n^{kG}(k, M) \quad \text{and} \quad H^n(G, M) = \operatorname{Ext}_{kG}^n(k, M)$$

for all n. Here, k is viewed as a (right and left) kG-module, by means of the trivial G-action. In particular, if

$$0 \longleftarrow k \xleftarrow{\varepsilon} P_*$$

is a resolution of k by projective right kG-modules, then

$$H_n(G, M) = H_n(P_* \otimes_{kG} M)$$

for all n. This definition doesn't depend upon the chosen resolution. More precisely, if

$$0 \longleftarrow k \xleftarrow{\varepsilon'} P'_*$$

is another kG-projective resolution of k, then there is a unique up to homotopy chain map $\varphi : P_* \longrightarrow P'_*$ satisfying $\varepsilon = \varepsilon' \varphi_0$, which is a homotopy equivalence and induces a canonical identification

$$H_n(P_* \otimes_{kG} M) \simeq H_n(P'_* \otimes_{kG} M)$$

for all n. Similarly, the cohomology groups $H^n(G, M)$ of G with coefficients in the left kG-module M can be computed by using a projective resolution Q_* of the trivial left kG-module k, as the cohomology groups of the cochain complex $\operatorname{Hom}_{kG}(Q_*, M)$.

In order to describe the so-called standard resolution of k, we let

$$S_n(G, k) = k[G^{n+1}] = \bigoplus \left\{ k \cdot (g_0, \ldots, g_n) : (g_0, \ldots, g_n) \in G^{n+1} \right\}$$

for all $n \geq 0$, with left (resp. right) kG-module structure induced by the left (resp. right) diagonal action of G on G^{n+1}; in particular, $S_0(G, k) = kG$. For all $n \geq 1$ and $i \in \{0, \ldots, n\}$ we define a k-linear map

$$\delta_i^n : S_n(G, k) \longrightarrow S_{n-1}(G, k) \,,$$

by letting

$$\delta_i^n(g_0, \ldots, g_n) = (g_0, \ldots, \widehat{g_i}, \ldots, g_n)$$

for any element $(g_0, \ldots, g_n) \in G^{n+1}$; here, the symbol $\,\widehat{}\,$ over an element denotes omission of that element. It is clear that the k-linear maps δ_i^n are, in fact, kG-linear for both left and right actions. We now define the operators

$$\delta_n, \delta_n' : S_n(G, k) \longrightarrow S_{n-1}(G, k) ,$$

by letting

$$\delta_n = \sum_{i=0}^{n} (-1)^i \delta_i^n \quad \text{and} \quad \delta_n' = \sum_{i=0}^{n-1} (-1)^i \delta_i^n$$

for all $n \geq 1$.

Proposition D.6 *Let k be a commutative ring and G a group.*
  (i) The chain complex

$$k \xleftarrow{\ \varepsilon\ } S_0(G, k) \xleftarrow{\ \delta_1\ } S_1(G, k) \xleftarrow{\ \delta_2\ } \cdots \xleftarrow{\ \delta_n\ } S_n(G, k) \xleftarrow{\ \delta_{n+1}\ } \cdots ,$$

where ε is the augmentation, is a free resolution $(S(G, k), \delta) = (S_*(G, k), \delta)$ of k as a trivial left or right kG-module.
  (ii) The chain complex

$$S_0(G, k) \xleftarrow{\ \delta_1'\ } S_1(G, k) \xleftarrow{\ \delta_2'\ } \cdots \xleftarrow{\ \delta_n'\ } S_n(G, k) \xleftarrow{\ \delta_{n+1}'\ } \cdots$$

is contractible as a complex of left or right kG-modules. In particular, it is acyclic. $\qquad\square$

In order to describe the functoriality of the (co-)homology groups, let us consider a group G and a homomorphism $f : M \longrightarrow M'$ of left kG-modules. Then, there are induced additive maps

$$f_* : H_n(G, M) \longrightarrow H_n(G, M') \quad \text{and} \quad f_* : H^n(G, M) \longrightarrow H^n(G, M')$$

for all n. On the other hand, let $\phi : G \longrightarrow G'$ be a group homomorphism. Then, ϕ induces a k-linear chain map

$$\widetilde{\phi} : S(G, k) \longrightarrow S(G', k) .$$

For any kG'-module M we denote by M_ϕ the kG-module obtained from M by restriction of scalars along ϕ. Then, the chain maps

$$\widetilde{\phi} \otimes 1 : S(G, k) \otimes_{kG} M_\phi \longrightarrow S(G', k) \otimes_{kG'} M$$

and

$$\mathrm{Hom}\left(\widetilde{\phi}, 1\right) : \mathrm{Hom}_{kG'}(S(G', k), M) \longrightarrow \mathrm{Hom}_{kG}(S(G, k), M_\phi)$$

induce additive maps

$$\phi_* : H_n(G, M_\phi) \longrightarrow H_n(G', M) \quad \text{and} \quad \phi^* : H^n(G', M) \longrightarrow H^n(G, M_\phi)$$

respectively for all n.

Remarks D.7 (i) Let k be a commutative ring, G a group and M a left kG-module. Then, the homology groups $H_n(G, M)$ and the cohomology groups $H^n(G, M)$ depend only upon the action of G on the abelian group M. In order to make this assertion precise, let us denote by M' the $\mathbf{Z}G$-module obtained from the kG-module M by restriction of scalars. Since

$$S_*(G, k) = S_*(G, \mathbf{Z}) \otimes_{\mathbf{Z}G} kG \quad \text{and} \quad S_*(G, k) = kG \otimes_{\mathbf{Z}G} S_*(G, \mathbf{Z})$$

as right and left kG-modules respectively, we have

$$\begin{aligned}
\operatorname{Tor}_n^{kG}(k, M) &= H_n(S_*(G, k) \otimes_{kG} M) \\
&= H_n(S_*(G, \mathbf{Z}) \otimes_{\mathbf{Z}G} M') \\
&= \operatorname{Tor}_n^{\mathbf{Z}G}(\mathbf{Z}, M')
\end{aligned}$$

and

$$\begin{aligned}
\operatorname{Ext}_{kG}^n(k, M) &= H^n(\operatorname{Hom}_{kG}(S_*(G, k), M)) \\
&= H^n(\operatorname{Hom}_{\mathbf{Z}G}(S_*(G, \mathbf{Z}), M')) \\
&= \operatorname{Ext}_{\mathbf{Z}G}^n(\mathbf{Z}, M'),
\end{aligned}$$

i.e. $H_n(G, M) = H_n(G, M')$ and $H^n(G, M) = H^n(G, M')$ for all n.

(ii) Let G be a group, M a left $\mathbf{Z}G$-module and $g \in G$ a fixed element. We consider the inner automorphism (conjugation) $I_g \in \operatorname{Aut}(G)$ and let $M_g = M_{I_g}$ be the $\mathbf{Z}G$-module obtained from M by restriction of scalars along I_g. We also consider the $\mathbf{Z}G$-linear map

$$\lambda_g : M \longrightarrow M_g ,$$

which is defined by $m \mapsto gm$, $m \in M$. Then, the compositions

$$H_n(G, M) \xrightarrow{(\lambda_g)_*} H_n(G, M_g) \xrightarrow{(I_g)_*} H_n(G, M)$$

and

$$H^n(G, M) \xrightarrow{(\lambda_g)_*} H^n(G, M_g) \xrightarrow{[(I_g)^*]^{-1}} H^n(G, M)$$

can be shown to be the identity operators for all n.

Examples D.8 (i) If G is a group and M a left $\mathbf{Z}G$-module, then

$$H_0(G, M) = \mathbf{Z} \otimes_{\mathbf{Z}G} M \simeq M/<gm - m : g \in G, m \in M>$$

is the coinvariance M_G of M (cf. Remark D.5(i)) and

$$H^0(G, M) = \operatorname{Hom}_{\mathbf{Z}G}(\mathbf{Z}, M) \simeq \{m \in M : gm = m \text{ for all } g \in G\}$$

is the invariance M^G of M (cf. Remark D.5(ii)).

(ii) Let k be a field of characteristic 0 and G a finite group. Then, in view of Maschke's theorem (Theorem 1.9), the groups $H_n(G, M)$ and $H^n(G, M)$ vanish for all $n > 0$ and all kG-modules M (cf. Remarks D.5(i),(ii)).

(iii) Let G be a finite cyclic group with generator τ. Then, the chain complex

$$\mathbf{Z} \xleftarrow{\varepsilon} \mathbf{Z}G \xleftarrow{1-\tau} \mathbf{Z}G \xleftarrow{N} \mathbf{Z}G \xleftarrow{1-\tau} \mathbf{Z}G \xleftarrow{N} \mathbf{Z}G \xleftarrow{1-\tau} \cdots ,$$

where $N = \sum \{t : t \in G\}$, is a $\mathbf{Z}G$-free resolution of $\mathbf{Z}$. In particular, for any $\mathbf{Z}G$-module M the homology groups of G with coefficients in M are computed as the homology groups of the chain complex

$$M \xleftarrow{1-\tau} M \xleftarrow{N} M \xleftarrow{1-\tau} M \xleftarrow{N} M \xleftarrow{1-\tau} \cdots$$

Similarly, the cohomology groups of G with coefficients in M are computed as the cohomology groups of the cochain complex

$$M \xrightarrow{1-\tau} M \xrightarrow{N} M \xrightarrow{1-\tau} M \xrightarrow{N} M \xrightarrow{1-\tau} \cdots$$

In particular, $H_n(G, M) = H_{n+2}(G, M)$ and $H^n(G, M) = H^{n+2}(G, M)$ for all $n > 0$.

Let k be a commutative ring and G a group. Then, G is said to have finite homological dimension over k if there exists an integer $n \geq 0$, such that $H_i(G, M) = 0$ for all $i > n$ and all kG-modules M. The smallest n with this property is the homological dimension $\mathrm{hd}_k G$ of G over k. In this way, Example D.8(ii) implies that $\mathrm{hd}_k G = 0$ if G is a finite group and k a field of characteristic 0.

Proposition D.9 *Let k be a commutative ring.*

(i) If G is a group of finite homological dimension over k and $H \leq G$ a subgroup, then H has finite homological dimension over k as well; in fact, $\mathrm{hd}_k H \leq \mathrm{hd}_k G$.

(ii) If $(G_i)_i$ is a family of groups of uniformly bounded homological dimension over k and $G = *_i G_i$ the corresponding free product, then G has finite homological dimension over k; in fact, $\mathrm{hd}_k G \leq \max\{1, \max_i \mathrm{hd}_k G_i\}$.

(iii) If $(G_i)_i$ is a directed system of groups of uniformly bounded homological dimension over k and $G = \varinjlim_i G_i$ the corresponding direct limit, then G has finite homological dimension over k; in fact, $\mathrm{hd}_k G \leq \max_i \mathrm{hd}_k G_i$. $\square$

D.2.2 H^2 and Extensions

Let G be a group and M a left $\mathbf{Z}G$-module. Then, an extension of G by M is a group X having M as a normal subgroup with $X/M \simeq G$, in such a way that conjugation induces the given action of G on M. Two extensions

$$1 \longrightarrow M \longrightarrow X \longrightarrow G \longrightarrow 1 \quad \text{and} \quad 1 \longrightarrow M \longrightarrow X' \longrightarrow G \longrightarrow 1$$

are called equivalent if there is a group homomorphism $f : X \longrightarrow X'$ that fits into a commutative diagram

$$
\begin{array}{ccccccccc}
1 & \longrightarrow & M & \longrightarrow & X & \longrightarrow & G & \longrightarrow & 1 \\
& & \| & & f\downarrow & & \| & & \\
1 & \longrightarrow & M & \longrightarrow & X' & \longrightarrow & G & \longrightarrow & 1
\end{array}
$$

There is a bijective correspondence between the set of equivalence classes of extensions of G by M and the cohomology group $H^2(G, M)$, such that the equivalence class of the semi-direct product $M \times G$ corresponds to the zero element of $H^2(G, M)$. In this way, the functorial behavior of H^2 corresponds to certain operations on extensions. In order to explicit these operations, let us consider an extension

$$1 \longrightarrow M \xrightarrow{\imath} X \xrightarrow{\pi} G \longrightarrow 1$$

and the corresponding cohomology class $\alpha \in H^2(G, M)$.

Functoriality in the coefficient module: Let M' be another $\mathbf{Z}G$-module and $f : M \longrightarrow M'$ a $\mathbf{Z}G$-linear map. We consider the $\mathbf{Z}X$-module M'_π obtained from M' by restricting its G-module structure along π and note that $\{(-f(m), \imath(m)) : m \in M\}$ is a normal subgroup of the semi-direct product $M'_\pi \times X$. The corresponding quotient group X' fits into the commutative diagram

$$
\begin{array}{ccccccccc}
1 & \longrightarrow & M & \xrightarrow{\imath} & X & \xrightarrow{\pi} & G & \longrightarrow & 1 \\
& & f\downarrow & & f'\downarrow & & \| & & \\
1 & \longrightarrow & M' & \xrightarrow{\imath'} & X' & \xrightarrow{\pi'} & G & \longrightarrow & 1
\end{array}
$$

where $\imath'$ and f' are obtained by composing the natural maps into $M' \times X$ with the projection onto X', whereas π' maps the class of an element (m', x) in X' onto $\pi(x) \in G$. Then, the G-module structure induced on M' by the extension in the bottom row of the above diagram is its original G-module structure and the element of $H^2(G, M')$ which classifies that extension is the image $f_*\alpha$ of α under the map

$$f_* : H^2(G, M) \longrightarrow H^2(G, M') \,.$$

Functoriality in the group: Let G' be another group and $\phi : G' \longrightarrow G$ a group homomorphism. We consider the subgroup $X'' = \{(x, g') \in X \times G' : \pi(x) = \phi(g')\}$ of the direct product $X \times G'$ and note that it fits into the commutative diagram

$$
\begin{array}{ccccccccc}
1 & \longrightarrow & M & \xrightarrow{\imath''} & X'' & \xrightarrow{\pi''} & G' & \longrightarrow & 1 \\
& & \| & & \phi''\downarrow & & \phi\downarrow & & \\
1 & \longrightarrow & M & \xrightarrow{\imath} & X & \xrightarrow{\pi} & G & \longrightarrow & 1
\end{array}
$$

where ϕ'' and π'' are the restrictions to X'' of the projections of $X \times G'$ onto X and G' respectively, whereas $\imath''$ maps $m \in M$ onto $(\imath(m), 1) \in X''$. Then, the G'-module structure induced on M by the extension in the top row of the above diagram is the one obtained by restricting its original G-module structure along ϕ, thereby identifying it with M_ϕ. Moreover, the element of $H^2(G', M_\phi)$ which classifies that extension is the image $\phi^* \alpha$ of α under the map

$$\phi^* : H^2(G, M) \longrightarrow H^2(G', M_\phi) \, .$$

Proposition D.10 *Consider a morphism of group extensions with abelian kernels*

$$
\begin{array}{ccccccccc}
1 & \longrightarrow & M & \longrightarrow & X & \longrightarrow & G & \longrightarrow & 1 \\
& & f \downarrow & & \downarrow & & \phi \downarrow & & \\
1 & \longrightarrow & M' & \longrightarrow & X' & \longrightarrow & G' & \longrightarrow & 1
\end{array}
$$

If $\alpha \in H^2(G, M)$ and $\alpha' \in H^2(G', M')$ are the cohomology classes classifying these extensions, then $f_* \alpha = \phi^* \alpha' \in H^2(G, M')$. □

D.2.3 Products

We fix a group G and consider the standard resolution $S_* = (S_*(G, \mathbf{Z}), \delta)$ of the trivial G-module $\mathbf{Z}$. We recall that $S_n = \mathbf{Z}[G^{n+1}]$ is projective both as a left and as a right $\mathbf{Z}G$-module. The chain complex $\mathrm{Tot}(S_* \otimes S_*)$ is a resolution of $\mathbf{Z} \otimes \mathbf{Z} = \mathbf{Z}$ (cf. Remark D.5(iii)), consisting of projective left (and right) $\mathbf{Z}G$-modules (the group G acts diagonally on the tensor product; cf. Lemma 1.8(ii)). Hence, there is a left $\mathbf{Z}G$-linear chain map

$$\Delta' : S_* \longrightarrow \mathrm{Tot}(S_* \otimes S_*) \, ,$$

which commutes with the augmentation maps to $\mathbf{Z}$ and is unique up to homotopy with this property. Similarly, there is a right $\mathbf{Z}G$-linear chain map

$$\Delta'' : S_* \longrightarrow \mathrm{Tot}(S_* \otimes S_*) \, ,$$

which commutes with the augmentation maps to $\mathbf{Z}$ and is unique up to homotopy with this property. Chain maps Δ' and Δ'' as above are homotopy equivalences and are referred to as diagonal approximations.

We now construct a specific diagonal approximation

$$\Delta : S_* \longrightarrow \mathrm{Tot}(S_* \otimes S_*) \, ,$$

called the Alexander-Whitney map. To that end, we define in degree n the additive map

$$\Delta_n : S_n \longrightarrow \bigoplus_{i=0}^n S_i \otimes S_{n-i} \, ,$$

by letting $\Delta_n(g_0, \ldots, g_n) = \sum_{i=0}^n (g_0, \ldots, g_i) \otimes (g_i, \ldots, g_n)$ for all $(n+1)$-tuples $(g_0, \ldots, g_n) \in G^{n+1}$.

Lemma D.11 *The Alexander-Whitney map* $\Delta = (\Delta_n)_n$ *is a chain map, which is* $\mathbf{Z}G$-*linear for both left and right actions and commutes with the augmentation maps to* $\mathbf{Z}$.   □

Using the Alexander-Whitney map, we define the cup- and cap-products.

Cup-products: Let M, N be left $\mathbf{Z}G$-modules and $M \otimes N$ the corresponding tensor product, viewed as a left $\mathbf{Z}G$-module with diagonal action. Then, there is a morphism of cochain complexes

$$\_ \cup \_ : \mathrm{Tot}(\mathrm{Hom}_{\mathbf{Z}G}(S_*, M) \otimes \mathrm{Hom}_{\mathbf{Z}G}(S_*, N)) \longrightarrow \mathrm{Hom}_{\mathbf{Z}G}(S_*, M \otimes N)$$

(cf. Example D.3(i)), which is defined as follows: For any $a \in \mathrm{Hom}_{\mathbf{Z}G}(S_i, M)$ and $b \in \mathrm{Hom}_{\mathbf{Z}G}(S_j, N)$, the element $a \cup b \in \mathrm{Hom}_{\mathbf{Z}G}(S_{i+j}, M \otimes N)$ maps any $x \in G^{i+j+1} \subseteq S_{i+j}$ with $\Delta_{i+j} x = \sum_{k+l=i+j} x_k \otimes x'_l \in \bigoplus_{k+l=i+j} S_k \otimes S_l$ onto $(-1)^{ij} a(x_i) \otimes b(x'_j) \in M \otimes N$. The induced additive maps

$$\_ \cup \_ : H^i(G, M) \otimes H^j(G, N) \longrightarrow H^{i+j}(G, M \otimes N), \quad i, j \geq 0,$$

are called *cup-product maps*. Some basic properties of them are summarized in the next result.

Proposition D.12 *The cup-product maps have the following properties:*

(i) (associativity) Let L, M and N be left $\mathbf{Z}G$-modules. Then, for any cohomology classes $\alpha \in H^i(G, L)$, $\beta \in H^j(G, M)$ and $\gamma \in H^k(G, N)$, we have $(\alpha \cup \beta) \cup \gamma = \alpha \cup (\beta \cup \gamma) \in H^{i+j+k}(G, L \otimes M \otimes N)$.

(ii) (graded-commutativity) Let M, N be left $\mathbf{Z}G$-modules. Then, for any cohomology classes $\alpha \in H^i(G, M)$ and $\beta \in H^j(G, N)$, we have $\alpha \cup \beta = (-1)^{ij} \tau_*(\beta \cup \alpha)$, where $\tau : N \otimes M \longrightarrow M \otimes N$ is the flip map.

(iii) (naturality with respect to the coefficient modules) Let M, N, M' and N' be left $\mathbf{Z}G$-modules. Then, for any $\mathbf{Z}G$-linear maps $f : M \longrightarrow M'$ and $h : N \longrightarrow N'$ there is a commutative diagram

$$
\begin{array}{ccc}
H^i(G, M) \otimes H^j(G, N) & \xrightarrow{\ \_ \cup \_\ } & H^{i+j}(G, M \otimes N) \\
{\scriptstyle f_* \otimes h_*} \downarrow & & \downarrow {\scriptstyle (f \otimes h)_*} \\
H^i(G, M') \otimes H^j(G, N') & \xrightarrow{\ \_ \cup \_\ } & H^{i+j}(G, M' \otimes N')
\end{array}
$$

for all $i, j \geq 0$.

(iv) (naturality with respect to the group) Let G' be another group and $\varphi : G' \longrightarrow G$ a group homomorphism. Then, for any $\mathbf{Z}G$-modules M, N there is a commutative diagram

$$
\begin{array}{ccc}
H^i(G, M) \otimes H^j(G, N) & \xrightarrow{\ \_ \cup \_\ } & H^{i+j}(G, M \otimes N) \\
{\scriptstyle \varphi^* \otimes \varphi^*} \downarrow & & \downarrow {\scriptstyle \varphi^*} \\
H^i(G', M') \otimes H^j(G', N') & \xrightarrow{\ \_ \cup \_\ } & H^{i+j}(G', M' \otimes N')
\end{array}
$$

for all $i, j \geq 0$. Here, $M' = M_\varphi$ and $N' = N_\varphi$ denote the $\mathbf{Z}G'$-modules obtained from M and N respectively by restriction of scalars along φ.   □

In particular, let us consider a commutative ring k, viewed as a trivial G-module. Then, the multiplication $k \otimes k \longrightarrow k$ enables one to consider the composition

$$H^i(G,k) \otimes H^j(G,k) \xrightarrow{\ \cup\ } H^{i+j}(G, k \otimes k) \longrightarrow H^{i+j}(G,k)$$

for all $i,j \geq 0$; these maps are referred to as cup-product maps as well.

Corollary D.13 *Let k be a commutative ring.*

(i) (k-algebra structure) The cup-product maps defined above endow the cohomology $H^\bullet(G,k) = \bigoplus_i H^i(G,k)$ with the structure of an associative and graded-commutative k-algebra.

(ii) (naturality with respect to the coefficient ring) Let K be another commutative ring and $f : k \longrightarrow K$ a ring homomorphism. Then, the induced map

$$f_* : H^\bullet(G,k) \longrightarrow H^\bullet(G,K)$$

is a ring homomorphism as well.

(iii) (naturality with respect to the group) Let G' be another group and $\varphi : G' \longrightarrow G$ a group homomorphism. Then, the induced map

$$\varphi^* : H^\bullet(G,k) \longrightarrow H^\bullet(G',k)$$

is a homomorphism of k-algebras. □

Cap-products: We consider again two left $\mathbf{Z}G$-modules M, N and their tensor product $M \otimes N$ (with diagonal G-action). Let $\alpha \in H^n(G, M)$ be a cohomology class, represented by a $\mathbf{Z}G$-linear map $a : S_n \longrightarrow M$. We denote by $S_*[n]$ the complex, which is given in degree i by S_{i-n} and whose differential is $(-1)^n \delta$, and consider the composition

$$S_* \otimes_{\mathbf{Z}G} N \xrightarrow{\Delta \otimes 1} \mathrm{Tot}(S_* \otimes S_*) \otimes_{\mathbf{Z}G} N \xrightarrow{(1 \otimes a \otimes 1)^{gr}} S_*[n] \otimes_{\mathbf{Z}G} (M \otimes N) \,,$$

where $(1 \otimes a \otimes 1)^{gr}$ maps an elementary tensor $(x_i \otimes x_j) \otimes y \in (S_i \otimes S_j) \otimes_{\mathbf{Z}G} N$ onto the tensor $(-1)^{ni} x_i \otimes (a(x_j) \otimes y) \in S_i \otimes_{\mathbf{Z}G} (M \otimes N)$ (resp. onto 0) if $j = n$ (resp. if $j \neq n$).[1] The induced additive maps

$$\alpha \cap \_ : H_i(G,N) \longrightarrow H_{i-n}(G, M \otimes N), \quad i \geq 0 \,,$$

depend only upon the cohomology class α and are called cap-product maps. Some basic properties of them are summarized in the next result.

---

[1] Here, we regard S_* and $\mathrm{Tot}(S_* \otimes S_*)$ as complexes of right $\mathbf{Z}G$-modules by letting any element $g \in G$ act as left multiplication by g^{-1}. As such, the complexes S_* and $\mathrm{Tot}(S_* \otimes S_*)$ provide us with projective resolutions of the trivial right $\mathbf{Z}G$-module $\mathbf{Z}$; cf. the discussion at the beginning of §D.2.4.

Proposition D.14 *The cap-product maps have the following properties:*

(i) (naturality with respect to the coefficient modules) Let M, N, M' and N' be left $\mathbf{Z}G$-modules. Then, for any $\mathbf{Z}G$-linear maps $f : M \longrightarrow M'$ and $h : N \longrightarrow N'$ and any cohomology class $\alpha \in H^n(G, M)$ there is a commutative diagram

$$
\begin{array}{ccc}
H_i(G, N) & \xrightarrow{\alpha \cap -} & H_{i-n}(G, M \otimes N) \\
{\scriptstyle h_*} \downarrow & & \downarrow {\scriptstyle (f \otimes h)_*} \\
H_i(G, N') & \xrightarrow{f_* \alpha \cap -} & H_{i-n}(G, M' \otimes N')
\end{array}
$$

for all $i \geq 0$.

(ii) (naturality with respect to the group) Let G' be another group and $\varphi : G' \longrightarrow G$ a group homomorphism. Then, for any left $\mathbf{Z}G$-modules M, N and any cohomology class $\alpha \in H^n(G, M)$ there is a commutative diagram

$$
\begin{array}{ccc}
H_i(G', N') & \xrightarrow{\varphi^* \alpha \cap -} & H_{i-n}(G', M' \otimes N') \\
{\scriptstyle \varphi_*} \downarrow & & \downarrow {\scriptstyle \varphi_*} \\
H_i(G, N) & \xrightarrow{\alpha \cap -} & H_{i-n}(G, M \otimes N)
\end{array}
$$

for all $i \geq 0$. Here, $M' = M_\varphi$ and $N' = N_\varphi$ denote the $\mathbf{Z}G'$-modules obtained from M and N respectively by restriction of scalars along φ.

(iii) (composition) Let L, M and N be left $\mathbf{Z}G$-modules. Then, for any cohomology classes $\alpha \in H^n(G, M)$ and $\beta \in H^m(G, L)$, the composition

$$
H_i(G, N) \xrightarrow{\alpha \cap -} H_{i-n}(G, M \otimes N) \xrightarrow{\beta \cap -} H_{i-n-m}(G, L \otimes M \otimes N)
$$

coincides with the cap-product map

$$
\gamma \cap - : H_i(G, N) \longrightarrow H_{i-n-m}(G, L \otimes M \otimes N) \,,
$$

where $\gamma = \beta \cup \alpha \in H^{n+m}(G, L \otimes M)$, for all $i \geq 0$. □

In particular, let k be a commutative ring (viewed as a trivial G-module) and $\alpha \in H^n(G, k)$ a cohomology class. Then, we may consider for all $i \geq 0$ the composition

$$
H_i(G, k) \xrightarrow{\alpha \cap -} H_{i-n}(G, k \otimes k) \longrightarrow H_{i-n}(G, k) \,,
$$

where the latter map is induced by the multiplication of k. These maps are referred to as cap-product maps as well.

Corollary D.15 *Let k be a commutative ring.*

(i) (naturality with respect to the coefficient ring) Let K be another commutative ring and $f : k \longrightarrow K$ a ring homomorphism. Then, for any cohomology class $\alpha \in H^n(G, k)$ there is a commutative diagram

$$
\begin{array}{ccc}
H_i(G, k) & \xrightarrow{\alpha \cap -} & H_{i-n}(G, k) \\
{\scriptstyle f_*} \downarrow & & \downarrow {\scriptstyle f_*} \\
H_i(G, K) & \xrightarrow{\alpha_K \cap -} & H_{i-n}(G, K)
\end{array}
$$

for all $i \geq 0$. Here, $\alpha_K = f_* \alpha$ is the image of α in the cohomology group $H^n(G, K)$.

(ii) (naturality with respect to the group) Let G' be another group and $\varphi : G' \longrightarrow G$ a group homomorphism. Then, for any cohomology class $\alpha \in H^n(G, k)$ there is a commutative diagram

$$
\begin{array}{ccc}
H_i(G', k) & \xrightarrow{\alpha' \cap -} & H_{i-n}(G', k) \\
\varphi_* \downarrow & & \downarrow \varphi_* \\
H_i(G, k) & \xrightarrow{\alpha \cap -} & H_{i-n}(G, k)
\end{array}
$$

for all $i \geq 0$. Here, $\alpha' = \varphi^* \alpha$ is the image of α in the cohomology group $H^n(G', k)$.

(iii) (composition) For any cohomology classes $\alpha \in H^n(G, k)$ and $\beta \in H^m(G, k)$, the composition

$$
H_i(G, k) \xrightarrow{\alpha \cap -} H_{i-n}(G, k) \xrightarrow{\beta \cap -} H_{i-n-m}(G, k)
$$

coincides with the cap-product map

$$
\gamma \cap - : H_i(G, k) \longrightarrow H_{i-n-m}(G, k) ,
$$

where $\gamma = \beta \cup \alpha \in H^{n+m}(G, k)$, for all $i \geq 0$. □

D.2.4 Duality

In this subsection, we briefly investigate the extent to which group cohomology is dual to group homology.

We fix a commutative ring k and a group G. First of all, we define the (co-)homology groups of G with coefficients in right kG-modules. To that end, we consider the involution

$$
\tau : kG \longrightarrow kG
$$

of the k-algebra kG, which is defined by letting $\tau(g) = g^{-1}$ for all $g \in G$. If U is a right kG-module, we let $\widetilde{U}$ be the left kG-module obtained from U by pulling back its right kG-module structure along τ. In other words, we define $g \cdot u = ug^{-1}$ for all $g \in G$ and $u \in U$. Then, the (co-)homology of G with coefficients in U is defined as the (co-)homology of G with coefficients in $\widetilde{U}$. We note that the right kG-module U is projective if and only if the left kG-module $\widetilde{U}$ is projective. Moreover, if U' is another right kG-module then a k-linear map $f : U \longrightarrow U'$ is a homomorphism of right kG-modules if and only if the map $f : \widetilde{U} \longrightarrow \widetilde{U'}$ is a homomorphism of left kG-modules. Therefore, we conclude that

$$
\operatorname{Hom}_{kG}(U, U') = \operatorname{Hom}_{kG}\left(\widetilde{U}, \widetilde{U'}\right) .
$$

It follows that the cohomology groups $H^n(G, U)$, $n \geq 0$, of G with coefficients in the right kG-module U can be computed using a resolution

$$0 \longleftarrow k \xleftarrow{\;\varepsilon\;} P_*$$

of k by projective right kG-modules, as the cohomology of the cochain complex $\operatorname{Hom}_{kG}(P_*, U)$. In a similar way, it turns out that the homology groups $H_n(G, U)$, $n \geq 0$, of G with coefficients in the right kG-module U can be computed using a resolution

$$0 \longleftarrow k \xleftarrow{\;\varepsilon\;} Q_*$$

of k by projective left kG-modules, as the homology of the chain complex $U \otimes_{kG} Q_*$.

We now consider a left kG-module M and a k-module J. Then, the k-module $U = \operatorname{Hom}_k(M, J)$ has a natural structure of a right kG-module, which is obtained by using the left G-action on M (cf. Exercise 1.3.1(i)). If

$$0 \longleftarrow k \xleftarrow{\;\varepsilon\;} P_*$$

is a resolution of k by projective right kG-modules, then there is a natural identification of cochain complexes

$$\operatorname{Hom}_{kG}(P_*, \operatorname{Hom}_k(M, J)) \xrightarrow{\;\sim\;} \operatorname{Hom}_k(P_* \otimes_{kG} M, J)$$

(cf. Exercise 1.3.1(iii)). In this way, we obtain k-linear maps

$$\theta_{G,M,J} : H^n(G, \operatorname{Hom}_k(M, J)) \longrightarrow \operatorname{Hom}_k(H_n(G, M), J)$$

for all $n \geq 0$ (cf. Example D.3(ii)).

Proposition D.16 *Let M be a left kG-module, J a k-module and consider the k-linear maps $\theta = \theta_{G,M,J}$ defined above.*

(i) For any cohomology class $\alpha \in H^n(G, \operatorname{Hom}_k(M, J))$ the k-linear map $\theta(\alpha) \in \operatorname{Hom}_k(H_n(G, M), J)$ is the composition

$$H_n(G, M) \xrightarrow{\;\alpha \cap\;} H_0\Big(G, \widetilde{\operatorname{Hom}_k(M, J)} \otimes M\Big) = \operatorname{Hom}_k(M, J) \otimes_{kG} M \xrightarrow{\;ev\;} J ,$$

where ev denotes the evaluation homomorphism.

(ii) If J is an injective k-module, then θ is an isomorphism for all n. □

Corollary D.17 *Assume that k is a field, viewed as a trivial G-module. Then, there is an isomorphism $H^n(G, k) \simeq \operatorname{Hom}_k(H_n(G, k), k)$, which identifies a cohomology class $\alpha \in H^n(G, k)$ with the cap-product map*

$$\alpha \cap {}_- : H_n(G, k) \longrightarrow H_0(G, k) = k$$

for all $n \geq 0$. □

D.2.5 The (co-)homology of an Extension

Let us now consider a group G, a normal subgroup $N \trianglelefteq G$ and the corresponding quotient $Q = G/N$. Our goal is to describe the relationship between the (co-)homology groups of G with coefficients in a $\mathbf{Z}G$-module M and certain (co-)homology groups of N and Q. This relationship can be properly described by using the notion of a spectral sequence. Instead of giving the details of the construction of the Lyndon-Hochschild-Serre spectral sequence, we adopt an ad hoc point of view and state a few results that make the techniques used in Chap. 4 intelligible.

Our first objective is to compute the homology groups $H_n(G, M)$, $n \geq 0$. We denote by M' the $\mathbf{Z}N$-module obtained from M by restriction of scalars and consider the homology groups $H_j(N, M')$, $j \geq 0$. For any $g \in G$ we let $I_g \in \mathrm{Aut}(N)$ be the conjugation by g and define $M'_g = M'_{I_g}$ to be the $\mathbf{Z}N$-module obtained from M' by restriction of scalars along I_g. If

$$\lambda_g : M' \longrightarrow M'_g$$

is the $\mathbf{Z}N$-linear map, which is defined by letting $\lambda_g(x) = gx$ for all $x \in M'$, then the composition

$$H_j(N, M') \xrightarrow{(\lambda_g)_*} H_j(N, M'_g) \xrightarrow{(I_g)_*} H_j(N, M')$$

is an additive endomorphism ϱ_g of the group $H_j(N, M')$. In this way, we obtain an action ϱ of G on the homology group $H_j(N, M')$. This action being trivial on N (cf. Remark D.7(ii)), we conclude that the group $H_j(N, M')$ admits a natural $\mathbf{Z}Q$-module structure for all $j \geq 0$. In particular, we may consider the homology groups $H_i(Q, H_j(N, M'))$ for all $i, j \geq 0$.

Theorem D.18 *Let G be a group, $N \trianglelefteq G$ a normal subgroup and $Q = G/N$ the corresponding quotient. We consider a $\mathbf{Z}G$-module M and let M' be the $\mathbf{Z}N$-module obtained from M by restriction of scalars. Then, for all $n \geq 0$ the homology group $H_n(G, M)$ admits a natural increasing filtration*

$$0 = F_{-1}H_n \subseteq F_0 H_n \subseteq F_1 H_n \subseteq \cdots \subseteq F_{n-1}H_n \subseteq F_n H_n = H_n(G, M) \,,$$

having the following properties:

(i) The group $F_p H_n / F_{p-1} H_n$ is a certain subquotient of the homology group $H_p(Q, H_{n-p}(N, M'))$ for all p, n.

(ii) The group $F_0 H_n$ is the image of the map

$$H_n(N, M') \longrightarrow H_n(G, M) \,,$$

which is induced by the inclusion $N \hookrightarrow G$.

(iii) The group $F_{n-1} H_n$ is the kernel of the map

$$H_n(G, M) \longrightarrow H_n(Q, M_N) \,,$$

which is induced by the natural maps $G \longrightarrow Q$ and $M \longrightarrow M_N$. □

Corollary D.19 *Let k be a commutative ring, G a group, $N \trianglelefteq G$ a normal subgroup and $Q = G/N$ the corresponding quotient. If the groups N, Q have finite homological dimension over k, then G has also finite homological dimension over k; in fact, $hd_k G \leq hd_k N + hd_k Q$.* □

There is a result analogous to Theorem D.18 for the cohomology groups of G with coefficients in M. As in the homology case, the conjugation action of G induces a natural $\mathbf{Z}Q$-module structure on the groups $H^j(N, M')$. More precisely, for any element $g \in G$ the action of $q = gN \in Q$ on $H^j(N, M')$ is given by the composition

$$H^j(N, M') \xrightarrow{(\lambda_g)_*} H^j\left(N, M'_g\right) \xrightarrow{[(I_g)^*]^{-1}} H^j(N, M') \,,$$

where M'_g, λ_g and I_g are defined as above.[2] In this way, we may consider the cohomology groups $H^i(Q, H^j(N, M'))$ for all $i, j \geq 0$.

Theorem D.20 *Let G be a group, $N \trianglelefteq G$ a normal subgroup and $Q = G/N$ the corresponding quotient. We consider a $\mathbf{Z}G$-module M and let M' be the $\mathbf{Z}N$-module obtained from M by restriction of scalars. Then, for all $n \geq 0$ the cohomology group $H^n(G, M)$ admits a natural decreasing filtration*

$$H^n(G, M) = F^0 H^n \supseteq F^1 H^n \supseteq \cdots \supseteq F^n H^n \supseteq F^{n+1} H^n = 0 \,,$$

having the following properties:

(i) The group $F^p H^n / F^{p+1} H^n$ is a certain subquotient of the cohomology group $H^p(Q, H^{n-p}(N, M'))$ for all p, n.

(ii) The group $F^1 H^n$ is the kernel of the map

$$H^n(G, M) \longrightarrow H^n(N, M') \,,$$

which is induced by the inclusion $N \hookrightarrow G$.

(iii) The group $F^n H^n$ is the image of the map

$$H^n\left(Q, M^N\right) \longrightarrow H^n(G, M) \,,$$

which is induced by the natural maps $G \longrightarrow Q$ and $M^N \hookrightarrow M$.

(iv) Assume that $M = k$ is a commutative ring, viewed as a trivial $\mathbf{Z}G$-module. Then, for any cohomology classes $\alpha \in F^p H^n$ and $\alpha' \in F^{p'} H^{n'}$ we have $\alpha \cup \alpha' \in F^{p+p'} H^{n+n'}$. □

We conclude our discussion with two results concerning the special cases of an extension as above, where the normal subgroup $N \trianglelefteq G$ is either finite or infinite cyclic.

---

[2] We note that the $\mathbf{Z}Q$-module structures defined on the (co-)homology groups of N with coefficients in M' are compatible with the duality maps θ; cf. Exercise D.3.6.

Proposition D.21 *Let k be a field of characteristic 0 and $\pi : G \longrightarrow Q$ a surjective group homomorphism with finite kernel. We consider a kQ-module V and let V_π be the kG-module obtained from V by restriction of scalars along π. Then, the natural maps*

$$\pi_* : H_n(G, V_\pi) \longrightarrow H_n(Q, V) \quad and \quad \pi^* : H^n(Q, V) \longrightarrow H^n(G, V_\pi)$$

are isomorphisms for all $n \geq 0$. $\square$

Proposition D.22 *Let*

$$1 \longrightarrow \mathbf{Z} \longrightarrow G \stackrel{\pi}{\longrightarrow} Q \longrightarrow 1$$

be a central extension and $\alpha \in H^2(Q, \mathbf{Z})$ the corresponding cohomology class.[3] We consider a $\mathbf{Z}Q$-module V and let V_π be the $\mathbf{Z}G$-module obtained from V by restriction of scalars along π. Then, there are exact sequences

$$H_n(G, V_\pi) \stackrel{\pi_*}{\longrightarrow} H_n(Q, V) \stackrel{\alpha \cap -}{\longrightarrow} H_{n-2}(Q, V) \longrightarrow H_{n-1}(G, V_\pi)$$

for all $n \geq 0$. $\square$

D.3 Exercises

1. Let G be an abelian group. The goal of this Exercise is to prove that G has finite homological dimension over $\mathbf{Q}$ if and only if it has finite rank.
   (i) If $T \subseteq G$ is a torsion subgroup, then show that $\mathrm{hd}_\mathbf{Q} T = 0$ and hence conclude that $\mathrm{hd}_\mathbf{Q} G \leq \mathrm{hd}_\mathbf{Q}(G/T)$.
   (ii) If $G = \mathbf{Z}^n$, show that $\mathrm{hd}_k G = n$ for any commutative ring k.
   (iii) If $G = \mathbf{Q}^n$, show that $\mathrm{hd}_k G = n$ for any commutative ring k.
   (iv) If G has finite rank, show that G has finite homological dimension over $\mathbf{Q}$.
   (v) If G is an abelian group of infinite rank, then show that G does not have finite homological dimension over any commutative ring k.
2. (i) Let (C, d), (C', d') and (C'', d'') be chain complexes of abelian groups. Show that the associativity isomorphisms

$$(C_i \otimes C_j') \otimes C_k'' \simeq C_i \otimes (C_j' \otimes C_k''), \quad i, j, k \geq 0 \,,$$

induce an isomorphism of chain complexes

$$\mathrm{Tot}(\mathrm{Tot}(C \otimes C') \otimes C'') \simeq \mathrm{Tot}(C \otimes \mathrm{Tot}(C' \otimes C'')) \,.$$

We view this isomorphism as an identification and denote the resulting chain complex by $\mathrm{Tot}(C \otimes C' \otimes C'')$.

---

[3] We note that Q acts trivially on $\mathbf{Z}$, since the extension is assumed to be central.

(ii) (coassociativity of the Alexander-Whitney map) Let G be a group and $S_* = (S_*(G, \mathbf{Z}), \delta)$ the standard resolution of the trivial G-module $\mathbf{Z}$. Show that the Alexander-Whitney map

$$\Delta : S_* \longrightarrow \mathrm{Tot}(S_* \otimes S_*)$$

induces a commutative diagram

$$
\begin{array}{ccc}
S_* & \xrightarrow{\Delta} & \mathrm{Tot}(S_* \otimes S_*) \\
{\scriptstyle \Delta} \downarrow & & \downarrow {\scriptstyle Tot(\Delta \otimes 1)} \\
\mathrm{Tot}(S_* \otimes S_*) & \xrightarrow{Tot(1 \otimes \Delta)} & \mathrm{Tot}(S_* \otimes S_* \otimes S_*)
\end{array}
$$

(iii) (existence of counit for the Alexander-Whitney map) Let G be a group and $S_* = (S_*(G, \mathbf{Z}), \delta)$ the standard resolution of the trivial G-module $\mathbf{Z}$. Note that $S_0 = \mathbf{Z}G$ and define

$$\varepsilon : S_* \longrightarrow \mathbf{Z}[0]$$

to be the chain map which is given by the usual augmentation in degree 0 and vanishes in positive degrees. Show that there is a commutative diagram

$$
\begin{array}{ccccc}
\mathrm{Tot}(S_* \otimes S_*) & \xleftarrow{\Delta} & S_* & \xrightarrow{\Delta} & \mathrm{Tot}(S_* \otimes S_*) \\
{\scriptstyle Tot(\varepsilon \otimes 1)} \downarrow & & \| & & \downarrow {\scriptstyle Tot(1 \otimes \varepsilon)} \\
\mathrm{Tot}(\mathbf{Z}[0] \otimes S_*) & \simeq & S_* & \simeq & \mathrm{Tot}(S_* \otimes \mathbf{Z}[0])
\end{array}
$$

where Δ is the Alexander-Whitney map.

3. Let G be a group and $S_* = (S_*(G, \mathbf{Z}), \delta)$ the standard resolution of the trivial G-module $\mathbf{Z}$. We consider three $\mathbf{Z}G$-modules L, M and N and two cohomology classes $\alpha \in H^n(G, M)$ and $\beta \in H^m(G, L)$. The goal of this Exercise is to prove Proposition D.14(iii). To that end, let us fix representatives $a : S_n \longrightarrow M$ and $b : S_m \longrightarrow L$ of α and β respectively and consider the representative $c = b \cup a : S_{n+m} \longrightarrow L \otimes M$ of $\gamma = \beta \cup \alpha \in H^{n+m}(G, L \otimes M)$.

(i) Show that the following diagram is commutative

$$
\begin{array}{ccc}
\mathrm{Tot}(S_* \otimes S_*) \otimes_{\mathbf{Z}G} N & \xrightarrow{(1 \otimes a \otimes 1)^{gr}} & S_*[n] \otimes_{\mathbf{Z}G} (M \otimes N) \\
{\scriptstyle Tot(\Delta \otimes 1) \otimes 1} \downarrow & & \downarrow {\scriptstyle \Delta \otimes 1 \otimes 1} \\
\mathrm{Tot}(S_* \otimes S_* \otimes S_*) \otimes_{\mathbf{Z}G} N & \xrightarrow{(1 \otimes 1 \otimes a \otimes 1)^{gr}} & \mathrm{Tot}(S_* \otimes S_*)[n] \otimes_{\mathbf{Z}G} (M \otimes N)
\end{array}
$$

Here, $(1 \otimes 1 \otimes a \otimes 1)^{gr}$ maps an elementary tensor $(x_i \otimes x_j \otimes x_k) \otimes y \in (S_i \otimes S_j \otimes S_k) \otimes_{\mathbf{Z}G} N$ onto the tensor $(-1)^{ni+nj}(x_i \otimes x_j) \otimes (a(x_k) \otimes y) \in (S_i \otimes S_j) \otimes_{\mathbf{Z}G} (M \otimes N)$ (resp. onto 0) if $k = n$ (resp. if $k \neq n$).

(ii) Show that the composition

$$H_i(G, N) \xrightarrow{\alpha \cap} H_{i-n}(G, M \otimes N) \xrightarrow{\beta \cap} H_{i-n-m}(G, L \otimes M \otimes N)$$

is induced by the composition

$$S_* \otimes_{\mathbf{Z}G} N \longrightarrow \text{Tot}(S_* \otimes S_* \otimes S_*) \otimes_{\mathbf{Z}G} N \longrightarrow S_*[n+m] \otimes_{\mathbf{Z}G} (L \otimes M \otimes N),$$

where the first chain map is $(\text{Tot}(\Delta \otimes 1) \circ \Delta) \otimes 1$ and the second one is $(1 \otimes b \otimes a \otimes 1)^{gr}$. Here, $(1 \otimes b \otimes a \otimes 1)^{gr}$ maps an elementary tensor $(x_i \otimes x_j \otimes x_k) \otimes y \in (S_i \otimes S_j \otimes S_k) \otimes_{\mathbf{Z}G} N$ onto $(-1)^{ni+nj+mi} x_i \otimes (b(x_j) \otimes a(x_k) \otimes y) \in S_i \otimes_{\mathbf{Z}G} (L \otimes M \otimes N)$ (resp. onto 0) if $j = m$ and $k = n$ (resp. if $j \neq m$ or $k \neq n$).

(iii) Show that the cap-product map

$$\gamma \cap \_ : H_i(G, N) \longrightarrow H_{i-n-m}(G, L \otimes M \otimes N)$$

is induced by the composition

$$S_* \otimes_{\mathbf{Z}G} N \longrightarrow \text{Tot}(S_* \otimes S_* \otimes S_*) \otimes_{\mathbf{Z}G} N \longrightarrow S_*[n+m] \otimes_{\mathbf{Z}G} (L \otimes M \otimes N),$$

where the first chain map is $(\text{Tot}(1 \otimes \Delta) \circ \Delta) \otimes 1$ and the second one is $(1 \otimes b \otimes a \otimes 1)^{gr}$.

(iv) Prove Proposition D.14(iii).

(*Hint:* Use the coassociativity of Δ; cf. Exercise 2(ii) above.)

4. Let G be a group and $S_* = (S_*(G, \mathbf{Z}), \delta)$ the standard resolution of the trivial G-module $\mathbf{Z}$. We consider a left $\mathbf{Z}G$-module M, an abelian group J and the right $\mathbf{Z}G$-module $U = \text{Hom}_{\mathbf{Z}}(M, J)$. The goal of this Exercise is to prove Proposition D.16(i), in the case where the coefficient ring k therein is that of integers (the proof for a general k is similar). To that end, let us fix a cohomology class $\alpha \in H^n(G, U)$, represented by a homomorphism $a : S_n \longrightarrow U$ of right $\mathbf{Z}G$-modules.

(i) Show that the map

$$(1 \otimes a \otimes 1)^{gr} : \text{Tot}(S_* \otimes S_*)_n \otimes_{\mathbf{Z}G} M \longrightarrow S_0 \otimes_{\mathbf{Z}G} \left(\widetilde{U} \otimes M \right),$$

followed by the natural quotient map

$$S_0 \otimes_{\mathbf{Z}G} \left(\widetilde{U} \otimes M \right) \simeq \widetilde{U} \otimes M \longrightarrow U \otimes_{\mathbf{Z}G} M,$$

coincides with the composition

$$\text{Tot}(S_* \otimes S_*)_n \otimes_{\mathbf{Z}G} M \xrightarrow{\widetilde{\varepsilon} \otimes 1} S_n \otimes_{\mathbf{Z}G} M \xrightarrow{a \otimes 1} U \otimes_{\mathbf{Z}G} M.$$

Here, $\widetilde{\varepsilon}$ denotes the map

$$\text{Tot}(\varepsilon \otimes 1) : \text{Tot}(S_* \otimes S_*)_n \longrightarrow \text{Tot}(\mathbf{Z}[0] \otimes S_*)_n \simeq S_n,$$

where ε is the counit defined in Exercise 2(iii) above.

(ii) Show that the cap-product map

$$\alpha \cap \_ : H_n(G, M) \longrightarrow H_0\left(G, \widetilde{U} \otimes M\right)$$

is induced by the additive map

$$a \otimes 1 : S_n \otimes_{\mathbf{Z}G} M \longrightarrow U \otimes_{\mathbf{Z}G} M = H_0\left(G, \widetilde{U} \otimes M\right).$$

(iii) Prove Proposition D.16(i), in the case where $k = \mathbf{Z}$.

5. The goal of this Exercise is to show that the duality homomorphisms $\theta_{G,M,J}$ of Proposition D.16 are natural with respect to the group G, the coefficient module M and the dualizing module J.

(i) (naturality with respect to the group) Let k be a commutative ring, $\phi : G' \longrightarrow G$ a group homomorphism, M a left kG-module and J a k-module. We denote by M' the left kG'-module obtained from M by restriction of scalars along ϕ and consider the right kG-module $U = \mathrm{Hom}_k(M, J)$. Then, the right kG'-module U' obtained from U by restriction of scalars along ϕ is identified with $\mathrm{Hom}_k(M', J)$. Show that the following diagram is commutative for all $n \geq 0$

$$\begin{array}{ccc} H^n(G, U) & \xrightarrow{\phi^*} & H^n(G', U') \\ \theta_{G,M,J} \downarrow & & \downarrow \theta_{G',M',J} \\ \mathrm{Hom}_k(H_n(G, M), J) & \xrightarrow{(\phi_*)^t} & \mathrm{Hom}_k(H_n(G', M'), J) \end{array}$$

Here, $(\phi_*)^t$ denotes the transpose of $\phi_* : H_n(G', M') \longrightarrow H_n(G, M)$.

(ii) (naturality with respect to the coefficient module) Let k be a commutative ring, G a group, $f : M' \longrightarrow M$ a homomorphism of left kG-modules and J a k-module. We consider the right kG-modules $U' = \mathrm{Hom}_k(M', J)$ and $U = \mathrm{Hom}_k(M, J)$ and note that the transpose $F = f^t : U \longrightarrow U'$ of f is kG-linear. Show that the following diagram is commutative for all $n \geq 0$

$$\begin{array}{ccc} H^n(G, U) & \xrightarrow{F_*} & H^n(G, U') \\ \theta_{G,M,J} \downarrow & & \downarrow \theta_{G,M',J} \\ \mathrm{Hom}_k(H_n(G, M), J) & \xrightarrow{(f_*)^t} & \mathrm{Hom}_k(H_n(G, M'), J) \end{array}$$

Here, $(f_*)^t$ denotes the transpose of $f_* : H_n(G, M') \longrightarrow H_n(G, M)$.

(iii) (naturality with respect to the dualizing module) Let k be a commutative ring, G a group, M a left kG-module and $\tau : J \longrightarrow J'$ a homomorphism of k-modules. We consider the right kG-modules $U = \mathrm{Hom}_k(M, J)$ and $U' = \mathrm{Hom}_k(M, J')$ and note that the induced map $T = \tau_* : U \longrightarrow U'$ is kG-linear. Show that the following diagram is commutative for all $n \geq 0$

$$\begin{array}{ccc} H^n(G, U) & \xrightarrow{T_*} & H^n(G, U') \\ \theta_{G,M,J} \downarrow & & \downarrow \theta_{G,M,J'} \\ \mathrm{Hom}_k(H_n(G, M), J) & \xrightarrow{\tau_*} & \mathrm{Hom}_k(H_n(G, M), J') \end{array}$$

Here, the map τ_* in the bottom row is that induced by τ between the Hom-groups.

6. Let G be a group, $N \trianglelefteq G$ a normal subgroup and $Q = G/N$. The goal of this Exercise is to show that the $\mathbf{Z}Q$-module structures defined in §D.2.5 on the (co-)homology groups of N with coefficients in restricted $\mathbf{Z}G$-modules are compatible with the duality homomorphisms θ of Proposition D.16. To that end, we consider a left $\mathbf{Z}G$-module M, an abelian group J and the right $\mathbf{Z}G$-module $U = \mathrm{Hom}_\mathbf{Z}(M, J)$. We fix an element $g \in G$ and let $q = gN \in Q$.

(i) Show that the left $\mathbf{Z}Q$-module structure on $H_n(N, M')$ defined in the text induces a right $\mathbf{Z}Q$-module structure on the abelian group $\mathrm{Hom}_\mathbf{Z}(H_n(N, M'), J)$, in such a way that $q \in Q$ acts as the composition of the transpose

$$((I_g)_*)^t : \mathrm{Hom}_\mathbf{Z}(H_n(N, M'), J) \longrightarrow \mathrm{Hom}_\mathbf{Z}\big(H_n\big(N, M'_g\big), J\big)$$

of $(I_g)_* : H_n\big(N, M'_g\big) \longrightarrow H_n(N, M')$, followed by the transpose

$$((\lambda_g)_*)^t : \mathrm{Hom}_\mathbf{Z}\big(H_n\big(N, M'_g\big), J\big) \longrightarrow \mathrm{Hom}_\mathbf{Z}(H_n(N, M'), J)$$

of $(\lambda_g)_* : H_n(N, M') \longrightarrow H_n\big(N, M'_g\big)$.

(ii) Show that the right $\mathbf{Z}N$-module U'_g obtained from the right $\mathbf{Z}N$-module $U' = \mathrm{Hom}_\mathbf{Z}(M', J)$ by restriction of scalars along the automorphism $I_g : N \longrightarrow N$ is identified with $\mathrm{Hom}_\mathbf{Z}\big(M'_g, J\big)$, whereas the transpose

$$\rho_g : U'_g \longrightarrow U'$$

of the map $\lambda_g : M' \longrightarrow M'_g$ defined in the text is a homomorphism of right $\mathbf{Z}G$-modules.

(iii) Show that the cohomology group $H^n(N, U')$ admits a right $\mathbf{Z}Q$-module structure, in such a way that $q \in Q$ acts as the composition

$$H^n(N, U') \xrightarrow{(I_g)^*} H^n\big(N, U'_g\big) \xrightarrow{(\rho_g)_*} H^n(N, U').$$

(iv) Show that the duality homomorphism

$$\theta_{N,M',J} : H^n(N, U') \longrightarrow \mathrm{Hom}_\mathbf{Z}(H_n(N, M'), J)$$

is a homomorphism of right $\mathbf{Z}Q$-modules for all $n \geq 0$.

(*Hint:* Use the naturality of θ with respect to the group and the coefficient module; cf. Exercise 5(i),(ii) above.)

E

Comparison of Projections

In this Appendix, we examine a few basic properties of the lattice of projections in a von Neumann algebra $\mathcal{N}$. We begin by reviewing the concepts of equivalence and weak ordering and then consider the notion of the central carrier of a projection. Our main goal is to prove the comparison theorem, which states that any two projections in $\mathcal{N}$ can be cut by a central projection into comparable sub-projections (for the precise statement, see Theorem E.7). This fact was used in a crucial way in §5.2.2, in order to prove the injectivity of the additive map t_*, which is induced in K-theory by the center-valued trace t on the von Neumann algebra of a group.

Of course, our presentation here is very limited, aiming only at those results that are necessary for the proof of the comparison theorem. For a more complete treatment of the structure theory of projections in a von Neumann algebra, the reader may consult specialized books on the subject, such as [18] and [36].

E.1 Equivalence and Weak Ordering

Let $\mathcal{B}(\mathcal{H})$ be the algebra of bounded linear operators on a Hilbert space $\mathcal{H}$. Any projection $e \in \mathcal{B}(\mathcal{H})$ is completely determined by its range $V = \operatorname{im} e$, which is a closed linear subspace of $\mathcal{H}$; indeed, e maps identically V onto itself and vanishes on the orthogonal complement $V^\perp$. Conversely, for any closed linear subspace $V \subseteq \mathcal{H}$ there is a (unique) projection in $\mathcal{B}(\mathcal{H})$ whose range is V; we denote that projection by p_V. Two projections e, f are orthogonal (i.e. $ef = 0$) if and only if $\operatorname{im} e \perp \operatorname{im} f$. If e, f are two commuting projections, then ef is the projection onto the subspace $\operatorname{im} e \cap \operatorname{im} f$. Given two projections e and f, we write $e \leq f$ if $ef = fe = e$ (cf. §1.1.1.II); in that case, the (closed) subspace $\operatorname{im} e$ is contained in $\operatorname{im} f$, whereas the operator $f - e$ is the orthogonal projection onto the subspace $(\operatorname{im} e)^\perp \cap \operatorname{im} f$. If e, f, c are projections in $\mathcal{B}(\mathcal{H})$, such that $e \leq f$ and c commutes with both e and f, then $ce \leq cf$. The correspondence $V \mapsto p_V$ defines an isomorphism of lattices between the lattice of closed subspaces of $\mathcal{H}$ and that of projections in $\mathcal{B}(\mathcal{H})$. For example, if $(e_i)_i$ is a family of projections and $V_i = \operatorname{im} e_i$ for all i, then the supremum $e = \sup_i e_i$ is the projection onto the closed linear subspace V of $\mathcal{H}$, which is generated by the V_i's (i.e. $V = \overline{\sum_i V_i}$). In the special case where the e_i's are orthogonal to each other, V is the orthogonal direct sum of the V_i's and $e = \sum_i e_i$.[1]

Recall that a linear operator $u \in \mathcal{B}(\mathcal{H})$ is called a partial isometry if there are closed linear subspaces $V, V' \subseteq \mathcal{H}$, such that u maps V isometrically onto V' and vanishes on the orthogonal complement $V^\perp$. In that case, the adjoint u^* maps V' isometrically onto V and vanishes on the orthogonal complement $V'^\perp$; therefore, we have $u^*u = p_V$ and $uu^* = p_{V'}$.[2] Let $e = p_V$ and $e' = p_{V'}$ be two projections in $\mathcal{B}(\mathcal{H})$. Then, the existence of a partial isometry u, such that $u^*u = e$ and $uu^* = e'$, is easily seen to be equivalent to the condition $\dim V = \dim V'$, where dim denotes the Hilbert space dimension.

We work with a fixed von Neumann algebra $\mathcal{N}$ of operators acting on the Hilbert space $\mathcal{H}$.

Lemma E.1 *The following conditions are equivalent for a closed subspace $V \subseteq \mathcal{H}$:*

  (i) $p_V \in \mathcal{N}$,
  (ii) the subspaces V and $V^\perp$ are $\mathcal{N}'$-invariant and
  (iii) the subspace V is $\mathcal{N}'$-invariant.

Proof. (i)→(ii): For any vector $\xi \in V$ and any operator $a \in \mathcal{N}'$ we have $a(\xi) = ap_V(\xi) = p_V a(\xi) \in \operatorname{im} p_V = V$; therefore, V is $\mathcal{N}'$-invariant. The same argument, applied to the complementary projection $1 - p_V$, shows that $V^\perp = \operatorname{im}(1 - p_V)$ is $\mathcal{N}'$-invariant as well.

(ii)→(iii): This is obvious.

(iii)→(i): In view of Lemma 1.17(ii), the $\mathcal{N}'$-invariance of V implies that the projection p_V is contained in $\mathcal{N}'' = \mathcal{N}$. □

Corollary E.2 *Let $(e_i)_i$ be a family of projections in $\mathcal{N}$. If $e \in \mathcal{B}(\mathcal{H})$ is the supremum of the e_i's, then $e \in \mathcal{N}$. In particular, if $e_i e_j = 0$ for $i \neq j$, then $\sum_i e_i \in \mathcal{N}$.*

Proof. Since $e_i \in \mathcal{N}$, Lemma E.1 implies that the subspace $V_i = \operatorname{im} e_i$ is $\mathcal{N}'$-invariant for all i. Then, the closed linear subspace $V = \overline{\sum_i V_i}$ is easily seen to be $\mathcal{N}'$-invariant as well. The result follows from another application of Lemma E.1, since $e = p_V$. □

We recall that two projections $e, f \in \mathcal{N}$ are called equivalent rel $\mathcal{N}$ if there is a partial isometry $u \in \mathcal{N}$, such that $e = u^*u$ and $f = uu^*$; in that case, we write $e \sim f$. Equivalently, $e \sim f$ if there is a partial isometry $u \in \mathcal{N}$, which maps $\operatorname{im} e$ isometrically onto $\operatorname{im} f$ and vanishes on the orthogonal complement

---

[1] Here, the infinite sum is understood as the SOT-limit of the net of finite sums.
[2] An algebraic description of partial isometries is provided in Exercise E.2.1.

$(\operatorname{im} e)^{\perp}$. It follows easily that the *equivalence rel* $\mathcal{N}$ is indeed an equivalence relation. We now state and prove a few basic properties of that relation, that are needed in the sequel.

Proposition E.3 *(i) Assume that $e, f \in \mathcal{N}$ are projections with $e \sim f$. Then, for any central projection $c \in \mathcal{N}$ we have $ce \sim cf$.*

(ii) Consider an operator $a \in \mathcal{N}$ and let V (resp. U) be the closure of the range $\operatorname{im} a$ of a (resp. of the range $\operatorname{im} a^*$ of a^*). Then, $p_V, p_U \in \mathcal{N}$ and $p_V \sim p_U$.

Proof. (i) Let $u \in \mathcal{N}$ be a partial isometry which maps $\operatorname{im} e$ isometrically onto $\operatorname{im} f$ and vanishes on the orthogonal complement $(\operatorname{im} e)^{\perp}$. We shall prove that $ce \sim cf$, by showing that $cu \in \mathcal{N}$ maps $\operatorname{im} ce$ isometrically onto $\operatorname{im} cf$ and vanishes on the orthogonal complement $(\operatorname{im} ce)^{\perp}$. Since $ue(\mathcal{H}) = u(\operatorname{im} e) = \operatorname{im} f = f(\mathcal{H})$, we have

$$cu(ce(\mathcal{H})) = cuce(\mathcal{H}) = c^2 ue(\mathcal{H}) = cue(\mathcal{H}) = cf(\mathcal{H}).$$

For any vector $\xi \in \operatorname{im} ce = \operatorname{im} c \cap \operatorname{im} e$ we have $cu(\xi) = uc(\xi) = u(\xi)$ and hence $\| cu(\xi) \| = \| u(\xi) \| = \| \xi \|$. It follows that cu maps the subspace $\operatorname{im} ce$ isometrically onto $\operatorname{im} cf$. Since u vanishes on $(\operatorname{im} e)^{\perp} = \operatorname{im}(1 - e)$, we have $u(1 - e) = 0$ and hence

$$cu(1 - ce) = cu - cuce = cu - c^2 ue = cu - cue = cu(1 - e) = 0 \,.$$

Therefore, cu vanishes on the subspace $\operatorname{im}(1 - ce) = (\operatorname{im} ce)^{\perp}$, as needed.

(ii) Since $a, a^* \in \mathcal{N}$, the subspaces $V = \overline{\operatorname{im} a}$ and $U = \overline{\operatorname{im} a^*}$ are easily seen to be $\mathcal{N}'$-invariant. Therefore, Lemma E.1 implies that $p_V, p_U \in \mathcal{N}$. If $a = u\,|a|$ is the polar decomposition of a (cf. Proposition 1.11), then u maps U isometrically onto V and vanishes on the orthogonal complement $U^{\perp}$. The proof is finished, since the partial isometry u is contained in the von Neumann algebra $\mathcal{N}$ (cf. Proposition 1.19). □

Proposition E.4 *Let $(V_i)_i$ and $(V_i')_i$ be two orthogonal families of closed subspaces of $\mathcal{H}$, which are such that $p_{V_i}, p_{V_i'} \in \mathcal{N}$ and $p_{V_i} \sim p_{V_i'}$ for all i. Then, $\sum_i p_{V_i} \sim \sum_i p_{V_i'}$.*

Proof. In view of Corollary E.2, we know that $\sum_i p_{V_i}, \sum_i p_{V_i'} \in \mathcal{N}$. For any index i there is a partial isometry $u_i \in \mathcal{N}$, which maps V_i isometrically onto V_i' and vanishes on the orthogonal complement $V_i^{\perp}$. Let V (resp. V') be the orthogonal direct sum of the family $(V_i)_i$ (resp. $(V_i')_i$). Then, any vector $\xi \in \mathcal{H}$ admits an orthogonal decomposition $\xi = \sum_i \xi_i + \eta$, where $\xi_i \in V_i$ for all i and $\eta \in V^{\perp}$. Then, we have

$$\sum_i \| u_i(\xi_i) \|^2 = \sum_i \| \xi_i \|^2 \le \| \xi \|^2 \,, \tag{E.1}$$

with equality if and only if $\eta = 0$ (i.e. if and only if $\xi \in V$). It follows that the sum $\sum_i u_i(\xi_i)$ is a well-defined element of V'. Hence, we may consider the linear map $u : \mathcal{H} \longrightarrow \mathcal{H}$, which is defined by mapping any element $\xi \in \mathcal{H}$ onto $\sum_i u_i(\xi_i)$, where $\xi = \sum_i \xi_i + \eta$ is the orthogonal decomposition of ξ as above. In view of (E.1), the operator u is bounded; in fact, u maps V isometrically onto V' and vanishes on the orthogonal complement $V^\perp$. Since $\sum_i p_{V_i} = p_V$ and $\sum_i p_{V_i'} = p_{V'}$, it only remains to prove that the partial isometry $u \in \mathcal{B}(\mathcal{H})$ is an element of the von Neumann algebra $\mathcal{N}$. To that end, we consider an operator a in the commutant $\mathcal{N}'$ and prove that $ua = au$. Since the projections p_V and p_{V_i} are contained in $\mathcal{N}$, Lemma E.1 implies that the subspaces $V^\perp$ and V_i are a-invariant for all i. We now fix a vector $\xi \in \mathcal{H}$ and consider the decomposition $\xi = \sum_i \xi_i + \eta$, where $\xi_i \in V_i$ for all i and $\eta \in V^\perp$. Then, $a(\xi) = \sum_i a(\xi_i) + a(\eta)$ is the associated decomposition of $a(\xi)$, since $a(\xi_i) \in V_i$ for all i and $a(\eta) \in V^\perp$. Hence,

$$
\begin{aligned}
ua(\xi) &= u\Big(\sum_i a(\xi_i) + a(\eta)\Big) \\
&= \sum_i u_i a(\xi_i) \\
&= \sum_i a u_i(\xi_i) \\
&= a\Big(\sum_i u_i(\xi_i)\Big) \\
&= au(\xi) \, ,
\end{aligned}
$$

where the third equality follows since the operator $u_i \in \mathcal{N}$ commutes with $a \in \mathcal{N}'$ for all i and the fourth one is a consequence of the continuity of a. This is the case for any vector $\xi \in \mathcal{H}$ and hence $ua = au$, as needed.   □

We now define the central carrier of a projection in $\mathcal{N}$; this notion will be our main technical tool in the proof of the comparison theorem.

Proposition E.5 *Let $e \in \mathcal{N}$ be a projection with range V (so that $e = p_V$) and consider the closed linear subspace $U = [\mathcal{N}V]^-$ of $\mathcal{H}$ generated by the set $\mathcal{N}V = \{a(\xi) : a \in \mathcal{N}, \xi \in V\}$. Then, the projection $c = p_U \in \mathcal{B}(\mathcal{H})$ has the following properties:*

(i) c is a projection in the center of the algebra $\mathcal{N}$,
(ii) $e \le c$ and
(iii) if c' is a projection in the center of $\mathcal{N}$ with $e \le c'$, then $c \le c'$.
The projection c is called the central carrier of e.

Proof. (i) In order to show that $c \in \mathcal{N} \cap \mathcal{N}'$, we shall prove that the closed linear subspace U is invariant under both $\mathcal{N}'$ and $\mathcal{N}'' = \mathcal{N}$ (cf. Lemma E.1). To that end, it suffices to show that the set $\mathcal{N}V$ is invariant under $\mathcal{N}'$ and $\mathcal{N}$. It is clear that $\mathcal{N}V$ is $\mathcal{N}$-invariant. On the other hand, $e = p_V$ is a projection in $\mathcal{N}$ and hence the subspace V is $\mathcal{N}'$-invariant (loc.cit.). It follows easily from this that $\mathcal{N}V$ is $\mathcal{N}'$-invariant as well.

(ii) The algebra $\mathcal{N}$ is unital and hence $V \subseteq \mathcal{N}V \subseteq [\mathcal{N}V]^- = U$; therefore, $e = p_V \le p_U = c$.

(iii) Let c' be a central projection in $\mathcal{N}$ with range U' and assume that $e \leq c'$. Since $c' \in \mathcal{N}'$, the subspace U' is invariant under $\mathcal{N}'' = \mathcal{N}$ (loc.cit.). Moreover, $e \leq c'$ and hence $V \subseteq U'$. It follows that

$$\mathcal{N}V = \{a(\xi) : a \in \mathcal{N}, \xi \in V\} \subseteq \{a(\xi) : a \in \mathcal{N}, \xi \in U'\} \subseteq U'$$

and hence $U = [\mathcal{N}V]^- \subseteq U'$. Therefore, $c = p_U \leq p_{U'} = c'$. $\qquad\square$

We now relate the orthogonality of the central carriers of two projections in $\mathcal{N}$ to the existence of non-zero equivalent sub-projections.

Proposition E.6 *The following conditions are equivalent for two projections* $e, f \in \mathcal{N}$:

(i) The projections e and f have no non-zero sub-projections, which are equivalent rel $\mathcal{N}$.

(ii) We have $eaf = 0$ for all $a \in \mathcal{N}$.

(iii) $c(e)c(f) = 0$, where $c(e)$ and $c(f)$ are the central carriers of e and f respectively.

Proof. (i)→(ii): Assume that there is an operator $a \in \mathcal{N}$ with $\underline{eaf \neq 0}$ and consider the projection p_V onto the closed linear subspace $V = \operatorname{im} eaf$; then, $0 \neq p_V \leq e$. The subspace V is easily seen to be $\mathcal{N}'$-invariant and hence Lemma E.1 implies that $p_V \in \mathcal{N}$. Since the operator $fa^*e = (eaf)^* \in \mathcal{N}$ is non-zero as well, the same $\underline{\text{argument}}$ shows that the projection p_U onto the closed linear subspace $U = \operatorname{im} fa^*e$ is a projection in $\mathcal{N}$ with $0 \neq p_U \leq f$. Therefore, p_V and p_U are non-zero sub-projections of e and f respectively, which are equivalent rel $\mathcal{N}$, in view of Proposition E.3(ii). It follows that if (i) holds then $eaf = 0$ for all $a \in \mathcal{N}$.

(ii)→(iii): Assume that $eaf = 0$ for all $a \in \mathcal{N}$. Then, for any $a, b \in \mathcal{N}$ we have $ea^*bf = 0$ and hence for any vectors $\xi, \eta \in \mathcal{H}$ we compute

$$< ae(\xi), bf(\eta) > = < \xi, ea^*bf(\eta) > = 0 .$$

It follows that the closed linear subspaces $[\mathcal{N}e(\mathcal{H})]^-$ and $[\mathcal{N}f(\mathcal{H})]^-$, which are generated by the sets $\mathcal{N}e(\mathcal{H}) = \{ae(\xi) : a \in \mathcal{N}, \xi \in \mathcal{H}\}$ and $\mathcal{N}f(\mathcal{H}) = \{bf(\eta) : b \in \mathcal{N}, \eta \in \mathcal{H}\}$ respectively, are orthogonal to each other. Since $c(e), c(f)$ are the orthogonal projections onto $[\mathcal{N}e(\mathcal{H})]^-$ and $[\mathcal{N}f(\mathcal{H})]^-$, it follows that $c(e)c(f) = 0$.

(iii)→(i): Assume that $c(e)c(f) = 0$ and consider two projections $e', f' \in \mathcal{N}$ with $e' \leq e$, $f' \leq f$ and $e' \sim f'$. Then, there is a partial isometry $u \in \mathcal{N}$, which maps $\operatorname{im} e'$ isometrically onto $\operatorname{im} f'$ and vanishes on the orthogonal complement $(\operatorname{im} e')^\perp$; in particular, we have $u = f'ue'$. Since $e' \leq e \leq c(e)$ (cf. Proposition E.5(ii)), we have $e' = e'c(e)$. Arguing similarly, we conclude that $f' = f'c(f)$. We now compute

$$u = f'ue' = f'c(f)ue'c(e) = f'ue'c(e)c(f) = 0 ,$$

where the third equality follows since $c(f)$ is central (cf. Proposition E.5(i)). Hence, we conclude that $e' = u^*u = 0$ and $f' = uu^* = 0$.   □

Recall that, given two projections $e, f \in \mathcal{N}$, we say that e is weaker than f rel $\mathcal{N}$ if there is a projection $e' \in \mathcal{N}$, such that $e \sim e'$ and $e' \leq f$; in that case, we write $e \preceq f$. The projection e is strictly weaker than f rel $\mathcal{N}$ if $e \preceq f$ and $e \not\sim f$; in that case, we write $e \prec f$.

We are now ready to state and prove the main result of this Appendix.

Theorem E.7 *(comparison theorem) For any two projections $e, f \in \mathcal{N}$ there is a central projection $c \in \mathcal{N}$, such that $ce \preceq cf$ and $(1-c)f \preceq (1-c)e$.*

Proof. We consider sets of ordered pairs $P = \{(e_i, f_i) : i \in I\}$, such that:
(i) e_i, f_i are projections in $\mathcal{N}$ with $e_i \sim f_i$ for all i,
(ii) $e_i \leq e$ and $f_i \leq f$ for all i and
(iii) $e_i e_j = 0$ and $f_i f_j = 0$ for all $i \neq j$.
Let $\mathcal{P}$ be the collection consisting of the sets P as above; then, $\mathcal{P} \neq \emptyset$, since $\{(0,0)\} \in \mathcal{P}$. Ordering the set $\mathcal{P}$ by inclusion, we note that it satisfies the assumptions of Zorn's lemma. Therefore, $\mathcal{P}$ has a maximal element $\overline{P} = \{(\overline{e_i}, \overline{f_i}) : i \in \overline{I}\}$. We now consider the projections $e_0 = e - \sum_i \overline{e_i}$ and $f_0 = f - \sum_i \overline{f_i}$, which are contained in $\mathcal{N}$ (cf. Corollary E.2). If $e', f' \in \mathcal{N}$ are non-zero sub-projections of e_0 and f_0 respectively with $e' \sim f'$, then $\overline{P} \cup \{(e', f')\}$ is an element of $\mathcal{P}$ which is strictly bigger than $\overline{P}$. In view of the maximality of $\overline{P}$, we conclude that the projections $e_0, f_0 \in \mathcal{N}$ have no non-zero equivalent sub-projections. Hence, Proposition E.6 implies that $c(e_0)c(f_0) = 0$, where $c(e_0)$ (resp. $c(f_0)$) is the central carrier of e_0 (resp. of f_0); since $e_0 = e_0 c(e_0)$ (cf. Proposition E.5(ii)), we have

$$c(f_0)e_0 = e_0 c(f_0) = e_0 c(e_0)c(f_0) = 0 .$$

Using Propositions E.3(i) and E.4, we conclude that

$$\begin{aligned}
c(f_0)e &= c(f_0)e_0 + c(f_0)\sum_i \overline{e_i} \\
&= c(f_0)\sum_i \overline{e_i} \\
&\sim c(f_0)\sum_i \overline{f_i} \\
&\leq c(f_0)f .
\end{aligned}$$

Since $(1 - c(f_0))f_0 = f_0 - c(f_0)f_0 = 0$ (cf. Proposition E.5(ii)), we also have

$$\begin{aligned}
(1 - c(f_0))f &= (1 - c(f_0))f_0 + (1 - c(f_0))\sum_i \overline{f_i} \\
&= (1 - c(f_0))\sum_i \overline{f_i} \\
&\sim (1 - c(f_0))\sum_i \overline{e_i} \\
&\leq (1 - c(f_0))e .
\end{aligned}$$

This finishes the proof, by letting $c = c(f_0)$.   □

Corollary E.8 *Let $e, f \in \mathcal{N}$ be two projections, which are not equivalent rel $\mathcal{N}$. Then, there is a central projection $c \in \mathcal{N}$, such that $ce \prec cf$ or $cf \prec ce$.*

Proof. In view of Theorem E.7, there is a central projection $c_0 \in \mathcal{N}$, such that $c_0 e \preceq c_0 f$ and $(1 - c_0)f \preceq (1 - c_0)e$. We shall finish the proof by showing that $c_0 e \prec c_0 f$ or $(1 - c_0)f \prec (1 - c_0)e$. To that end, we argue by contradiction and assume that $c_0 e \sim c_0 f$ and $(1 - c_0)f \sim (1 - c_0)e$. In that case, we may invoke Proposition E.4 and conclude that

$$ e = c_0 e + (1 - c_0)e \sim c_0 f + (1 - c_0)f = f \,, $$

contradicting our assumption that $e \not\sim f$. $\qquad\square$

E.2 Exercises

1. Let $\mathcal{H}$ be a Hilbert space and $u \in \mathcal{B}(\mathcal{H})$ a bounded linear operator. Show that the following conditions are equivalent:
   (i) u is a partial isometry,
   (i)' u^* is a partial isometry,
   (ii) $u = uu^*u$,
   (ii)' $u^* = u^*uu^*$,
   (iii) u^*u is a projection and
   (iii)' uu^* is a projection.
2. Let $\mathcal{N}$ be a von Neumann algebra of operators acting on the Hilbert space $\mathcal{H}$. Two projections $e, f \in \mathcal{N}$ are called unitarily equivalent in $\mathcal{N}$ if there is a unitary operator $u \in \mathcal{N}$, such that $f = u^*eu$. Show that the relation of unitary equivalence in $\mathcal{N}$ implies that of equivalence rel $\mathcal{N}$.
3. Let $\mathcal{N}$ be a von Neumann algebra of operators acting on the Hilbert space $\mathcal{H}$. Then, $\mathcal{N}$ is called finite if there is no projection $e \in \mathcal{N}$ with $e \neq 1$, which is equivalent to 1 rel $\mathcal{N}$.
   (i) Assume that $\mathcal{N}$ is finite and let $e, f \in \mathcal{N}$ be two projections, such that $e \leq f$ and $e \sim f$. Then, show that $e = f$.
   (*Hint:* Use Proposition E.4.)
   (ii) Assume that two projections in $\mathcal{N}$ are equivalent rel $\mathcal{N}$ if and only if they are unitarily equivalent in $\mathcal{N}$ (cf. Exercise 2 above). Then, show that $\mathcal{N}$ is finite.
   (iii) Let G be a group and $\mathcal{N}G$ the associated von Neumann algebra. Show that the von Neumann algebra $\mathbf{M}_n(\mathcal{N}G)$ of $n \times n$ matrices with entries in $\mathcal{N}G$ is finite for all $n \geq 1$.
   (*Hint:* Use Proposition 5.43(i),(ii).)
4. Let $\mathcal{N}$ be a finite von Neumann algebra of operators acting on the Hilbert space $\mathcal{H}$ (cf. Exercise 3 above). The goal of this Exercise is to show that two projections in $\mathcal{N}$ are equivalent rel $\mathcal{N}$ if and only if they are unitarily

equivalent in $\mathcal{N}$ (cf. Exercise 2 above).[3] To that end, let $e, f \in \mathcal{N}$ be two projections with $e \sim f$.

(i) Show that $1 - e \sim 1 - f$.

(*Hint:* Argue by contradiction, using Corollary E.8.)

(ii) Show that e and f are unitarily equivalent in $\mathcal{N}$.

---

[3] In view of Exercise 3(ii) above, it follows that a von Neumann algebra $\mathcal{N}$ is finite if and only if the relation *equivalence rel* $\mathcal{N}$ coincides with *unitary equivalence in* $\mathcal{N}$.

References

1. Adem, A., Karoubi, M.: Periodic cyclic cohomology of group rings. C. R. Acad. Sci. Paris Ser. I Math. **326**, 13–17 (1998)
2. Alexander, G.M.: Semisimple elements in group algebras. Comm. Algebra **21**, 2417–2435 (1993)
3. Atiyah, M.F., Macdonald, I.G.: Introduction to Commutative Algebra. Addison-Wesley 1969
4. Bass, H.: Euler characteristics and characters of discrete groups. Invent. Math. **35**, 155–196 (1976)
5. Bass, H.: Traces and Euler characteristics. Homological Group Theory (Ed. C.T.C. Wall), London Math. Soc. Lecture Notes Series **36**, 1–26 (1979)
6. Baum, P., Connes, A.: Geometric K-theory for Lie groups and foliations. IHES preprint 1982
7. Borel, A., Serre, J.-P.: Le Théorème de Riemann-Roch. Bull. Soc. Math. de France **86**, 97–136 (1958)
8. Bourbaki, N.: Éléments de mathématique. Algèbre Commutative. Paris: Masson 1985
9. Brown, K.S.: Cohomology of groups. (Grad. Texts Math. **87**) Berlin Heidelberg New York: Springer 1982
10. Burghelea, D.: The cyclic homology of the group rings. Comment. Math. Helv. **60**, 354–365 (1985)
11. Chadha, G.K., Passi, I.B.S.: Centralizers and homological dimension. Comm. Alg. **22**(14), 5703–5708 (1994)
12. Cliff, G.H.: Zero divisors and idempotents in group rings. Can. J. Math. **32**, 596–602 (1980)
13. Cliff, G.H.: Ranks of projective modules over group rings. Comm. Alg. **13**(5), 1115–1130 (1985)
14. Cliff, G.H., Sehgal, S.K.: On the trace of an idempotent in a group ring. Proc. Amer. Math. Soc. **62**, 11–14 (1977)
15. Connes, A.: Non-commutative differential geometry. Publ. Math. IHES **62**, 41–144 (1985)
16. Connes, A., Moscovici, H.: The L^2-index theorem for homogeneous spaces of Lie groups. Ann. of Math. **115**, 291–330 (1982)
17. Cuntz, J.: K-theoretic amenability for discrete groups. J. für Reine und Angew. Math. **344**, 180–195 (1983)

18. Dixmier, J.: Les Algèbres d' Opérateurs dans l'Espace Hilbertien. Gauthier-Villars, Paris 1969
19. Eckmann, B.: Cyclic homology of groups and the Bass conjecture. Comment. Math. Helv. **61**, 193–202 (1986)
20. Eckmann, B.: Projective and Hilbert modules over group algebras, and finitely dominated spaces. Comment. Math. Helv. **71**, 453–462 (1996). Addendum. Comment. Math. Helv. **72**, 329 (1997)
21. Eckmann, B.: Idempotents in a complex group algebra, projective modules, and the group von Neumann algebra. Arch. Math. (Basel) **76**, 241–249 (2001)
22. Emmanouil, I.: On a class of groups satisfying Bass' conjecture. Invent. Math. **132**, 307–330 (1998)
23. Emmanouil, I.: Projective modules, augmentation and idempotents in group algebras. J. Pure Appl. Alg. **158**, 151–160 (2001)
24. Emmanouil, I.: Traces and idempotents in group algebras. Math. Z. **245**, 293–307 (2003)
25. Emmanouil, I.: Induced projective modules over group von Neumann algebras. (to appear in K-Theory)
26. Emmanouil, I.: On the trace of idempotent matrices over group algebras. (preprint)
27. Emmanouil, I., Passi, I.B.S.: A contribution to Bass' conjecture. J. Group Theory **7**, 409–429 (2004)
28. Farrell, F.T., Linnell, P.A.: Whitehead groups and the Bass conjecture. Math. Ann. **326**, 723–757 (2003)
29. Formanek, E.: Idempotents in Noetherian group rings. Can. J. Math. **25**, 366–369 (1973)
30. Hall, P.: On the finiteness of certain soluble groups. Proc. London Math. Soc. (3) **9**, 595–622 (1959)
31. Hartshorne, R.: Algebraic Geometry. (Grad. Texts Math. **52**) Berlin Heidelberg New York: Springer 1977
32. Hattori, A: Rank element of a projective module. Nagoya J. Math. **25**, 113–120 (1965)
33. Hewitt, E., Ross, K.A.: Abstract Harmonic Analysis (vol. 1). Berlin Heidelberg New York: Springer 1963
34. Hilton, P., Stammbach, U.: A Course in Homological Algebra. (Grad. Texts Math. **4**) Berlin Heidelberg New York: Springer 1976
35. Ji, R.: Nilpotency of Connes' periodicity operator and the idempotent conjectures. K-Theory **9**, 59–76 (1995)
36. Kadison, R.V., Ringrose, J.R.: Fundamentals of the Theory of Operator Algebras. (2 volumes) San Diego Orlando: Academic Press 1983, 1986
37. Kaplansky, I.: Projective modules. Ann. Math. **68**, 372–377 (1958)
38. Kaplansky, I.: Fields and Rings. University of Chicago Press, Chicago 1969
39. Karoubi, M.: K-Theory: An Introduction. (Grundlehren der Math. Wissenschaften **226**) Berlin Heidelberg New York: Springer 1978
40. Karoubi, M., Villamayor, O.: Homologie cyclique d'algèbres de groupes. C.R. Acad. Sci. Paris **311**, 1–3 (1990)
41. Lam, T.Y.: A First Course in Noncommutative Rings. (Grad. Texts Math. **131**) Berlin Heidelberg New York: Springer 1991
42. Lang, S.: Algebra. Addison-Wesley 1993
43. Linnell, P.A.: Decomposition of augmentation ideals and relation modules. Proc. London Math. Soc. (3) **47**, 83–127 (1983)

44. Loday, J.L.: Cyclic homology. (Grundl. Math. Wiss. **301**) Berlin Heidelberg New York: Springer 1992
45. Loday J.L., Quillen, D.: Cyclic homology and the Lie algebra homology of matrices. Comment. Math. Helv. **59**, 565–591 (1984)
46. Lück, W.: The relation between the Baum-Connes conjecture and the trace conjecture. Invent. Math. **149**, 123–152 (2002)
47. Mac Carthy, R.: Morita equivalence and cyclic homology. C.R. Acad. Sci. Paris **307**, 211–215 (1988)
48. Mac Lane, S.: Homology. (Grundlehren der mathematischen Wissenschaften **114**) Berlin Heidelberg New York: Springer 1995
49. Malcev, A.I.: On faithful representations of infinite groups my matrices. Math. Sb. **8**(50), 405–422 (1940)
50. Marciniak, Z.: Cyclic homology and idempotents in group rings. Springer Lect. Notes in Math. **1217**, 253–257 (1985)
51. Marciniak, Z.: Cyclic homology of group rings. Banach Center Publications **18**, 305–312 (1986)
52. Matsumura, H.: Commutative ring theory. Cambridge Stud. Adv. Math. **8**, Cambridge University Press 1992
53. Moody, J.A.: Ph.D. thesis. Columbia University, 1986
54. Passi, I.B.S., Passman, D.S.: Algebraic elements in group rings. Proc. Amer. Math. Soc. **108**, 871–877 (1990)
55. Passman, D.S.: The algebraic structure of group rings. (Pure Appl. Math.) Wiley-Interscience, New York 1977
56. Pedersen, G.K.: Analysis Now. (Grad. Texts Math. **118**) Berlin Heidelberg New York: Springer 1989
57. Pimsner, M., Voiculescu, D.: K-groups of reduced crossed products by free groups. J. Oper. Theory **8**, 131–156 (1982)
58. Rosenberg, J.: Algebraic K-Theory and its Applications. (Grad. Texts Math. **147**) Berlin Heidelberg New York: Springer 1994
59. Roy, R.: The Trace Conjecture – A Counterexample. K-Theory **17**, 209–213 (1999)
60. Rudin, W.: Functional Analysis. McGraw-Hill 1973
61. Rudin, W.: Real and Complex Analysis. McGraw-Hill 1987
62. Schafer, J.A.: Traces and the Bass conjecture. Michigan Math. J. **38**, 103–109 (1991)
63. Schafer, J.A.: Relative cyclic homology and the Bass conjecture. Comment. Math. Helv. **67**, 214–225 (1992)
64. Schafer, J.A.: The Bass conjecture and group von Neumann algebras. K-Theory **19**, 211–217 (2000)
65. Serre, J.P.: Arbres, Amalgames, SL_2. Astérisque **46**, 1977
66. Stallings, J.: Centerless groups – an algebraic formulation of Gottlieb's theorem. Topology **4**, 129–134 (1965)
67. Stevenhagen, P. and Lenstra, H.W.Jr.: Chebotarëv and his Density Theorem. Math. Intelligencer **18**, 26–37 (1996)
68. Strebel, R.: Homological methods applied to the derived series of groups. Comment. Math. Helv. **49**, 302–332 (1974)
69. Strojnowski, A.: Idempotents and zero divisors in group rings. Comm. Alg. **14**(7), 1171–1185 (1986)
70. Swan, R.G.: Induced representations and projective modules. Ann. Math. **71**, 552–578 (1960)

71. Swan, R.G.: Vector bundles and projective modules. Trans. Amer. Math. Soc. **105**, 264–277 (1962)
72. Wall, C.T.C.: Finiteness conditions for CW-complexes I. Ann. of Math. **81**, 56–69 (1965); Finiteness conditions for CW-complexes II. Proc. Royal Soc. Ser. A **295**, 129–139 (1966)
73. Weibel, C.: An introduction to algebraic K-theory. (book in progress, http://math.rutgers.edu/~weibel/Kbook.html)
74. Weiss, A.: Idempotents in group rings. J. Pure Appl. Algebra **16**, 207–213 (1980)
75. Zaleskii, A.E.: On a problem of Kaplansky. Soviet. Math. **13**, 449–452 (1972)

Index

Universitext